Sustainable Construction through Utilization of Optimization Tools and Experimental Methods

Sustainable Construction through Utilization of Optimization Tools and Experimental Methods

Uroš Klanšek
Tomaž Žula

MDPI

Basel • Beijing • Wuhan • Barcelona • Belgrade • Novi Sad • Cluj • Manchester

Editors

Uroš Klanšek
Faculty of Civil Engineering,
Transportation Engineering
and Architecture
University of Maribor
Maribor
Slovenia

Tomaž Žula
Faculty of Civil Engineering,
Transportation Engineering
and Architecture
University of Maribor
Maribor
Slovenia

Editorial Office
MDPI AG
Grosspeteranlage 5
4052 Basel, Switzerland

This is a reprint of articles from the Special Issue published online in the open access journal *Sustainability* (ISSN 2071-1050) (available at: www.mdpi.com/journal/sustainability/special_issues/W7229CS0HZ).

For citation purposes, cite each article independently as indicated on the article page online and as indicated below:

Lastname, A.A.; Lastname, B.B. Article Title. *Journal Name* **Year**, *Volume Number*, Page Range.

ISBN 978-3-7258-2338-3 (Hbk)
ISBN 978-3-7258-2337-6 (PDF)
doi.org/10.3390/books978-3-7258-2337-6

Contents

sustainability

Editorial

Sustainable Construction through Utilization of Optimization Tools and Experimental Methods—An Editorial

Uroš Klanšek * **and Tomaž Žula**

Faculty of Civil Engineering, Transportation Engineering and Architecture, University of Maribor, Smetanova ulica 17, 2000 Maribor, Slovenia; tomaz.zula@um.si
* Correspondence: uros.klansek@um.si

1. Introduction

It is widely recognized that the adverse impacts of the construction industry on economic, social, and environmental sustainability are staggering. With growing global awareness of the importance of the above, the construction industry has been forced to explore innovative techniques to ensure that its processes and products are sustainable. Here, optimization-based engineering can pave the way for sustainable construction practices in various areas from the utilization of natural resources and the organization of business processes and production through the reduction of energy demands, costs, and environmental footprints all the way to ensuring health and safety conditions, among other areas.

For the engineering problem at hand, optimization can identify the best solution in the set of all possible solutions. In this way, optimization lies at the very core of the basic mission of engineering, i.e., to develop new, better, more-efficient, and sustainable systems and improve the functioning of existing ones. The potential of optimization to determine the best solution without actually testing all possible solutions comes from the use of advanced mathematical methods and is only realized after performing iterative numerical calculations that follow clearly defined logical procedures or computational algorithms implemented in computer software and after consuming the necessary processor time.

This Special Issue is dedicated to the latest developments in the field of sustainable construction through the utilization of optimization tools and experimental methods. On the one hand, this Special Issue aims to publish influential and innovative articles that deal with the challenges of construction sustainability through the use of optimization tools that involve either exact mathematical programming or meta/hyper/bio-heuristics, alternative hybrid approaches, or multi-criteria decision-making techniques. On the other hand, field or laboratory research on materials, products, structures, objects, or operations often provides key data for developing optimization models. Therefore, research engaging with experimental methods to address current issues associated with sustainable construction is also covered by this volume. Moreover, scientific works combining optimization tools and experimental methods to reach synergistic effects in favor of sustainable construction are included within this Special Issue as well.

check for updates

Citation: Klanšek, U.; Žula, T. Sustainable Construction through Utilization of Optimization Tools and Experimental Methods—An Editorial. *Sustainability* **2024**, *16*, 7666. https://doi.org/10.3390/su16177666

Received: 29 August 2024
Accepted: 2 September 2024
Published: 4 September 2024

2. Key Insights of This Volume

This thematic volume consists of eleven original articles. The paper by Rogulj et al. (Contribution 1) deals with flat-roof renovation planning on public buildings using fuzzy multi-criteria analysis. The study establishes the priority of the renovation of the flat roofs of public buildings, employing the multi-criteria method PROMETHEE II in fuzzy logic form. The proposed approach demonstrates the ability to transform the uncertain and vague information received from an expert into a fuzzy number. This makes it possible to obtain an objective outcome, remove the criteria conflict, and enable the ranking of alternatives, as well as mutual comparison. Based on the defined goals and criteria, the roofs are evaluated and ranked according to the priority for renovation. The suggested approach is validated on an actual example. Accordingly, it is argued that when it comes to flat roof renovation decision-making, a multi-criteria approach can support and assist building owners or facility managers in making informed decisions based on a set of indicators beyond just cost or aesthetics. The article also states that this can lead to more-informed decisions involving the full range of factors affecting the performance, durability, and sustainability of a flat roof system.

The study by Premrov and Kozem Šilih (Contribution 2) presents a numerical analysis of the racking behavior of multi-story timber-framed buildings considering the load-bearing function of double-skin façade elements (DSF). Based on the research performed, the use of the DSF elements as load-bearing structural elements to increase the racking load-bearing capacity of the whole structure proves to be reasonable. It is claimed that the development of racking-resistant timber DSF elements can open new perspectives in designing contemporary multi-story timber buildings located in seismic areas with strong winds and with a strong asymmetrical position of the transparent glass areas. However, this may decrease the energy demand for heating and provides more daylight, contributing to better living comfort within the building. The paper indicates the potentially beneficial socioeconomic effects of the research outcomes, as they can encourage the construction of multi-story timber buildings, which enable better use of forested areas and contribute to reducing the impact of buildings on the environment.

The article by Bekdaş et al. (Contribution 3) addresses the optimal dimensions of post-tensioned concrete cylindrical walls using harmony search and ensemble learning with SHapley Additive exPlanations (SHAP) methodology. The research is motivated by the fact that the optimal design of prestressed concrete cylindrical walls has a rather beneficial effect on the economic and environmental aspects of construction. Here, the unit costs of concrete and steel, the specific weight of the stored fluid, and the height of the cylindrical wall are the input parameters, whereas the optimum thicknesses of the wall with and without post-tensioning are the output variables. Founded on this database, advanced ensemble learning techniques like the extreme gradient boosting, light gradient boosting machine, categorical gradient boosting, and random forest algorithms are trained, while the impacts of various input features on the predictions of distinct machine learning models are analyzed via the SHAP approach. The paper infers that the gained predictive equations could be included in the engineering design process to facilitate structural optimization.

The research by Jelušič and Žula (Contribution 4) addresses the sustainable design of circular reinforced concrete column sections via multi-objective optimization. With this purpose, a mixed-integer nonlinear programming optimization model is developed, incorporating the requirements of Eurocodes for structural design. This model is created in MATLAB software, where a genetic algorithm (GA) is used for optimization. On these bases, parametric structural optimization is conducted separately for various combinations of applied loads and the objective functions of material costs and CO_2 emissions generated during the production of reinforced concrete. Multi-objective optimization is executed to identify a range of structural design solutions that offer optimal balances between economic and environmental objectives, i.e., minimum material cost and CO_2 emissions. The article highlights that the model is developed in a general form and can provide the optimal solution for various structural design parameters, including different concrete strength properties.

The investigation by Jelušič et al. (Contribution 5) reveals the potential of using waste materials in flexible pavement structures identified via and optimization design approach. The novelty of this work is the optimization model, which may provide an optimal pavement structure design and is able to consider various material properties affected by the inclusion of waste materials, both in terms of minimum construction cost and CO_2 emissions. GA is used to minimize construction cost and CO_2 emissions in accordance with pavement design guidelines. In order to show the effects of the material properties of the asphalt, base layer, sub-base layer, and subgrade, as affected by the inclusion of waste materials, a parametric study is conducted to establish the optimal design of flexible pavement structures. The paper points out that by applying the proposed optimization model, engineers can aim for a pavement structure that is both structurally sound and environmentally sustainable while minimizing construction costs.

The contribution by Galjanić et al. (Contribution 6) tackles the identification of key factors for project performance within a multi-stakeholder environment, the definition of a performance measurement framework for construction investments, and the constitution of a link between performance measurement and project performance prediction. Here, the performance management fields are derived from formerly executed, comprehensive bibliographical research analysis and multiple case-studies that consider predefined performance measures, including their outcomes in selected construction investment projects. The suggested framework for creating a pattern from which to predict the project success is based on actual construction investment projects and is thoroughly analyzed using a multiple-case-study method followed by semi-structured interviews with identified stakeholders. The study draws attention to the significance of multi-stakeholder factors on performance measurement and project success, even if the contractor is a single company.

The work by Ruá et al. (Contribution 7) evaluates the viability of different roof rehabilitation systems and identifies the best one based on a multi-criteria analysis, which comprises environmental, economic, and performance factors. Simulation tools are applied to assess the energy savings, payback periods, and environmental effects for a selected building in the survey area. The obtained results are mapped to a neighborhood level. The research outcomes highlight the value of considering parameters like weight, cost, and user preferences when choosing appropriate solutions for refurbishment. In addition to assessing the potential energy savings and carbon emission reductions in the area, the findings also emphasize the importance of roof refurbishments for extending a building's life span, thus contributing to sustainable construction.

The report by Roux et al. (Contribution 8) concentrates on geopolymer (GP) compositions designed for structural use with the objective of minimizing the environmental impact. A life cycle assessment model for metakaolin– and potassium–silicate-solution-based GP mortars is developed. The model is applied to optimize the GP matrix as well as the granular skeleton, beginning from an existing formulation, throughout the decrease in greenhouse gas (GHG) emissions. The findings demonstrate that 3D-printed GP formulations do not yet constitute a mature technology, and their short-term applications are not inevitably environmentally favorable. Despite that, GPs contain the potential to reduce GHG emissions with conceivable innovations. The material designed shows adequate extrudability and buildability such that it could be utilized in high-performance applications owing to its high compressive strength. Regardless of its low share of aggregates, the dealt material holds an environmental value since it contains a substantial portion of earth that is broadly at hand and frequently also treated as waste.

The text by Simón-Portela et al. (Contribution 9) deals with the optimization of glulam roof structures, aiming to enhance sustainable construction. The study assesses the impact of glulam strength classes on the design of roof structures made of timber double-tapered beams and purlins, taking into account the requirements of Eurocode 5. A GA-based tool is developed for the purpose of optimization, and based on the results, various equations are suggested to identify the optimal geometry of structural elements, including their spatial configuration given the roof length, span, snow load, and strength class. In addition, a general equation is derived to anticipate the optimal volume needed for the roof structure, bearing in mind various strength classes, which may promote the efficient usage of resources and economic benefits while meeting structural and safety requirements.

The writing by Pennacchia et al. (Contribution 10) gives a catalog of optimized and sustainable solutions for strengthening the energy efficiency of the most-common types of roofs that distinguish the national residential building's heritage, focusing particularly on the principles of standardization and prefabrication. The methodological approach includes the identification, study, and classification of roofings according to their construction period.

Standardized optimal solutions are identified in terms of the essential energy retrofitting of deteriorated residential building stock. The performance assessments of the obtained results allowed for the implementation of a matrix, which may be a valuable support for designers in selecting optimal solutions for the building heritage in question based on energy efficiency and sustainability criteria.

Finally, the treatise by Villar-García et al. (Contribution 11) concerns both the static and kinetic friction coefficients of chestnut timber. The purpose of the research is to spread the application of Castanea sativa in the design of structures involving frictional forces, encouraging construction sustainability with the use of less-exploited materials, which entails the diversification of the species utilized in construction, thereby mitigating the demand for those more commonly exploited. The experimental program considers the orthotropic nature of material by assessing different wood orientations, engaging both wooden frictional pairs and wood against a steel plate. The effect of the moisture content is also taken into account in this study. The outcomes of the experimental program form a comprehensive database that is useful as an input for engineering optimizations, which may then be reflected in the more-prudent use of the considered natural resource.

3. Conclusions

The present Special Issue of *Sustainability* addresses various pressing topics related to sustainable construction and demonstrates diverse optimization, multi-criteria decision-making, and experimental approaches that represent the best available solutions to actual engineering problems occurring in the field. Established authors of reputable research institutions from all over the world—from Croatia, the Czech Republic, France, Italy, the Republic of Korea, Slovenia, Spain, Turkey, and the USA—responded to the invitation and contributed their studies, giving the content of this Special Issue its particular value. From the foundation built by the contributing authors, this Special Issue is intended to promote and disseminate the knowledge required to advance sustainable construction practices.

Author Contributions: Conceptualization, U.K. and T.Ž.; methodology, U.K. and T.Ž.; validation, U.K. and T.Ž.; writing—original draft preparation, U.K. and T.Ž.; writing—review and editing, U.K. and T.Ž.; funding acquisition, U.K. and T.Ž. All authors have read and agreed to the published version of the manuscript.

Funding: The authors acknowledge the financial support from the Slovenian Research and Innovation Agency (Research Core Funding No. P2-0129).

Conflicts of Interest: The authors declare no conflicts of interest.

List of Contributions

1. Rogulj, K.; Jajac, N.; Batinić, K. Flat Roofs Renovation Planning on Public Buildings Using Fuzzy Multi-Criteria Analysis. *Sustainability* **2023**, *15*, 6280.
2. Premrov, M.; Kozem Šilih, E. Numerical Analysis of the Racking Behaviour of Multi-Storey Timber-Framed Buildings Considering Load-Bearing Function of Double-Skin Façade Elements. *Sustainability* **2023**, *15*, 6379.
3. Bekdaş, G.; Cakiroglu, C.; Kim, S.; Geem, Z.W. Optimal Dimensions of Post-Tensioned Concrete Cylindrical Walls Using Harmony Search and Ensemble Learning with SHAP. *Sustainability* **2023**, *15*, 7890.
4. Jelušič, P.; Žula, T. Sustainable Design of Circular Reinforced Concrete Column Sections via Multi-Objective Optimization. *Sustainability* **2023**, *15*, 11689.
5. Jelušič, P.; Gücek, S.; Žlender, B.; Gürer, C.; Varga, R.; Bračko, T.; Taciroğlu, M.V.; Korkmaz, B.E.; Yarcı, Ş.; Macuh, B. Potential of Using Waste Materials in Flexible Pavement Structures Identified by Optimization Design Approach. *Sustainability* **2023**, *15*, 13141.
6. Galjanić, K.; Marović, I.; Hanak, T. Performance Measurement Framework for Prediction and Management of Construction Investments. *Sustainability* **2023**, *15*, 13617.
7. Ruá, M.J.; Pitarch, Á.M.; Arín, I.; Reig, L. A Roof Refurbishment Strategy to Improve the Sustainability of Building Stock: A Case Study. *Sustainability* **2024**, *16*, 2028.

8. Roux, C.; Archez, J.; Le Gall, C.; Saadé, M.; Féraille, A.; Caron, J.-F. Towards Sustainable Material: Optimizing Geopolymer Mortar Formulations for 3D Printing: A Life Cycle Assessment Approach. *Sustainability* **2024**, *16*, 3328
9. Simón-Portela, M.; Villar-García, J.R.; Vidal-López, P.; Rodríguez-Robles, D. Enhancing Sustainable Construction: Optimization Tool for Glulam Roof Structures According to Eurocode 5. *Sustainability* **2024**, *16*, 3514
10. Pennacchia, E.; Romeo, C.; Zylka, C. Towards High-Efficiency Buildings for Sustainable Energy Transition: Standardized Prefabricated Solutions for Roof Retrofitting. *Sustainability* **2024**, *16*, 3850.
11. Villar-García, J.R.; Moya Ignacio, M.; Vidal-López, P.; Rodríguez-Robles, D. Frictional Behavior of Chestnut (*Castanea sativa* Mill.) Sawn Timber for Carpentry and Mechanical Joints in Service Class 2. *Sustainability* **2024**, *16*, 3886.

sustainability MDPI

Article

Flat Roofs Renovation Planning on Public Buildings Using Fuzzy Multi-Criteria Analysis

Katarina Rogulj *, **Nikša Jajac** and **Katja Batinić**

Faculty of Civil Engineering, Architecture and Geodesy, University of Split, 21000 Split, Croatia; njajac@gradst.hr (N.J.)
* Correspondence: katarina.rogulj@gradst.hr

Abstract: Renovation of flat roofs typically involves repairing or replacing the existing roof to improve its performance and extend its lifespan. The renovation process may include a range of tasks depending on the condition of the roof, such as repairing leaks, replacing damaged or deteriorated materials, adding insulation, or upgrading drainage systems. This research aim was to establish the priority of renovation of flat roofs of the public building based on the principles of multi-criteria analysis and fuzzy set theory, using the multi-criteria method PROMETHEE II in fuzzy logic form (F-PROMETHEE II). The proposed approach is adequate due to its ability to transform the uncertain and vague information received from an expert into a fuzzy number. This way, the objective outcome can be obtained, the criteria conflict removed and the alternatives ranking and mutual comparison enabled. It was necessary to analyze the existing literature, the flat roofs of a public building in terms of their current condition, and define the main goals and criteria for the roof renovation project. Based on the defined goals and criteria, the roofs are evaluated and ranked according to the priority for renovation. The planning process of renovation of flat roofs was carried out specifically on the building of the Faculty of Civil Engineering, Architecture and Geodesy in Split.

Keywords: renovation planning; flat roof; multi-criteria analysis; fuzzy logic; F-PROMETHEE II

Citation: Rogulj, K.; Jajac, N.; Batinić, K. Flat Roofs Renovation Planning on Public Buildings Using Fuzzy Multi-Criteria Analysis. *Sustainability* **2023**, *15*, 6280. https://doi.org/10.3390/su15076280

Academic Editors: Uroš Klanšek and Tomaž Žula

Received: 10 March 2023
Revised: 1 April 2023
Accepted: 4 April 2023
Published: 6 April 2023

1. Introduction

Roofs are one of the most important elements of any building, which are exposed to all weather conditions and difficulties. The condition of the roofs, whether they are old, dilapidated or renovated, tells us about the maintenance of the building and, thus, also about the users of that building. A proper and high-quality roof is of great importance for the building because it provides safety and comfort to users and protects against external climatic conditions and temperature differences. Along with the correct way of performing the roof, its maintenance is also important.

Flat roofs are subject to wear and gradual loss of protective properties and are particularly prone to this type of damage due to their geometry and use as coverings for some materials. The first minor defects may appear several years after the operation. Damage can be removed with timely roof repair. Simple measures to restore the roof will help to extend its life, and what is important, will not require higher costs.

The need for renovation of a flat roof depends on a variety of factors, such as the age of the roof, the materials used in its construction, the quality of the initial installation, the level of maintenance it has received, and the climate it is exposed to, among others. If a flat roof has been well-maintained, constructed with quality materials, and is relatively new, then it may not require renovation. However, if a flat roof has been poorly maintained, has suffered damage due to weather or other external factors, or is approaching the end of its lifespan, then renovation may be necessary. Renovation of a flat roof may involve repairing or replacing damaged areas, reinforcing the structure to improve its load-bearing capacity, and improving the waterproofing system to prevent leaks and water damage. It may also involve upgrading the insulation to improve energy efficiency and reduce heating and cooling costs. In general, it is important to regularly inspect and maintain flat roofs to detect any problems early on and prevent the need for costly renovations.

Flat roof renovation typically involves a series of phases to repair, restore or replace the roof system to improve its structural integrity and waterproofing properties. The specific steps involved in flat roof renovation will depend on the extent of the damage or deterioration and the type of flat roof system in place. Flat roof renovation can be a complex process that requires the expertise of experienced professionals. It is important to choose a reputable contractor who can provide a comprehensive renovation plan and deliver high-quality workmanship.

When renovating roofs, it is important to take care of finances. It may involve significant funds and project and technical documentation preparation. From all this, it can be seen that it is a complex problem that needs to be looked at from several angles. In addition to concluding that this assessment should be done using several methods of evaluation. So, it is a multi-criteria problem. The need to include important stakeholders in the renovation planning process also contributes to this, as they have different views on the planning process itself, i.e., the priorities and importance of the criteria, and thus the final ranking, may be different for them. When it comes to flat roof renovation decision-making, a multi-criteria approach can support and assist building owners or facility managers to make informed decisions based on a range of factors beyond just cost or aesthetics. This can lead to more informed decisions that take into account the full range of factors that can impact the performance, durability, and sustainability of a flat roof system.

This research analyzes the renovation of flat roofs of public buildings, i.e., priorities are established for the undertaking of flat roof renovation activities, specifically on the example of the building of the Faculty of Civil Engineering, Architecture and Geodesy in Split. In order to achieve the stated goal, an approach of multi-criteria analysis based on fuzzy set theory and using the PROMETHEE II method in the fuzzy logic form (F-PROMETHEE II) is carried out. The main advantage of the approach is that the uncertain and vogue information received from stakeholders or an expert can be transformed into fuzzy numbers and thus obtain an objective outcome. Using fuzzy multi-criteria analysis, the criteria conflict can be removed, enabling alternative ranking and mutual comparison. Defuzzification of net flows can easily be calculated using the centroid method. Preference usual and linear functions are used to define relationships between alternatives according to the selected criteria. Euclidian distance is proposed to use to determine the distance between alternatives.

The proposed approach is considered relevant and effective because it takes into account several aspects of the problem analysis. For the purposes of the mentioned method, it is necessary to establish a set of criteria, identify relevant stakeholders and determine the weights of the criteria that express the attitude of the involved stakeholders and their compromise attitude.

Various studies used multi-criteria analysis to solve problems with the repairs or renovations of buildings or their elements, among which are different types of roofs. Miniotaite [1] investigated thermal and technological methods of renovation alternatives of flat roofs using multi-criteria analysis, considering used materials, the structure of the roof and roofing technologies. Gagliano et al. [2], in their study, developed an analysis of comparison of three typologies of the roof, standard, cool and green roof. The roofs were numerically compared by their environmental and energy capabilities. Petrakova and Grznar [3] used multi-criteria decision-making to find an optimal solution for roof envelope repair. Goncalves et al. [4] investigated waterproofing options applied on the flat roofs. The life cycle of each option is compared regarding economic and environmental impact, using data from various European and generic databases. Aguacil et al. [5] presented a multi-criteria assessment of different scenarios of photovoltaic options in the renovation of buildings to achieve savings in energy costs and materials and reduce fossil fuels and greenhouse gas emissions. The renovations are also applied to the roofs of the buildings.

Rosasco and Perini [6] developed a sustainability approach based on the multi-criteria analysis for the selection of green roof systems, considering the parameters that have an influence on the experts' and decision makers' decisions. Kuznecov and Šaparauskas [7] studied sloping-pitched roofs regarding renovation and reconstruction technologies and the efficiency and applicability of each technology. Firstly, the main characteristics of each roof are analyzed, as well as damaged and defective causes of roofs, and then suggestions and routines for roofs' performance improvement are proposed. All mentioned was achieved by distinguishing three technologies of reconstruction. These technologies were mutually compared by multi-criteria methods, additive radio assessment and complex proportional assessment. Using a multi-criteria approach shows that polyurethane foam is the best selection for the reconstruction of a roof. Kalibatas and Kovalitis [8] used the multi-criteria method SAW and three game theory rules to select the most appropriate and effective membrane for the analyzed type of roof. Golić, Kosorić and Furundžić [9] proposed an approach for the renovation of the residential building by integrating solar water heating systems. The approach is applied through a few stages to extenuate problem-solving and to provide optimization of the different designs. Studying all the stages, the multi-criteria analysis is shown to be the most appropriate methodology to obtain a compromised ranking after the assessment of design alternatives and for selecting the best solar water heating system. Krstić Furundžić, Kosorić and Golić [10], in their other study, presented all the advantages of solar water heating systems regarding fuel oil-, gas- and electricity-based heating systems. Considering that solar thermal collectors were proposed and analyzed, and using a multi-criteria approach, collectors' alternatives were evaluated and mutually compared to obtain their priority ranking. For the comparison process, various aspects are discussed, such are ecological, functional, economic, technical and esthetic aspects. Sangkakool et al. [11] defined and analyzed the main aspects that affect the adoption of green roofs by using the mixed method design. These aspects are the definition of qualitative analysis, structuring by internal and external factors, and numerical evaluation using the analytical hierarchy process under experts' opinions. Turskis, Morkunaite and Kutut [12] developed a hybrid model to prioritize the ranking of heritage buildings according to the activity of renovation, considering the main elements of the buildings, such are facades and roofs. The model is based on two multi-criteria decision-making methods analytic hierarchy process and evaluation based on distance from average solution, identifying relevant criteria concerning the cultural heritage buildings renovation. Naing, Nitivattananon and Shipin [13] presented a methodology for green roof retrofitting on the campus of the Asian Institute of Technology. Using multi-criteria decision making, visual inspection, questionnaires survey literature review as inputs, evaluation of potential alternatives and their ranking were generated. Gonzales-Dominguez, Sanchez-Barroso and Garcia-Sanz-Calcedo [14], in their research, proposed an optimization process of the flat roofs maintenance periodicity. The process is applied in twelve hospitals in Spain to increase the usefulness, reliability and stability of the flat roofs. The Markov Chain approach is used, and three types of flat roofs were analyzed: elastomeric, bitumen and PVC. Wu et al. [15] developed a methodology of fusing fuzzy logic, multi-criteria analysis and multi-objective programming to select the best large-scale rooftop photovoltaic portfolio project. Furthermore, the criteria weights are defined using an analytic hierarchy process, and the PROMETHEE II method is applied to rank the large-scale rooftop photovoltaic portfolio alternatives.

The core problems addressed in the studies listed are related to flat roof renovation and the selection of the most appropriate renovation technologies, waterproofing membranes, and roof systems. Other core problems include the environmental and economic impact of roofing solutions, potential energy savings, CO_2 emissions reduction, and integration of solar water heating systems in building refurbishment. In addition, some papers address the prospect of green roofs in urban areas, multi-criteria decision-making in the choice of alternative solutions, and preventive maintenance optimization of accessible flat roofs in healthcare centers. Overall, the papers aim to provide innovative solutions to address the core problems in flat roof renovation, which include sustainability, cost-effectiveness, energy efficiency, and environmental impact. Examining these studies, the proposed approach has several differences and main contributions compared with previous studies. None of the studies considered the problem in a holistic manner, taking into account simultaneously technological, economic, environmental and sociological aspects of the problem to achieve the main goal. Moreover, the approach presents the multi-criteria analysis fused with fuzzy set theory, applied in the assessment process of the flat roofs renovation priority. To be precise, the fuzzy PROMETHEE II method, shown to be the most suitable methodology for problems with a high degree of uncertainty and complicatedness among the criteria, is used. The proposed methodology is applied to flat roof renovation planning to prove the applicability and efficiency of the participatory decision-making framework, which was not the case in other studies.

As for the fuzzy PROMETHEE II method, there were applications on various case studies and research such are: tourism environmental impact [16], supplier performance evaluation [17], selection of inland nuclear power plant [18], evaluation of solar power plant location [19], exploration of the landslide susceptibility mapping [20], evaluation and ranking of alternative energy exploitation schemes of a low temperature geothermal field [21], concepts assessment and ranking for a new product development process [22], machine tools selection for the company's productivity and capabilities [23], quality of the passenger services in the railway [24], improvement of the emergency department and hospital resources [25], selection and ranking of the performance indicators for municipal solid waste management systems [26], evaluation of the power plants for the investment projects [27], evaluation of the irrigation methods for the two plains for cotton production in Turkey [28], assessment of the nuclear medicine imaging devices [29], evaluation of the outsourcing for the information systems field [30], risk ranking for the highway construction projects [31], evaluation of the solution for the cancer treatment [32] and selection of the facility location of new organization [33].

2. Materials and Methods

This section gives more about flat roofs, their main characteristics, types and renovation. Moreover, the fuzzy multi-criteria analysis is explained, as well as criteria for the evaluation of the flat roofs.

2.1. Flat Roofs

A flat roof is a structural element that consists of a load-bearing structure, a covering and a roof soffit. Roofs, especially flat roofs, are among the most critical parts of the building. The load-bearing part of the flat roof is, therefore, constructed as a mezzanine structure. The roof must protect the building from external influences such as precipitation, heat, cold and wind. With proper execution and the use of quality materials, a flat roof is an economical construction that provides significantly greater freedom in composing volume and space [34]. The external influences are water, snow, groats, sudden summer temperature changes (after storms), very high temperatures in summer, very low winter temperatures, sound penetration, moisture that occurs during construction, atmospheric water that arrives during construction, and during the process of binding and drying different materials, small cracks appear in the materials. The internal influences are desirable constant heat inside the rooms, large temperature differences inside/outside, increased indoor air humidity due to use, etc. [35]

The advantages of the flat roofs are lower price compared to a pitched roof, a flat roof has a smaller area, so the consumption of materials is lower. The installation and further maintenance are simplified, the arrangement of a flat construction is simpler than an inclined one, and it is easier to move along a horizontal plane than along an inclined one. There is a possibility of obtaining additional usable space; the surface of the flat roof can easily be used as a terrace, a walking platform, a garden with a lawn, etc. [35].

There are several types of roofs; they differ in layers, slope and purpose. Each one is specific in its own way, both in terms of appearance and functionality. Any type of flat roof must protect the structure from external influences. Every flat roof must be of high quality, mechanically resistant, waterproof, must have UV protection and protect the structure from condensation. Great importance is attached to the energy efficiency of the flat roof [34]. According to the slope, there are flat roofs with a small slope of 0.5–1%, with a normal slope of 1–2.5%, a higher slope of 4–15% and pitched roofs (15–40%). Types of flat roofs according to the arrangement of layers are classic (cold), reverse (warm) and ventilating, while according to purpose or accessibility, they are passable, impassable and green (intensive or extensive). Types of flat roofs according to the method of rainwater drainage are drainage of rainwater along the outer edges and drainage of stormwater using drains.

The function of the insulated roof structure primarily includes the protection of the structure and the space under the roof from climatic influences and the optimization of the energy efficiency of the space under the roof. The main factors when designing a roof are primarily physical construction. Then there are economic and environmental factors, of course, both in the function of the desired architectural solutions.

Among the most important construction physical influences on flat roofs, on which technical solutions depend, are considered [35]:

- Fire—in the event of a fire, the roof structure is the first structural system that is exposed to extremely high temperatures. The use of non-combustible materials prevents the spread of fire and the release of harmful gases or burning dripping particles, which are often the cause of numerous deaths.
- Temperature—temperatures on the surface of flat roofs, depending on the season and the geographical location of the building, can range from −25 °C to +75 °C. Often, in certain weather conditions, the temperature range in one day exceeds 80 °C, which directly affects the final layer or roof membrane. This certainly affects the layer of thermal insulation and the mode of passage of water vapor through the roof structure. By choosing an adequate thickness of flat roof thermal insulation, we reduce the building's heat losses in the winter and prevent overheating of the premises in the summer. It is recommended that the thermal insulation does not change its dimensions with changes in temperature because, in this way, unwanted stretching is caused, which thereby burdens the roof membrane or causes unwanted thermal bridges at the contact of the insulation panels.
- Wind—wind loads at tall buildings are definitely significantly higher than at low ones. With the static calculation of the roof structure, we must also calculate the impact of wind load, which directly determines the method of attachment and the arrangement of dowels for attaching the thermal insulation and the final layer of the roof to the load-bearing structure. An important factor, in this case, is the weight of the thermal insulation layer itself, which is affected by the negative wind pressure. In this case, heavier materials, such as stone wool, have an advantage over light organic insulation.
- Mechanical load—static loads of a flat roof, such as the weight of its own structure and the weight of snow, affect the dimensioning of the load-bearing structure of the roof, while the dimensioning of the thermal insulation layer is affected by occasional mechanical loads of the roof during the installation phase or later maintenance of the roof surface. The key quality factor is not only the surface load-bearing capacity of the insulation layer but also the resistance to individual point loads. They occur at the places of mechanical fixing of the roof foil and also when walking on the roof. Occasional walking on the roof cannot be avoided because sometimes the roof needs to be maintained, and there may be subsequent works that require temporary use of that surface.
- Moisture and vapor pressure—rain, snow and ice directly load the roof membrane from the outside. In addition, moisture also stresses the roof structure from the inside. The mode of vapor passage through the building envelope must be such that moisture

does not stop in the thermal insulation layer. Excessive ingress of moisture into the thermal insulation layer of a flat roof is prevented by installing a vapor barrier before the insulation layer on the inside.

- Noise—when designing buildings in an urban environment, we are often faced with the requirement of lowering the noise level from the surroundings to the interior of the building. Stone wool, with its interwoven structure of stone fibers, absorbs noise and vibrations from the environment much better than insulation materials of organic origin. Sound energy is broken down in the fibrous structure of stone wool and converted into other forms of energy.

2.2. Maintenance of Flat Roof

Maintaining a flat roof is an important segment of roof and building durability. The maintenance of the flat roof, and especially the maintenance of the waterproofing protection, should be approached immediately at the beginning of the exploitation of the flat roof and the building. The reason for this is to remove possible hidden defects that occurred during the execution of works and appeared during exploitation. Localized deformations and damages should be removed immediately after they appear.

In the process of maintaining a flat roof, the important elements that should be taken into account are as follows:

- not to stress the roof surface, more specifically, the waterproofing protection with unforeseen subsequent loads;
- not to damage and penetrate the waterproofing protection;
- not to use the roof for purposes for which it was not intended (drying laundry, sunbathing in the summer);
- to maintain and clean drains and bays (from leaves, paper, bags);
- when cleaning heavy snow so as not to damage the waterproofing protection.

The above procedures and actions can and should be performed only with the knowledge and in the presence of an expert who knows how to evaluate and decide on actions that are planned, that is, that are undertaken on the roof. Flat roof maintenance also includes smaller repairs on the roof. Such repairs should be carried out immediately after they are noticed. If they are not removed as soon as possible, i.e., not immediately, they can cause new major deformations and damage. Then, their removal will be of greater scope, both physical and financial. Such minor procedures that should be carried out immediately are as follows:

- repair of separate places of putty on the eaves sheets;
- repair of possible wind damage on the eaves sheets;
- repair of the drainage slits between the sheets;
- if the wind has moved the layer of gravel, it should be evenly distributed again;
- if the light protection on impenetrable roofs is washed or shed, repaint the surface with suitable paint for roofs.

In addition to those mentioned, other, smaller alterations should be carried out, which aim to extend the durability of the waterproofing protection and the flat roof. All these works should be performed by a professional contractor specializing in this type of work. A regular inspection should be performed on a flat roof twice a year. In this way, even for a long time, the occurrence of further harmful phenomena will be prevented.

2.3. Renovation of Flat Roof

The condition of roofs depends on the wear and tear of the cover, the type of cover, maintenance of the cover over time and other influences. Roofs must be properly maintained, which prolongs their life. Periodic inspections of the roofs are definitely recommended, and if it is necessary to clean the roofs of leaves and other impurities through annual plans and programs, depending on the needs of the building itself. Bigger problems on roofs are caused by accumulated dirt in drainage pipes or gutters.

Any reconstruction of a flat roof begins with identifying the problem. In most cases, the roof leaks, and it is precisely this circumstance that leads to the realization of the need for repair. The roof itself has three main functions: insulation, waterproofing and aesthetics.

- Aesthetics—in this case, classic general construction or repair work is carried out related to the replacement of the existing finish coating or the decoration of the parapet or the installation of equipment, etc. In the event of a violation of the thermal or waterproofing of a flat roof, the repair will be significantly more difficult both from an engineering point of view and with regard to the intensity of work and, as a rule, considering the price of work.
- Repair of thermal insulation—a roof leak is not always related to a breach of waterproofing. For example, the roof can simply "float" if the vapor barrier is not installed. A related but different case is insufficient roof insulation. In this case, the dew point is not located in the insulation, as it should be, but directly on the floor plate, thus causing abundant condensation. The recipe for "healing" this roof is quite effective—it needs to be re-insulated. If this is not feasible from above due to the already finished finish or for other reasons, it is possible to do it from the inside. To do this, it is necessary to attach an additional layer of insulation under the floorboard to the ceiling and install a vapor barrier. The simplest option is something made of self-adhesive tape. After that, the thickness of the insulation will be sufficient; the dew point will be in the insulation, a vapor barrier will be created and the roof will stop floating. Again, it is unprofitable to do it from the top because of the roof disassembly, but to do it from the bottom, under the floorboard, is quite simple and convenient.
- Repair of waterproofing—the most difficult case from the point of view of repairing a flat forge is a violation of waterproofing (its damage or leakage due to natural wear). If the roof waterproofing is open, that is, if the final layer is not laid on the roof, the repair is quite simple—first of all, you need to find holes and patch them. It is recommended to repair a flat roof using modern waterproofing materials. It is necessary to make waterproofing over the entire surface of the waterproofing because if it is a bitumen roof that has really lasted its term, then it has been destroyed in many places. If stone wool insulation was used and it got wet, in that case, the roof would have to be completely dismantled, the insulation replaced and a new waterproofing layer installed. In the event that the waterproofing is covered with gravel, tiles or some other finish, the repair will be a rather difficult task, as the search for damage may take a long time. In this case, the works are carried out with the complete removal of the roof, and depending on the condition of the waterproofing and insulation, either their replacement is performed or the waterproofing is repaired.
- Repair of local damage (cuts, punctures)—mechanical damage can occur as a result of snow removal in winter, city encroachment and human movement. Other potential reasons are base deformation and initial imperfections allowed during coating application. Local defects are removed by placing a patch on the damaged area. The patch should be made of the same material as the base cover. Its edges are usually rounded to reduce the probability of laying, and the size is made to cover the defect by 10–15 cm in all directions.

Therefore, repairing a flat roof is a complex and expensive task, usually associated with its disassembly completely or in separate sections. All linings and roof coverings are not eternal; even the most modern of them are subject to wear and gradual loss of protective properties. Under the influence of precipitation, temperature changes and mechanical factors, the roofing material collapses, cracks, bubbles, defects, etc., appear on it, and then leakage is just one step further.

2.4. Fuzzy Multi-Criteria Analysis

Multi-criteria analysis refers to structuring, planning and problem-solving. The analysis is approached only if it is determined with certainty that the problem is characterized by a number of alternative solutions. Therefore, if a multi-criteria analysis is used to solve the problem, it is necessary to define all possible solutions to the problem. The aim of the multi-criteria analysis is to support the decision maker when there is a wide choice of alternatives when solving a problem. When using multi-criteria decision-making, it is necessary to use the personal wishes of the decision-maker. A problem for solution may have the best of the alternatives offered, the best alternative or a small set of good alternatives.

The characteristics of multi-criteria analysis are a large number of criteria, inconsistencies between criteria, incomparable measurement units of criteria, the choice of the best solution and the ranking of alternatives.

Fuzzy multi-criteria analysis combines multi-criteria methods with fuzzy logic, in this case, is PROMETHEE II method. Some of the aspects of fuzzy set theory are given in the following:

Definition 1. *[36] Let a be a triangular fuzzy number defined as* $a = (a_L, a_M, a_U)$*, where* $0 < a_L < a_M < a_U < 1$*, then the membership function can be written as*

$$f_a(x) = \begin{Bmatrix} \frac{x-a_L}{a_L-a_M} & a_L \leq x \leq a_M \\ \frac{a_U-x}{a_U-a_M} & a_M \leq x \leq a_U \\ 0 & else \end{Bmatrix} \quad (1)$$

2.5. The Approach for Flat Roofs Ranking Using F-PROMETHEE II Method

In this research, the PORMETHEE II method based on fuzzy set theory is used for ranking flat roofs for the activity of renovation. Fusing fuzzy theory with multi-criteria methods has proven to be more effective in numerous studies than traditional methods when dealing with uncertain and vague data and information. Compared to traditional decision-making methods, the F-PROMETHEE II approach presented here has several advantages and innovations, such as

1. The fuzzy logic component of the method allows for the handling of imprecise and uncertain data. This is particularly useful when data are incomplete or subjective, as the fuzzy logic can provide a more nuanced evaluation of different options.
2. The multi-criteria method allows for the evaluation of multiple criteria simultaneously, which can provide a more comprehensive evaluation of different options. The fuzzy logic component of the method allows the inclusion of subjective or qualitative criteria, which may not be easily quantifiable.
3. The F-PROMETHEE II method provides a clear and easy-to-interpret ranking of different options. The use of decision matrices and preference functions can help stakeholders understand and compare the relative strengths and weaknesses of different options.
4. The method is flexible and can be adapted to different decision-making contexts and criteria. This makes it useful in a wide range of applications, including flat roof renovation decision-making.

Overall, the biggest innovation and advantage of the F-PROMETHEE II method is its ability to handle imprecise and uncertain data while incorporating multiple criteria and providing a clear and robust outcome of different options. This can lead to more informed and transparent decision-making processes, particularly when dealing with incomplete or subjective data.

The F-PROMETHEE method is composed of three phases. In the first phase, stakeholders are selected and gathered together to discuss the problem and generate alternatives. In the second phase, adequate criteria are defined and their weights are assigned, while in the third phase, alternatives are evaluated according to criteria. The approach for flat roofs ranking using the F-PROMETHEE II method is presented in Figure 1 and in the following steps:

Step 1. Gathering of DMs and defining their importance.

Let D be a set of decision-makers, where $D = \{DM_1, \ldots, DM_l\}$, then the weight of each DM $w_{DM} = \{w_{DM_1}, \ldots, w_{DM_l}\}$.

Step 2. Definition of the linguistic variables and their triangular fuzzy numbers.

Linguistic variables in the form of scale are defined with their triangular fuzzy numbers. These variables are used to define criteria weights and evaluations of the alternatives.

Step 3. Fuzzy criteria weights determination.

Each DM defined criteria weights using linguistic variables and their triangular fuzzy numbers. Aggregated weights are calculated as follows:

$$a_L = \frac{1}{n}\sum_{n=1}^{n} a_{L_n}, a_M = \frac{1}{n}\sum_{n=1}^{n} a_{M_n}, a_U = \frac{1}{n}\sum_{n=1}^{n} a_{U_n} \quad (2)$$

Step 4. Evaluation of the alternatives using linguistic values and triangular fuzzy numbers.

Alternatives are evaluated according to each criterion by each DM, then the aggregation of the alternatives evaluation is calculated. Finally, the normalized fuzzy decision matrix is determined as follows [37]:

For maximized criteria:

$$x_{ij} = \frac{a_{L-j}}{a_{U_{ij}}}, x_{ij} = \frac{a_{L-j}}{a_{M_{ij}}}, x_{ij} = \frac{a_{L-j}}{a_{L_{ij}}} j \in E\ a_{L-j} = \min_i a_{L_{ij}} \quad (3)$$

For minimized criteria:

$$x_{ij} = \frac{a_{L_{ij}}}{a_{U^+{}_j}}, x_{ij} = \frac{a_{M_{ij}}}{a_{U^+{}_j}}, x_{ij} = \frac{a_{U_{ij}}}{a_{U^+{}_j}} j \in F\ a_{U^+{}_j} \max_i a_{U_{ij}} \quad (4)$$

While normalized fuzzy decision matrix is

$$X = \left[x_{ij}\right]_{m \times n} i = 1, \ldots, m\ j = 1, \ldots, n \quad (5)$$

where $x_{ij} = \left(t_{ij}, v_{ij}, y_{ij}\right)$ is an evaluation by DMs based on jth criterion for ith alternative which. u_{ij}, v_{ij} and y_{ij} denote the low boundary, medium boundary and up boundary of the triangular fuzzy number, respectively.

Step 5. Assign the preference function $P_j = \left(x_{ij}, x_{kj}\right)$ for each criterion.

According to [38], there are six types of preference functions. Hereby, the authors, together with stakeholders, decided to use the usual type and linear priority type, as they were the most appropriate way to describe the type of criteria.

Usual preference function:

$$P(d) = \begin{cases} 1, & d > 0 \\ 0, & d \leq 0 \end{cases} \quad (6)$$

Linear priority preference function:

$$P(d) = \begin{cases} 1, & d > p \\ \frac{d}{p}, & d \leq p \end{cases} \quad (7)$$

where p is the threshold value between the indifferent and strict preference areas, and d presents the Euclidean distance [37] between alternatives based on jth criterion. The value of d is expressed for the maximized and minimized criteria as follows:

For maximized criteria:

$$d(x_{ij}, x_{kj}) = \begin{cases} \sqrt{\frac{(t_{ij}-t_{kj})^2+(v_{ij}-v_{kj})^2+(y_{ij}-y_{kj})^2}{2}}, & x_{ij} > x_{kj} \\ 0, & else \end{cases} \quad (8)$$

For minimized criteria:

$$d(x_{ij}, x_{kj}) = \begin{cases} \sqrt{\frac{(t_{ij}-t_{kj})^2+(v_{ij}-v_{kj})^2+(y_{ij}-y_{kj})^2}{2}}, & x_{ij} < x_{kj} \\ 0, & else \end{cases} \tag{9}$$

where $i \neq k$ and $i, k = 1, 2, \ldots, m$.

Step 6. Preference index calculation.

By using the preference index, it is proved that the degree of the ith alternative is higher than the jth alternative, which indicates the rank of the ith alternative. The index is expressed as follows:

$$\pi(x_i, x_k) = \sum_{j=1}^{n} w_j \times P_j\left(x_{ij}, x_{kj}\right) = \left(t_{ij}^{\pi}, v_{ij}^{\pi}, y_{ij}^{\pi}\right)\ i \neq k \text{ and } i, k = 1, 2, \ldots, m \tag{10}$$

Moreover, the criteria weight is denoted with $w_{ij} = \{w_1, \ldots w_m\}$.

Step 7. Calculation of positive (leaving) and negative (entering) flow for each alternative.

The positive flow $\Phi^+(x_i)$ defines the strength of the ith alternative, while the entering flow $\Phi^-(x_i)$ measures its weakness. Following are the expressions of the positive and negative flows:

Positive flow:

$$\Phi^+(x_i) = \sum_{k=1}^{m} \pi(x_i, x_k) = \left(t_i^{\Phi^+}, v_i^{\Phi^+}, y_i^{\Phi^+}\right)\ i \neq k \text{ and } i, k = 1, 2, \ldots, m \tag{11}$$

Negative flow:

$$\Phi^-(x_i) = \sum_{k=1}^{m} \pi(x_i, x_k) = \left(t_i^{\Phi^-}, v_i^{\Phi^-}, y_i^{\Phi^-}\right)\ i \neq k \text{ and } i, k = 1, 2, \ldots, m \tag{12}$$

Step 8. Definition of the net flow for each alternative.

An alternative with a higher net flow will have a higher ranking. The net flow of ith alternative is expressed as

$$\Phi(x_i) = \Phi^+(x_i) - \Phi^-(x_i) = \left(t_i^{\Phi}, v_i^{\Phi}, y_i^{\Phi}\right)\ i = 1, 2, \ldots, m \tag{13}$$

Step 9. Net flow defuzzification for final ranking.

For the defuzzification of the alternatives evaluations defined with net flows, the centroid method is used as follows [39]:

$$\bar{\Phi}(x_i) = \frac{t_i^{\Phi} + v_i^{\Phi} + y_i^{\Phi}}{3}\ i = 1, 2, \ldots, m \tag{14}$$

If $\bar{\Phi}(x_1) > \bar{\Phi}(x_2)$, it means that alternative 1 is higher ranked than alternative 2.

2.6. Criteria Definition

Multi-criteria decision-making includes four components: alternatives definition, criteria determination, criteria weights assignment and decision maker's attitude. The decision-maker needs to rank the alternatives or choose between them, while criteria serve to evaluate and compare alternatives. In order to better carry out the assessment of the alternative, it is necessary to define measurement scales. By defining requirements for problem-solving, the goals of multi-criteria problems are established. In order for the goals to be defined as effectively as possible, the views of all stakeholders in the decision-making process must be taken into account. Furthermore, in Table 1, criteria are presented with a description, preference function and type regarding minimum or maximum. All criteria are defined by decision-makers, which were the project manager, construction engineer and user. Criteria C1, C2 and C3 are defined from the technical aspects of the problem, C4, C5 and C6 from the economical and C7, C8 and C9 from the ecological aspect. All stakeholders were gathered, and throughout the discussion, criteria were defined, considering the importance of all stakeholders equally. Criteria from C1 to C6 are defined with linear priority preference function, while criteria from C7 to C9 with usual. These functions were chosen as they were the most appropriate to describe each criterion preference.

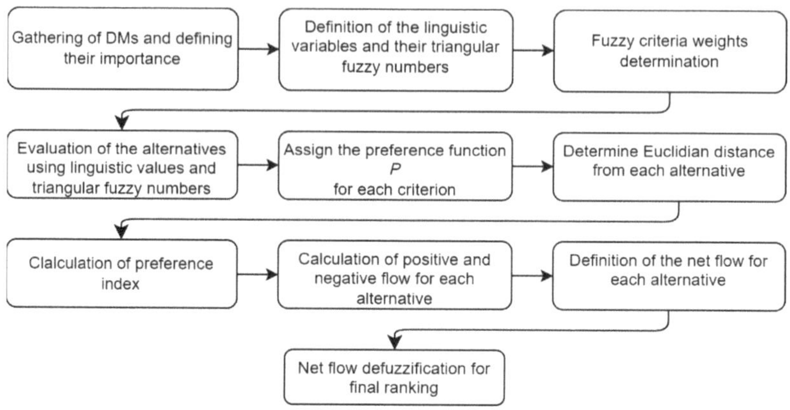

Figure 1. The approach for the flat roofs renovation ranking.

Table 1. Defined criteria, with preference functions, for renovation planning of flat roofs.

Label	Criteria Name	Description	Preference Function	Min/Max
C1	Condition of the roof	The current state of the roof is regarding mechanical stability and resistance.	Linear priority	min
C2	Selected material	Quality and durability of the material.	Linear priority	max
C3	The simplicity of renovation	It refers to the simplicity of roof renovation.	Linear	max
C4	Renovation cost	It refers to the price of materials for the roof construction and the costs of carrying out the work.	Linear priority	min
C5	Maintenance cost	Includes roof maintenance costs after renovation.	Linear priority	min
C6	Renovation duration	It includes the time required for the renovation of the roof structure and rainwater drainage.	Linear priority	min
C7	Thermal efficiency	Meeting the minimum requirements regarding the improvement of the thermal protection properties.	Usual	max
C8	Sound efficiency	Attenuation from noise and external sounds.	Usual	max
C9	Waterproofing	It includes protection against water penetration into the interior of the building.	Usual	max

3. Results

In this section, the effectiveness and applicability of the defined criteria and F-PROMETHEE II method are verified in the assessment of the flat roofs for the activity of renovation.

3.1. Case Study

The FCEAG's building consists of three interconnected parts. Firstly, they were constructed separately, but today they represent a physical unit, sharing a common entrance, communication areas and other service areas. The building serves as a higher education institution. The analyzed flat roofs are located on three construction plots of the building, meaning building A at lot no. 6474/4, building B at lot no. 6474/8, and building C, at lot no. 6474/5. Building A consists of a basement, ground floor, three floors, roof and superstructure. Building B consists of a basement, ground floor, five floors, roof and superstructure. Building C consists of a basement, a ground floor, three floors and a roof. The roofs are presented in Figure 2.

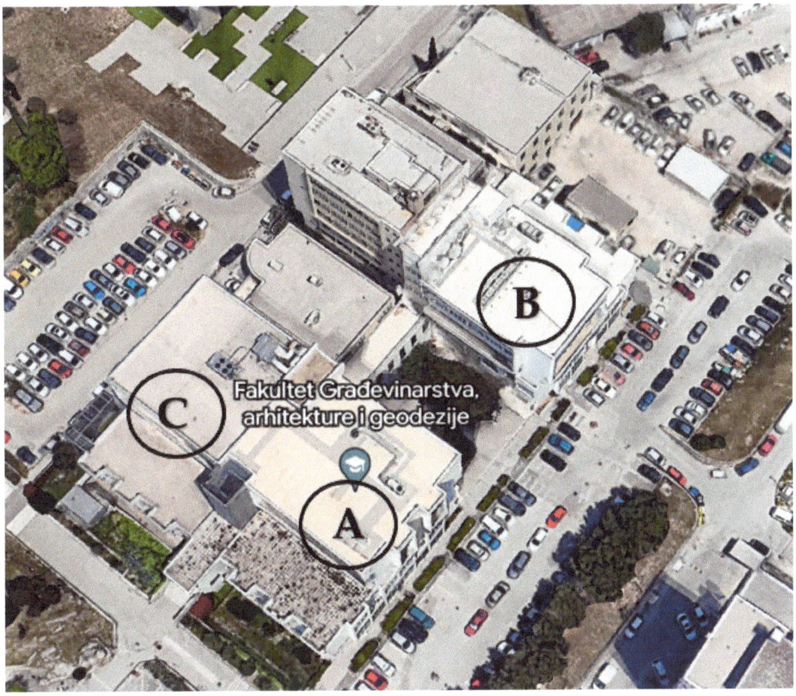

Figure 2. The FCEAG's building with A, B and C buildings.

Building A was completed in 1976, and building B in 1978. The first of the reconstructions was in 2005 when building C was completed. After the construction of building C, the outer envelope of building A and building B was reconstructed, including all the outer walls, i.e., facades, flat roofs and openings. The construction of the building consists of a system of load-bearing reinforced concrete beams and monolithic reinforced concrete slabs. The roof is solved as a flat structure, and the facades are in good condition. The roof of building A, located above the third floor, was renovated in 2021 when old layers were completely removed with new ones. The roof of building A above the first floor has not been renovated since the construction of the building in 1976, and considering that it has no thermal insulation at all, as determined from the existing project documentation, it does not meet the standards of thermal resistance of the building's outer envelope. Based on field inspection, the roof is in an unmaintained state. Indentations on the roof slab, cracked concrete slabs and weeds are visible.

Building B consists of two flat roofs on the first and fifth floors, both of which were renovated in 2019. Moreover, in building B old layers were removed and new ones were constructed on the existing concrete. Gravel was placed on the roof above the first floor, while concrete slabs were placed on the roof above the fifth floor.

The roof of building C was built in 2005 and has not been renovated since then. Although it is visibly in satisfactory condition, it does not meet the thermal resistance standards of the building's outer envelope and requires renovation [19].

Overall, seven roofs have been evaluated, which are as follows:

- building A: roof above the 1st flour (A1), roof A above the 4th flour (A2);
- building B: roof above the 1st flour (B1), roof B above the 5th flour (B2);

- roof above the corridor (H);
- building C: roof above the 1st flour (K1), and roof above the 4th flour (K2).

3.2. The Ranking of Flat Roofs

In this section, the ranking of seven flat roofs is obtained by the methodology presented in Section 2.4.

There are three decision makers (DM) with many years of experience in structural engineering. All three DMs are equally important, and for that reason, their weights are all defined with a crisp value of 1.

Furthermore, in Table 2, linguistic values with corresponding triangular fuzzy numbers are given. By using these values, DMs defined criteria weights and by Equation (2), aggregated fuzzy weight is calculated, as is presented in Table 3.

Table 2. The linguistic variables and corresponding fuzzy numbers.

Linguistic Variable	Fuzzy Numbers
Very Low/Poor (VL/VP)	(0.0, 0.0, 0.1)
Low/Poor (L/P)	(0.0, 0.1, 0.25)
Medium Low/Poor (ML/MP)	(0.1, 0.25, 0.4)
Fair (F)	(0.25, 0.4, 0.6)
Medium-High/Good (MH/MG)	(0.4, 0.6, 0.8)
High/Good (H/G)	(0.6, 0.8, 1.0)
Very High/Good (VH/VG)	(0.8, 1.0, 1.0)

Table 3. Criteria fuzzy weight and aggregated fuzzy weight.

Criteria	DM1	DM2	DM3	Aggregated Fuzzy Weight
C1	VH	VH	VH	(0.8, 9.97, 1.0)
C2	VH	H	H	(0.67, 0.87, 1.0)
C3	H	MH	H	(0.53, 0.73, 0.93)
C4	MH	F	MH	(0.35, 0.53, 0.73)
C5	MH	F	MH	(0.35, 0.53, 0.73)
C6	H	MH	H	(0.53, 0.73, 0.93)
C7	H	VH	H	(0.67, 0.87, 1.0)
C8	VH	H	VH	(0.73, 0.93, 1.0)
C9	VH	H	VH	(0.73, 0.93, 1.0)

Likewise, the evaluations of alternatives are given by DMs according to the values in Table 2, and aggregated fuzzy evaluations are calculated. The normalized fuzzy decision matrix is calculated using Equations (3) and (4), respectively. Table 4 presents the evaluation of each alternative according to each criterion by each DM, their aggregated evaluations and the normalized decision matrix.

Table 4. The evaluation of each alternative and aggregated fuzzy evaluation.

Criteria	Roof	DM1	DM2	DM3	Aggregated Fuzzy Evaluation	Normalized Fuzzy Decision Matrix
C1	A1	P	P	P	(0.1, 0.25, 0.4)	(0.25, 0.4, 1.0)
	A2	H	VH	H	(0.67, 0.87, 1.0)	(0.1, 0.12, 0.15)
	B1	H	VH	VH	(0.73, 0.93, 1.0)	(0.1, 0.11, 0.14)
	B2	H	VH	VH	(0.73, 0.93, 1.0)	(0.1, 0.11, 0.14)
	H	H	H	VH	(0.67, 0.87, 1.0)	(0.1, 0.12, 0.15)
	K1	F	H	MH	(0.42, 0.6, 0.8)	(0.13, 0.17, 0.24)
	K2	F	H	MH	(0.42, 0.6, 0.8)	(0.13, 0.17, 0.24)

Table 4. *Cont.*

Criteria	Roof	DM1	DM2	DM3	Aggregated Fuzzy Evaluation	Normalized Fuz Decision Matrix
C2	A1	H	H	H	(0.6, 0.8, 1.0)	(0.6, 0.8, 1.0)
	A2	MH	MH	MH	(0.4, 0.6, 0.8)	(0.4, 0.6, 0.8)
	B1	MH	MH	MH	(0.4, 0.6, 0.8)	(0.4, 0.6, 0.8)
	B2	MH	MH	MH	(0.4, 0.6, 0.8)	(0.4, 0.6, 0.8)
	H	MH	MH	F	(0.35, 0.53, 0.73)	(0.35, 0.53, 0.73)
	K1	MH	F	MH	(0.35, 0.53, 0.73)	(0.35, 0.53, 0.73)
	K2	MH	F	MH	(0.35, 0.53, 0.73)	(0.35, 0.53, 0.73)
C3	A1	P	MP	MP	(0.1, 0.25, 0.4)	(0.1, 0.25, 0.4)
	A2	MH	MH	MH	(0.4, 0.6, 0.8)	(0.4, 0.6, 0.8)
	B1	MH	MH	MH	(0.4, 0.6, 0.8)	(0.4, 0.6, 0.8)
	B2	MH	MH	MH	(0.4, 0.6, 0.8)	(0.4, 0.6, 0.8)
	H	MH	MH	MH	(0.4, 0.6, 0.8)	(0.4, 0.6, 0.8)
	K1	MH	MH	F	(0.35, 0.53, 0.73)	(0.35, 0.53, 0.73)
	K2	F	MH	F	(0.3, 0.47, 0.67)	(0.3, 0.47, 0.67)
C4	A1	H	H	H	(0.6, 0.8, 1.0)	(0.1, 0.13, 0.17)
	A2	MH	MH	MH	(0.4, 0.6, 0.8)	(0.13, 0.17, 0.25)
	B1	F	F	F	(0.25, 0.4, 0.6)	(0.17, 0.25, 0.4)
	B2	F	F	F	(0.25, 0.4, 0.6)	(0.17, 0.25, 0.4)
	H	MP	MP	MP	(0.1, 0.25, 0.4)	(0.25, 0.4, 1.0)
	K1	VH	VH	VH	(0.8, 1.0, 1.0)	(0.1, 0.1, 0.13)
	K2	VH	VH	VH	(0.8, 1.0, 1.0)	(0.1, 0.1, 0.13)
C5	A1	H	H	H	(0.6, 0.8, 1.0)	(0.1, 0.13, 0.17)
	A2	F	F	F	(0.25, 0.4, 0.6)	(0.17, 0.25, 0.4)
	B1	MP	MP	MP	(0.1, 0.25, 0.4)	(0.25, 0.4, 1.0)
	B2	F	F	F	(0.25, 0.4, 0.6)	(0.17, 0.25, 0.4)
	H	MP	MP	MP	(0.1, 0.25, 0.4)	(0.25, 0.4, 1.0)
	K1	MH	MH	MH	(0.4, 0.6, 0.8)	(0.13, 0.17, 0.25)
	K2	VH	VH	VH	(0.8, 1.0, 1.0)	(0.1, 0.1, 0.13)
C6	A1	VH	VH	VH	(0.67, 0.87, 0.93)	(0.11, 0.12, 0.15)
	A2	F	F	F	(0.25, 0.4, 0.6)	(0.17, 0.25, 0.4)
	B1	MP	MP	MP	(0.1, 0.25, 0.4)	(0.25, 0.4, 1.0)
	B2	MP	MP	MP	(0.1, 0.25, 0.4)	(0.25, 0.4, 1.0)
	H	MP	MP	MP	(0.1, 0.25, 0.4)	(0.25, 0.4, 1.0)
	K1	H	H	H	(0.6, 0.8, 1.0)	(0.1, 0.13, 0.17)
	K2	H	H	H	(0.6, 0.8, 1.0)	(0.1, 0.13, 0.17)
C7	A1	VP	VP	VP	(0.0, 0.0, 0.1)	(0.0, 0.0, 0.1)
	A2	VH	VH	VH	(0.8, 1.0, 1.0)	(0.8, 1.0, 1.0)
	B1	VH	VH	VH	(0.8, 1.0, 1.0)	(0.8, 1.0, 1.0)
	B2	VH	VH	VH	(0.8, 1.0, 1.0)	(0.8, 1.0, 1.0)
	H	VH	VH	VH	(0.8, 1.0, 1.0)	(0.8, 1.0, 1.0)
	K1	VP	VP	VP	(0.0, 0.0, 0.1)	(0.0, 0.0, 0.1)
	K2	VP	VP	VP	(0.0, 0.0, 0.1)	(0.0, 0.0, 0.1)
C8	A1	VP	VP	VP	(0.0, 0.0, 0.1)	(0.0, 0.0, 0.1)
	A2	VH	VH	VH	(0.8, 1.0, 1.0)	(0.8, 1.0, 1.0)
	B1	VH	VH	VH	(0.8, 1.0, 1.0)	(0.8, 1.0, 1.0)
	B2	VH	VH	VH	(0.8, 1.0, 1.0)	(0.8, 1.0, 1.0)
	H	VH	VH	VH	(0.8, 1.0, 1.0)	(0.8, 1.0, 1.0)
	K1	VP	VP	VP	(0.0, 0.0, 0.1)	(0.0, 0.0, 0.1)
	K2	VP	VP	VP	(0.0, 0.0, 0.1)	(0.0, 0.0, 0.1)
C9	A1	VP	VP	VP	(0.0, 0.0, 0.1)	(0.0, 0.0, 0.1)
	A2	VH	VH	VH	(0.8, 1.0, 1.0)	(0.8, 1.0, 1.0)
	B1	VH	VH	VH	(0.8, 1.0, 1.0)	(0.8, 1.0, 1.0)
	B2	VH	VH	VH	(0.8, 1.0, 1.0)	(0.8, 1.0, 1.0)
	H	VH	VH	VH	(0.8, 1.0, 1.0)	(0.8, 1.0, 1.0)
	K1	VP	VP	VP	(0.0, 0.0, 0.1)	(0.0, 0.0, 0.1)
	K2	VP	VP	VP	(0.0, 0.0, 0.1)	(0.0, 0.0, 0.1)

The preference intensity P is determined according to Equations (6) and (7), where d is Euclidian distance calculated for the maximized and minimized criteria as defined in Equations (8) and (9), respectively. The P values are shown in Table 5. Consequently, shown in Table 6, the preference index for each alternative is calculated by Equation (10).

Table 5. The preference intensity of each alternative by each criterion.

	C1	C2	C3	C4	C5	C6	C7	C8	C9
P(A1,A2)	0.64	0.00	0.53	0.00	0.00	0.00	1.00	1.00	1.00
P(A1,B1)	0.65	0.00	0.53	0.00	0.00	0.00	1.00	1.00	1.00
P(A1,B2)	0.65	0.00	0.53	0.00	0.00	0.00	1.00	1.00	1.00
P(A1,H)	0.64	0.00	0.53	0.00	0.00	0.00	1.00	1.00	1.00
P(A1,K1)	0.57	0.00	0.46	0.00	0.00	0.01	0.00	0.00	0.00
P(A1,K2)	0.57	0.00	0.41	0.00	0.00	0.01	0.00	0.00	0.00
P(A2,A1)	0.00	0.24	0.00	0.11	0.19	0.21	0.00	0.00	0.00
P(A2,B1)	0.01	0.00	0.00	0.00	0.00	0.00	0.00	0.00	0.00
P(A2,B2)	0.01	0.00	0.00	0.00	0.00	0.00	0.00	0.00	0.00
P(A2,H)	0.00	0.00	0.00	0.00	0.00	0.00	0.00	0.00	0.00
P(A2,K1)	0.00	0.00	0.00	0.12	0.12	0.19	0.00	0.00	0.00
P(A2,K2)	0.00	0.00	0.00	0.12	0.23	0.19	0.00	0.00	0.00
P(B1,A1)	0.00	0.24	0.00	0.22	0.63	0.64	0.00	0.00	0.00
P(B1,A2)	0.00	0.00	0.00	0.20	0.44	0.44	0.00	0.00	0.00
P(B1,B2)	0.00	0.00	0.00	0.00	0.44	0.00	0.00	0.00	0.00
P(B1,H)	0.00	0.00	0.00	0.00	0.00	0.00	0.00	0.00	0.00
P(B1,K1)	0.00	0.00	0.00	0.59	0.56	0.63	0.00	0.00	0.00
P(B1,K2)	0.00	0.00	0.00	0.24	0.66	0.63	0.00	0.00	0.00
P(B2,A1)	0.00	0.24	0.00	0.22	0.19	0.64	0.00	0.00	0.00
P(B2,A2)	0.00	0.00	0.00	0.20	0.00	0.44	0.00	0.00	0.00
P(B2,B1)	0.00	0.00	0.00	0.00	0.00	0.00	0.00	0.00	0.00
P(B2,H)	0.00	0.00	0.00	0.00	0.00	0.00	0.00	0.00	0.00
P(B2,K1)	0.00	0.00	0.00	0.59	0.12	0.63	0.00	0.00	0.00
P(B2,K2)	0.00	0.00	0.00	0.24	0.23	0.63	0.00	0.00	0.00
P(H,A1)	0.00	0.32	0.00	0.66	0.63	0.64	0.00	0.00	0.00
P(H,A2)	0.00	0.08	0.00	0.62	0.44	0.44	0.00	0.00	0.00
P(H,B1)	0.01	0.08	0.00	0.57	0.00	0.00	0.00	0.00	0.00
P(H,B2)	0.01	0.08	0.00	0.57	0.44	0.00	0.00	0.00	0.00
P(H,K1)	0.00	0.00	0.00	0.68	0.56	0.63	0.00	0.00	0.00
P(H,K2)	0.00	0.00	0.00	0.68	0.66	0.63	0.00	0.00	0.00
P(K1,A1)	0.00	0.32	0.00	0.00	0.07	0.00	0.00	0.00	0.00
P(K1,A2)	0.08	0.08	0.35	0.00	0.00	0.00	0.00	0.00	0.00
P(K1,B1)	0.09	0.08	0.35	0.00	0.00	0.00	0.00	0.00	0.00
P(K1,B2)	0.09	0.08	0.35	0.00	0.00	0.00	0.00	0.00	0.00
P(K1,H)	0.08	0.00	0.35	0.00	0.00	0.00	0.00	0.00	0.00
P(K1,K2)	0.00	0.00	0.00	0.00	0.00	0.00	0.00	0.00	0.00
P(K2,A1)	0.00	0.32	0.35	0.00	0.00	0.00	0.00	0.00	0.00
P(K2,A2)	0.08	0.08	0.35	0.00	0.00	0.00	1.00	1.00	1.00
P(K2,B1)	0.09	0.08	0.35	0.00	0.00	0.00	1.00	1.00	1.00
P(K2,B2)	0.09	0.08	0.35	0.00	0.00	0.00	1.00	1.00	1.00
P(K2,H)	0.08	0.00	0.33	0.00	0.00	0.00	1.00	1.00	1.00
P(K2,K1)	0.00	0.00	0.34	0.00	0.00	0.00	0.00	0.00	0.00

Finally, in Table 7, the fuzzy positive and negative flows and fuzzy net flow are defined by Equations (11)–(13), while in Table 8, defuzzified net flow, defined by centroid method (14), and ranking positions of flat roofs are presented. Centroid method values are also presented in Table 7, and the final ranking of alternatives is obtained.

Table 6. Preference index for each alternative.

Π	TFN	Π	TFN
Π(A1,A2)	(0.33, 0.42, 0.46)	Π(B2,H)	(0.0, 0.0, 0.0)
Π(A1,B1)	(0.33, 0.42, 0.46)	Π(B2,K1)	(0.07, 0.09, 0.12)
Π(A1,B2)	(0.33, 0.42, 0.46)	Π(B2,K2)	(0.06, 0.08, 0.1)
Π(A1,H)	(0.33, 0.42, 0.46)	Π(H,A1)	(0.11, 0.16, 0.21)
Π(A1,K1)	(0.08, 0.1, 0.11)	Π(H,A2)	(0.07, 0.11, 0.14)
Π(A1,K2)	(0.08, 0.1, 0.11)	Π(H,B1)	(0.03, 0.04, 0.06)
Π(A2,A1)	(0.04, 0.06, 0.07)	Π(H,B2)	(0.05, 0.07, 0.09)
Π(A2,B1)	(0.0, 0.0, 0.0)	Π(H,K1)	(0.9, 0.12, 0.17)
Π(A2,B2)	(0.0, 0.0, 0.0)	Π(H,K2)	(0.9, 0.13, 0.17)
Π(A2,H)	(0.0, 0.0, 0.0)	Π(K1,A1)	(0.03, 0.03, 0.04)
Π(A2,K1)	(0.02, 0.03, 0.04)	Π(K1,A2)	(0.03, 0.04, 0.05)
Π(A2,K2)	(0.02, 0.03, 0.04)	Π(K1,B1)	(0.03, 0.04, 0.05)
Π(B1,A1)	(0.09, 0.13, 0.16)	Π(K1,B2)	(0.03, 0.04, 0.05)
Π(B1,A2)	(0.05, 0.07, 0.1)	Π(K1,H)	(0.03, 0.04, 0.04)
Π(B1,B2)	(0.02, 0.03, 0.04)	Π(K1,K2)	(0, 0.01, 0.01)
Π(B1,H)	(0.0, 0.0, 0.0)	Π(K2,A1)	(0.04, 0.06, 0.07)
Π(B1,K1)	(0.08, 0.12, 0.16)	Π(K2,A2)	(0.27, 0.35, 0.39)
Π(B1,K2)	(0.07, 0.1, 0.14)	Π(K2,B1)	(0.27, 0.35, 0.39)
Π(B2,A1)	(0.07, 0.1, 0.13)	Π(K2,B2)	(0.27, 0.35, 0.39)
Π(B2,A2)	(0.03, 0.05, 0.06)	Π(K2,H)	(0.26, 0.34, 0.38)
Π(B2,B1)	(0.0, 0.0, 0.0)	Π(K2,K1)	(0.02, 0.03, 0.03)

Table 7. The positive and negative flows.

Roof	$\Phi^{+}x_i$	$\Phi^{-}x_i$	$\Phi(x_i)$
A1	(1.46, 1.86, 2.06)	(0.39, 0.54, 0.68)	(1.07, 1.32, 1.38)
A2	(0.09, 0.13, 0.16)	(0.79, 1.03, 1.20)	(−0.7, −0.91, −1.03)
B1	(0.31, 0.45, 0.59)	(0.66, 0.85, 0.96)	(−0.35, −0.4, −0.36)
B2	(0.23, 0.32, 0.42)	(0.69, 0.91, 1.03)	(−0.47, −0.58, −0.61)
H	(0.43, 0.63, 0.84)	(0.62, 0.79, 0.88)	(−0.18, −0.16, −0.04)
K1	(0.16, 0.21, 0.25)	(0.35, 0.5, 0.64)	(−0.19, −0.29, −0.38)
K2	(1.14, 1.47, 1.64)	(0.32, 0.45, 0.58)	(0.82, 1.02, 1.06)

Table 8. The defuzzified net flows and rank of the flat roofs.

Roof	$\bar{\Phi}(x_i)$	Rank
A1	1.26	1
A2	−0.88	7
B1	−0.37	5
B2	−0.56	6
H	−0.13	3
K1	−0.29	4
K2	0.97	2

4. Conclusions

This research presented theoretical features related to flat roofs, and a fuzzy multi-criteria analysis of the renovation problem of flat roofs on public buildings was carried out. The fuzzy multi-criteria approach is developed for the activity of renovation of flat roofs, using the F-PROMETHEE II method. Relevant stakeholders and their attitudes were identified, as well as their inclusion in the planning process of renovation activities was achieved. The approach is validated on the flat roofs of the Faculty of Civil Engineering, Architecture and Geodesy in Split. A total of seven roofs were analyzed, and an evaluation of each of them according to all defined criteria was carried out.

The final ranking is achieved when roof A1 is ranked first. This roof is ranked as the most priority for renovation, with a net flow value of $\bar{\Phi}(x_i) = 1.26$, and the least priority is roof A2, with $\bar{\Phi}(x_i) = -0.88$.

Considering the obtained results, which were presented to the final decision makers and accepted by them, the following can be concluded about the proposed model:

- The proposed fuzzy multi-criteria approach is useful and effective for evaluating different flat roof renovation options, particularly in contexts where there is a high degree of uncertainty and imprecision.
- The use of a decision-making process where different stakeholders are involved in the evaluation and ranking of flat roofs ensures that all relevant factors are considered and that the final decision is widely accepted.
- The criteria are selected based on their relevance to the flat roof renovation context and are weighted appropriately based on their relative importance.
- The use of fuzzy set theory allowed the handling of imprecise and uncertain data, which is particularly important in the context of flat roof renovation where data are incomplete or unreliable.
- The F-PROMETHEE II method is a particularly effective method for flat roof renovation decision-making, as it provides a clear and easy-to-interpret ranking of different options based on the selected criteria.

While the proposed approach has shown to be useful and effective in the decision-making process for flat roof renovation, there may be several potential gaps in research and analysis that should be considered, including the identification of criteria. The selection of appropriate criteria for evaluating flat roofs for renovation can be challenging, particularly when considering the varying needs and preferences of different stakeholders. The proposed approach is designed to be used in a participatory decision-making process, where different stakeholders are involved in the evaluation and ranking of flat roofs for the activity of renovation. However, the involvement of stakeholders can be challenging, particularly when dealing with complex technical issues. Moreover, flat roof renovation decision-making often involves a high degree of uncertainty, particularly when considering the long-term performance and durability of different roofing materials and technologies. For this reason, it is important to handle a high uncertainty. Finally, for some non-experts, the proposed design might be complex and difficult to understand.

The future study will consider the wider application of the methodology on all public buildings in the town of Split, Croatia. Considering all the variants regarding green, standard and cool roofs, more research will be implemented to identify the most important criteria for different flat roof renovation contexts and to determine how to weigh and combine these criteria in the evaluation process. To handle missing or incomplete data, the most appropriate methods for collecting and analyzing data for flat roof renovation decision-making will be determined. Moreover, the methodology based on mixed integer programming will be used to obtain managers' approaches regarding the financial cycle with defined required constraints. This way, the managing process of roof renovation will be more systematic, as each year, it will be known which roofs are going to be reconstructed. Moreover, an approach for predicting the future condition of roofs using fuzzy neural networks will be proposed to enable managers to maintain roofs for longer periods properly.

Author Contributions: Conceptualization, K.R. and N.J.; methodology, K.R. and N.J.; software, K.R., N.J. and K.B.; validation, K.R., N.J. and K.B.; formal analysis, K.R. and N.J.; investigation, K.R., N.J. and K.B.; resources, K.R., N.J. and K.B.; data curation, K.R., N.J. and K.B.; writing—original draft preparation, K.R., N.J. and K.B.; writing—review and editing, K.R. and N.J.; visualization, K.R., N.J. and K.B.; supervision, K.R. and N.J. The principal authors of the paper are K.R. and N.J. All authors have read and agreed to the published version of the manuscript.

Funding: This research received no external funding.

Institutional Review Board Statement: Not applicable.

Informed Consent Statement: Not applicable.

Data Availability Statement: Data available on request due to restrictions e.g., privacy or ethical. The data presented in this study are available on request from the corresponding author. The data are not publicly available due to further research to be published.

Acknowledgments: This research is partially supported through project KK.01.1.1.02.0027, a project co-financed by the Croatian Government and the European Union through the European Regional Development Fund—the Competitiveness and Cohesion Operational Programme.

Conflicts of Interest: The authors declare no conflict of interest.

References

1. Miniotaite, R. Optimization of flat roof thermal renovation technologies. In *Interaction between Theory and Practice in Civil Engineering and Construction*; Komurlu, R., Gurgun, A.P., Singh, A., Yazdani, S., Eds.; ISEC Press: North Fargo, ND, USA, 2016; pp. 311–316.
2. Gagliano, A.; Detommaso, M.; Nocera, F.; Evola, G. A multi-criteria methodology for comparing the energy and environmental behavior of cool, green and traditional roofs. *Build. Environ.* **2015**, *90*, 71–81. [CrossRef]
3. Petrakova, Z.; Grznar, M. Methods of multi-criteria decision-making in the choice of an alternative solution in the reconstruction process of a flat roof. *Slovak J. Civ. Eng.* **2004**, *3*, 1–11.
4. Gonçalves, M.; Silvestre, J.D.; de Brito, J.; Gomes, R. Environmental and economic comparison of the life cycle of waterproofing solutions for flat roofs. *J. Build. Eng.* **2019**, *24*, 100710. [CrossRef]
5. Aguacil Moreno, S.; Lufkin, S.; Rey, E. Architectural design scenarios with building-integrated photovoltaic solutions in renovation processes: Case study in Neuchâtel (Switzerland). In Proceedings of the PLEA 2016–32nd International Conference on Passive and Low Energy Architecture—Cities, Buildings, People: Towards Regenerative Environments, Los Angeles, CA, USA, 24 October 2016; pp. 486–492.
6. Rosasco, P.; Perini, K. Selection of (green) roof systems: A sustainability-based multi-criteria analysis. *Buildings* **2019**, *9*, 134. [CrossRef]
7. Kuznecov, R.; Šaparauskas, J. Multi-criteria assessment of pitched roof reconstruction technologies. In Proceedings of the 13th International Conference: Modern Building Materials, Structures and Techniques, Vilnius, Lithuania, 16–17 May 2019.
8. Kalibatas, D.; Kovaitis, V. Selecting the most effective alternative of waterproofing membranes for multifunctional inverted flat roofs. *J. Civ. Eng. Manag.* **2017**, *23*, 650–660. [CrossRef]
9. Golić, K.; Kosorić, V.; Furundžić, A.K. General model of solar water heating system integration in residential building refurbishment—Potential energy savings and environmental impact. *Renew. Sustain. Energy Rev.* **2011**, *15*, 1533–1544. [CrossRef]
10. Furundzic, A.K.; Kosoric, V.; Golic, K. Potential for reduction of CO2 emissions by integration of solar water heating systems on student dormitories through building refurbishment. *Sustain. Cities Soc.* **2012**, *2*, 50–62. [CrossRef]
11. Sangkakool, T.; Techato, K.; Zaman, R.; Brudermann, T. Prospects of green roofs in urban Thailand–a multi-criteria decision analysis. *J. Clean. Prod.* **2018**, *196*, 400–410. [CrossRef]
12. Turskis, Z.; Morkunaite, Z.; Kutut, V. A hybrid multiple criteria evaluation method of ranking of cultural heritage structures for renovation projects. *Int. J. Strateg. Prop. Manag.* **2017**, *21*, 318–329. [CrossRef]
13. Naing, Y.M.; Nitivattananon, V.; Shipin, O.V. Green roof retrofitting: Potential assessment in an academic campus. *Eng. J.* **2017**, *21*, 57–74. [CrossRef]
14. Gonzalez-Dominguez, J.; Sanchez-Barroso, G.; Garcia-Sanz-Calcedo, J. Preventive maintenance optimisation of accessible flat roofs in healthcare centres using the Markov chain. *J. Build. Eng.* **2020**, *32*, 101775. [CrossRef]
15. Wu, Y.; Xu, C.; Ke, Y.; Chen, K.; Sun, X. An intuitionistic fuzzy multi-criteria framework for large-scale rooftop PV project portfolio selection: Case study in Zhejiang, China. *Energy* **2017**, *143*, 295–309. [CrossRef]
16. Tian, C.; Peng, J.; Zhang, W.; Zhang, S.; Wang, J. Tourism environmental impact assessment based on improved AHP and picture fuzzy PROMETHEE II methods. *Technol. Econ. Dev. Econ.* **2020**, *26*, 355–378. [CrossRef]
17. Agrawal, N. Multi-criteria decision-making toward supplier selection: Exploration of PROMETHEE II method. *Benchmarking Int. J.* **2022**, *29*, 2122–2146. [CrossRef]
18. Wu, Y.; Liu, F.; Huang, Y.; Xu, C.; Zhang, B.; Ke, Y.; Jia, W. A two-stage decision framework for inland nuclear power plant site selection based on GIS and type-2 fuzzy PROMETHEE II: Case study in China. *Energy Sci. Eng.* **2020**, *8*, 1941–1961. [CrossRef]
19. Samanlioglu, F.; Ayağ, Z. A fuzzy AHP-PROMETHEE II approach for evaluation of solar power plant location alternatives in Turkey. *J. Intell. Fuzzy Syst.* **2017**, *33*, 859–871. [CrossRef]
20. Roodposhti, M.S.; Rahimi, S.; Beglou, M.J. PROMETHEE II and fuzzy AHP: An enhanced GIS-based landslide susceptibility mapping. *Nat. Hazards* **2014**, *73*, 77–95. [CrossRef]
21. Goumas, M.; Lygerou, V. An extension of the PROMETHEE method for decision making in fuzzy environment: Ranking of alternative energy exploitation projects. *Eur. J. Oper. Res.* **2000**, *123*, 606–613. [CrossRef]
22. Samanlioglu, F.; Ayağ, Z. Concept selection with hesitant fuzzy ANP-PROMETHEE II. *J. Ind. Prod. Eng.* **2021**, *38*, 547–560. [CrossRef]

23. Samanlioglu, F.; Ayağ, Z. Fuzzy ANP-based PROMETHEE II approach for evaluation of machine tool alternatives. *J. Intell. Fuzzy Syst.* **2016**, *30*, 2223–2235. [CrossRef]
24. Liu, P.; Guan, Z. Evaluation Research on the Quality of the Railway Passenger Service Based on the Linguistic Variables and the Improved PROMETHEE-II Method. *J. Comput.* **2009**, *4*, 265–270. [CrossRef]
25. Amaral, T.M.; Costa, A.P. Improving decision-making and management of hospital resources: An application of the PROMETHEE II method in an Emergency Department. *Oper. Res. Health Care* **2014**, *3*, 1–6. [CrossRef]
26. AlHumid, H.A.; Haider, H.; AlSaleem, S.S.; Shafiquzamman, M.; Sadiq, R. Performance indicators for municipal solid waste management systems in Saudi Arabia: Selection and ranking using fuzzy AHP and PROMETHEE II. *Arab. J. Geosci.* **2019**, *12*, 1–23. [CrossRef]
27. Tabaraee, E.; Ebrahimnejad, S.; Bamdad, S. Evaluation of power plants to prioritise the investment projects using fuzzy PROMETHEE method. *Int. J. Sustain. Energy* **2018**, *37*, 941–955. [CrossRef]
28. Burak, S.; Samanlioglu, F.; Ülker, D. Evaluation of irrigation methods in Söke Plain with HF-AHP-PROMETHEE II hybrid MCDM method. *Agric. Water Manag.* **2022**, *271*, 107810. [CrossRef]
29. Ozsahin, D.U.; Uzun, B.; Musa, M.S.; Şentürk, N.; Nurçin, F.V.; Ozsahin, I. Evaluating nuclear medicine imaging devices using fuzzy PROMETHEE method. *Procedia Comput. Sci.* **2017**, *120*, 699–705. [CrossRef]
30. Wang, T.C.; Chen, L.Y.; Chen, Y.H. Applying fuzzy PROMETHEE method for evaluating IS outsourcing suppliers. In Proceedings of the Fifth International Conference on Fuzzy Systems and Knowledge Discovery, Jinan, China, 18–20 October 2008; Volume 3, pp. 361–365.
31. Moradpour, S.; Ebrahimnejad, S.; Mehdizadeh, E.; Mohamadi, A. Using hybrid fuzzy PROMETHEE II and fuzzy binary goal programming for risk ranking: A case study of highway construction projects. *J. Optim. Ind. Eng.* **2011**, *9*, 47–55.
32. Ozsahin, D.U.; Uzun, B.; Musa, M.S.; Helwan, A.; Wilsona, C.N.; Nurçina, F.V.; Senturka, N.; Ozsahin, I. Evaluating cancer treatment alternatives using fuzzy PROMETHEE method. *Int. J. Adv. Comput. Sci. Appl.* **2017**, *8*, 177–182.
33. Athawale, V.M.; Chatterjee, P.; Chakraborty, S. Decision making for facility location selection using PROMETHEE II method. *Int. J. Ind. Syst. Eng.* **2012**, *11*, 16–30. [CrossRef]
34. Zdolec, M. Flat Roofs. Master's Thesi, University of North, Koprivnica, Croatia, 2020.
35. Đurić, D. *Flat Roof: Physical Characteristics of Reinforced Concrete Roofs*; University of Zagreb, Faculty of Architecture: Zagreb, Croatia, 2020.
36. Zadeh, L.A. Fuzzy sets. *Inf. Control.* **1965**, *8*, 338–353. [CrossRef]
37. Tong, L.; Pu, Z.; Chen, K.; Yi, J. Sustainable maintenance supplier performance evaluation based on an extend fuzzy promethee II approach in petrochemical industry. *J. Clean. Prod.* **2020**, *273*, 122771. [CrossRef]
38. Brans, J.P.; Vincke, P. A Preference Ranking Organisation Method: (The PROMETHEE Method for Multiple Criteria. Decision-Making). *Manag. Sci.* **1985**, *31*, 647–656. [CrossRef]
39. Rogulj, K.; Kilić Pamuković, J.; Jajac, N. Knowledge-Based Fuzzy Expert System to the Condition Assessment of Historic Road Bridges. *Appl. Sci.* **2021**, *11*, 1021. [CrossRef]

sustainability

Article

Numerical Analysis of the Racking Behaviour of Multi-Storey Timber-Framed Buildings Considering Load-Bearing Function of Double-Skin Façade Elements

Miroslav Premrov and Erika Kozem Šilih *

Faculty of Civil Engineering, Transportation Engineering and Architecture, University of Maribor, 2000 Maribor, Slovenia
* Correspondence: erika.kozem@um.si

Abstract: The paper presents an innovative approach in the modelling of multi-storey timber-framed buildings, where double-skin façade elements (DSF) are additionally considered as load-bearing wall elements against a horizontal load impact. The mathematical model with a fictive diagonal element developed for timber-framed wall elements with classical oriented strand boards (OSB) or fibre–plaster sheathing boards (FPB) is upgraded for DSF elements. The diameter of the fictive diagonal is determined with either experimental results or numerically obtained results using the time-consuming FEM model with elastic spring elements, which simulates the bonding line between the timber frame and both glazing panes. In the second part of the study, the numerical analysis of a specially selected three-storey timber-framed building was performed using the developed mathematical model with fictive diagonal elements. Two alternative calculations were performed with the DSF elements as non-resisting and racking-resisting wall elements. It was demonstrated on the selected case that the racking resistance (R) of a building can essentially increase up to 35% if DSF elements are considered as resisting wall elements. As a secondary goal of the study, it is also important to point out that by using DSF elements as racking-resisting elements, the distortion in the first floor essentially decreased. It is demonstrated on the selected numerical example that this torsional influence decreased notably (by almost 18%) when the load-bearing DSF elements were used for seismic excitation in the X direction. Therefore, such an approach can open new perspectives in designing multi-storey timber-framed buildings with a more attractive and dynamic floor plan and structure.

Keywords: sustainability; timber; structures; multi-storey; numerical analysis; DSF; racking resistance

Citation: Premrov, M.; Kozem Šilih, E. Numerical Analysis of the Racking Behaviour of Multi-Storey Timber-Framed Buildings Considering Load-Bearing Function of Double-Skin Façade Elements. *Sustainability* **2023**, *15*, 6379. https://doi.org/10.3390/su15086379

Academic Editor: Syed Minhaj Saleem Kazmi

Received: 6 March 2023
Revised: 30 March 2023
Accepted: 6 April 2023
Published: 7 April 2023

1. Introduction

In a sense, decreasing CO_2 emissions and designing energy-efficient buildings has been a topic of research since 1970. If the green house gas (GHG) emissions will continue to grow as they currently are, the Earth's average temperature will increase by at least 1.50 °C until the end of 21st century [1]. Consequently, a new strategy to design buildings with net zero emissions has to be adopted not only for new buildings, but also for a wide range of building renovations [2,3], integrating a life-cycle approach as well [4]. Timber, as a natural raw material with the potential to store CO_2, has the capacity to rapidly decrease GHG emissions, and seems to be the best possible solution to this problem.

A similar increasing trend may be observed in the construction of new multi-storey (MSTB) and high-rise timber buildings (HRTB). However, a "high-rise" building is mainly considered as such when surpassing 25 m [5,6] or having more than ten storeys [7,8]. On the other hand, the term "multi-storey" tends to be a lot more defined and refers to buildings of four storeys or more, which is why it will be used in this paper. Despite the increase and advantages shown by numerous studies in recent years, the global spread of multi-storey timber construction is still relatively low compared to massive steel construction with essentially higher lateral resistance compared to timber structures. An additional aspect is the eventual non-regularity in a multi-storey building floor-plan, which can essentially increase distortion effects in each storey and amounts to an additional limitation of multi-storey timber buildings. As a result, buildings that could be built from timber are still built from reinforced concrete or steel, which leads to an environmental performance that is poorer than if this housing stock had been built with timber—even the new timber structural systems that have been recently developed, such as those in [9].

On the other hand, the use of glazing in buildings has always contributed to openness, visual comfort, and a better daylight situation. Over the years, manufacturers have improved the thermal insulation and strength of glass [10], which enabled not only the internal illumination of buildings with large glass surfaces which were primarily south oriented, but also solar energy heating with increased solar heating gains through the transparent areas. On the other hand, the installed glazing areas that are non-load bearing in their planes in terms of assuming horizontal loads further aggravate the problem of required horizontal load-bearing capacity from a structural perspective.

Therefore, it is crucial in such cases to develop a load-bearing timber–glass wall element, which can significantly contribute to resisting the increased horizontal load impact. With their racking stiffness, such elements can increase the horizontal stiffness of the whole building and consequently decrease the torsional effects of seismic forces. In this sense, so-called single-skin timber–glass wall elements were initially developed [11], where the single-layer or thermal insulated two- or three-layered glass pane is rigidly connected to the timber frame with the bonding line [12]. It was concluded by many experimental [12–14] and numerical studies [15–17] that by using only single-skin timber–glass wall elements, the racking stiffness in particular did not increase in the expected manner and was not in the same range as the timber-framed walls with the classical sheathing boards, such as OSB or fibre–plaster boards. Therefore, special double-skin façade (DSF) timber–glass wall elements were further developed, supported by experimental [18] and further numerical studies [19]. The DSF elements were first developed primarily to be used for the energy and structural renovation of existing old buildings but can be used in new multi-storey timber buildings as well, especially in cases of a strong asymmetric position of transparent areas on the building envelope.

However, this numerical study was performed with a very time-consuming finite element model (FE) of a composite timber–glass wall element using elastic springs for simulating the bonding line between the timber frame and both glazing panes. Therefore, such an FE model is not appropriate, in practice, for any engineering application in the static or dynamical stability of multi-storey timber buildings and can be performed only in theoretical studies and analyses of a single wall element.

To avoid such time-consuming calculations in the presented study, the developed DSF timber element was integrated into a quite simple mathematical model with the fictive diagonal for simulating the racking stiffness of the bracing timber–glass wall element (Section 3). The FE model with the fictive diagonal was already developed in [20], but only for the classical fibre–plaster and OSB sheathing boards, where the sheets are stapled to the timber frame and not continuously bonded such as in the case of DSF elements. In Section 4, such modelled DSF elements are integrated into the static model of a specially selected four-storey timber-framed building with an asymmetrical position of transparent DSF elements. The influence on seismic behaviour of this selected timber building is numerically studied using an FE calculation programme SAP 2000 [21], with a special impact dedicated to the influence of the developed DSF elements to increase the racking resistance and stiffness of the building. Special attention is paid to a decrease in torsional effects caused by the seismic force if the building is analysed without and with racking resisting DSF elements.

2. Structural Stability of Multi-Storey Timber-Framed Buildings

2.1. Structural Design of Multi-Storey Timber Buildings

In the last decades, new timber products (cross-laminated timber, for instance, at the beginning of the 1990s) changed the form and especially the maximum possible height of timber buildings. In this sense, timber structures became competitive with other structures built with conventional and commonly used structural materials. Furthermore, combining timber structural elements with other commonly used building materials (brick, concrete, steel, and lately also glass) can open new perspectives on the attractive architectural forms of such hybrid timber buildings. Therefore, a combination with load-bearing glazing will be presented at the end of this section.

There are many different basic structural systems which are commonly used in multi-storey timber buildings schematically presented in Figure 1. They also differ from each other in a horizontal load transfer, and therefore allow different limits in the maximal height of timber structures. For instance, it is recommended to build timber frame-and-panel structures with up to four storeys, and a solid timber construction with CLT elements with up to ten storeys. Higher high-rise timber structures are constructed in a frame construction (14-storey Tree building in Bergen with the maximal height of 52.8 m) or even as hybrid structures combining timber primarily with a reinforced-concrete (RC) core (24-storey HoHo timber building in Vienna with the maximal height of 84 m). The limits in the height of timber buildings also strongly depend on the location of the building and the subsequent horizontal load impact (wind or earthquake).

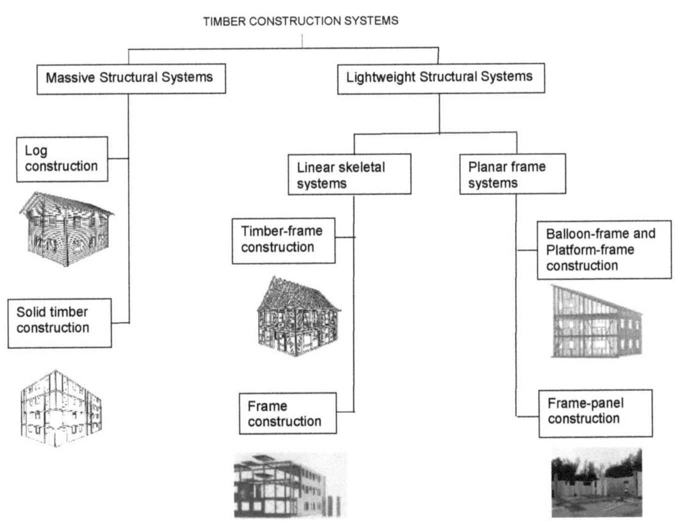

Figure 1. Classification of timber structural systems according to their load-bearing function.

In most cases, the more problematic of these two horizontal load actions is an earthquake, which subjects a building to a high-intensity dynamic load often resulting in catastrophic consequences. One of the basic principles when designing a building to resist seismic loads is trying to avoid plan irregularity, clearly described in, and prescribed by, the Eurocode 8 (2005) standard [22]. This means that the centre of gravity (M), where the resulting seismic force (F_b) acts, and the centre of rigidity/stiffness (R) of a building with a resulting horizontal resisting force should be as close to each other as possible. Unfortunately, this is an issue for energy efficient buildings with large glazing areas predominantly placed on one side of the structure, resulting in an uneven stiffness of their floor plan and an important dislocation between the centre of gravity (M) and the centre of rigidity (R), as schematically presented in Figure 2. To avoid this distortion, it is important to consider the most external walls on the south façade as racking-resisting load-bearing elements. This means that walls with fixed glazing areas (but not windows) should also be treated as racking-resisting bracing elements and will be treated as composite elements in the timber frame-and-panel system of a timber frame and a glass sheathing, which can transmit a considerable share of horizontal forces to the basement. With such an approach, the racking resistance and stiffness of the whole analysed building can be increased, and new limits in the maximal height of multi-storey buildings constructed in a timber-framed structural system can be set. The influence of such an approach will be numerically analysed on a specially selected four-storey timber-framed building in Section 4.

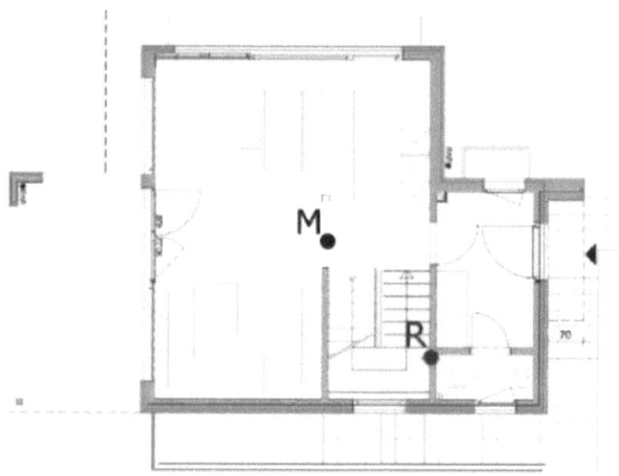

Figure 2. An example of dislocation between the centre of gravity (M) and the centre of rigidity (R), [16].

It must be emphasised, however, that the final structural design of a multi-storey timber-framed building additionally depends on its micro-location and height. It is generally known that according to Eurocode 1 [23], wind loads exponentially increase only from a certain height of a building upwards, while earthquake loads according to Eurocode 8 [22] increase almost linearly with the height of a building as schematically shown in Figure 3 for cases when higher loads are caused by earthquake loads. Consequently, a high horizontal load impact is particularly significant in the first storey of the building where a maximal load-bearing capacity has to be reached with all resisting wall elements. In lower stories of middle-rise or high-rise timber buildings, it is therefore of the utmost importance to consider all wall elements with fixed glass panes as resisting bracing elements, which can contribute to the overall racking resistance and stiffness of the whole building in the first storey.

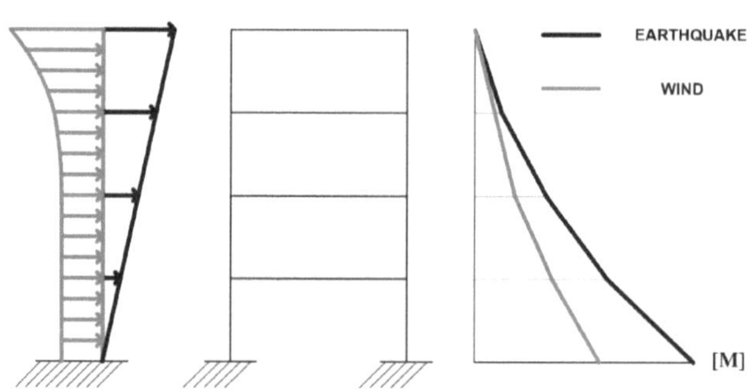

Figure 3. Display of how wind and earthquake load increase with the height of a building at a certain location.

2.2. Load-Bearing Timber–Glass Wall Elements

As mentioned before, transparent glass areas in energy-efficient timber houses are mainly installed on the south façade of the building envelope. To decrease the torsional effects of seismic forces, it is crucial to develop load-bearing timber–glass wall elements, which can contribute to the overall racking resistance and stiffness of the whole building and decrease the torsional effects. In such timber–glass wall elements, a conventional sheathing board (fibre–plaster or OSB, Figure 4b) is replaced with a glass pane (Figure 4c). The main concept of such timber–glass wall elements is that according to the basic horizontal force distribution (Figure 4a), the bonding line between the glass pane and the timber frame will take over the shear flow and the glass pane will take over the diagonal tensile force transmission (Figure 4c).

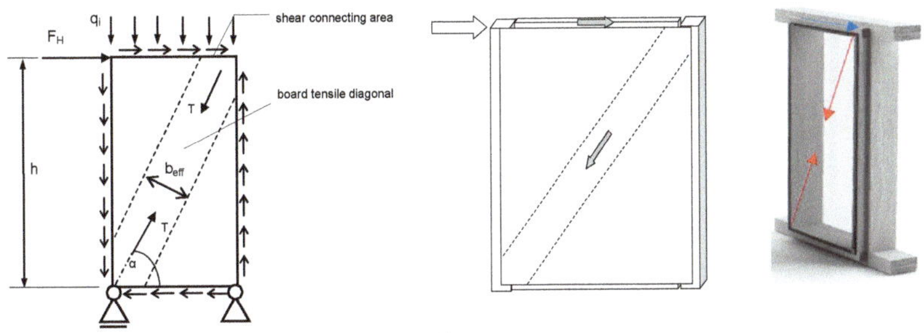

(a) (b) (c)

Figure 4. (**a**) Horizontal point load distribution in timber-framed wall element; (**b**) timber-framed wall with a conventional sheathing board; (**c**) timber–glass resisting wall element for a horizontal load impact (shear force (blue arrow), tensile diagonal force (red arrow)) [16].

In the case of timber–glass wall elements, the horizontal point load acting at the top of the element is consequently transferred to the supports in the same manner as presented in Figure 4a for any timber-framed wall elements for any sheathing boards:

- The adhesive takes over the shear stress in the gluing line;
- The tensile diagonal of the glass pane shifts the force to the support.

Timber–glass wall elements will be further separately treated as single-skin façades (SSF) with a single-glass panel and double-skin façade (DSF) elements with two panes of glazing.

2.2.1. Single-Skin Façade (SSF) Timber–Glass Wall Elements

The problem with single-skin glazing in timber-framed walls acting as racking resisting elements actually began with substantial research work in [11,13,24]. In these studies, a single-skin timber-framed load-bearing wall element with one-pane non-insulating glazing was developed using experimental and numerical tests. In [24], a special substructure was used to connect the glass pane to the timber frame. The most important technological advantage of such a type of connection is a relatively simple replacement of the glazing if it breaks. In the study of Blyberg [25], a shear wall element intended to be used as a load-bearing façade element was designed. In contradiction to [24] in this case, a non-insulating glass one-pane was rigidly bonded to a timber frame using different types of adhesives.

However, all such load-bearing timber–glass elements actually do not have any thermal insulating function and cannot be used as building envelope elements at all. Therefore, in the Wood Wisdom international research project [12], load-bearing timber–glass wall elements were further developed, where a double- or even a three-layered thermal insulating single-skin glazing was rigidly bonded directly to the timber frame structure. Such elements can be treated as single-skin façade elements. Since various parameters (such as the type and thickness of the adhesive, the type and thickness of the glass pane, wall element dimensions, etc.) can significantly affect the racking resistance and stiffness of such timber–glass wall elements, a unique mathematical model was further developed using special spring elements to simulate the slip in the bonding line between the glass pane and the timber frame [15]. A major parametrical numerical study varying the most influential parameters stated above was further performed. It was demonstrated in many experimental and numerical studies that triple-insulating glazing can foster higher racking resistance and stiffness compared with single non-insulating glazing. However, many challenges still lie ahead to improve the type of the bonding line, the type of the used adhesive, the position of the glazing, etc., to enhance the horizontal resistance and stiffness of timber–glass wall elements, improving the structural stability of the whole timber building. The racking stiffness with a polyurethane or silicone adhesive is not in the same range as the compared timber-framed walls with conventional sheathing boards, such as OSB or fibre–plaster boards, which are prescribed by standards as primary load-bearing racking-resistant structural wall elements.

2.2.2. Double-Skin Façade (DSF) Timber–Glass Wall Elements

As presented in Section 2.2.1, many experimental and numerical studies highlight that the racking resistance obtained with the developed timber single-skin façade (SSF) elements is not sufficient to improve the structural behaviour of a whole building under a horizontal load impact, especially if the building is exposed to a more significant horizontal load impact as schematically presented in Figure 3. Consequently, it is crucial to develop a new load-bearing timber–glass wall element as a resisting bracing element in another way, not by using only two- or three-layer insulating single-skin (SSF) glazing, but by adding a structurally important external non-insulating single-layer glass pane. Such façade elements are treated as double-skin façade elements (DSF) and will be further presented. The introduction of innovative solutions by applying the developed DSF elements would thus expand the range of multi-storey timber building design options on account of said structural advantages. In practice, this would mean that buildings with a slightly more complex design could be built (a higher degree of floor plan or façade asymmetry, more storeys, etc., are allowed) within the scope of the same boundary conditions, and better energy efficiency could be achieved.

The schematic presentation of such a load-bearing DSF element is provided in Figure 5. It is important to point out that the thermal-insulating three-layered float glazing is placed on the internal side of the façade element and a single-layer non-insulating fully-tempered glazing on the external side. Solar shading systems can be integrated within the cavity, and the width of the cavity can vary from 200 mm to even more than 2 m. The installation of any ventilation devices, which would also be optimal for the building, is not suitable for load-bearing DSF elements. Ventilation requires openings in the load-bearing elements of the DSF system, which significantly affect the horizontal load-bearing capacity of such a DSF wall element as presented in our previous study on timber-framed wall elements with fibre–plaster sheathing boards [26]. Within the scope of the Home+ development project, the developed timber DSF wall elements constitute an additional potential of transparent areas in multi-storey timber construction [18]. In addition to the foreseen structural advantages, they provide better sound insulation of the building envelope and better energy performance as compared to the regular triple-insulating glazing [27–35].

Recently, many studies have analysed the thermal and acoustic performance of DSF elements, but almost none of them have analysed their structural behaviour, especially in terms of determining their racking resistance. Such façade elements were in the past primarily developed with the goal of essentially improving the thermal and acoustic resistance of the building envelope [30–32]. Therefore, DSF elements have been proposed as a promising passive building technology to enhance energy efficiency and improve indoor thermal comfort [33]. Such a constructed envelope DSF element demonstrates better acoustic resistance [29] in comparison with the widely used and previously described single-skin façade (SSF) elements and can, therefore, be suitable for high-noise areas where a high level of sound insulation is required [34,35]. In wide research in Pomponi and D'Amico [28], a structural approach with a timber DSF was studied, but only for the vertical load impact. Consequently, a load-bearing DSF timber element must also be developed for a horizontal load impact to increase the possibility of also using larger transparent glass areas in multi-storey timber-framed buildings and consequently to increase the potential of wood as a natural and eco-friendly material in a variety of multi-storey residential buildings.

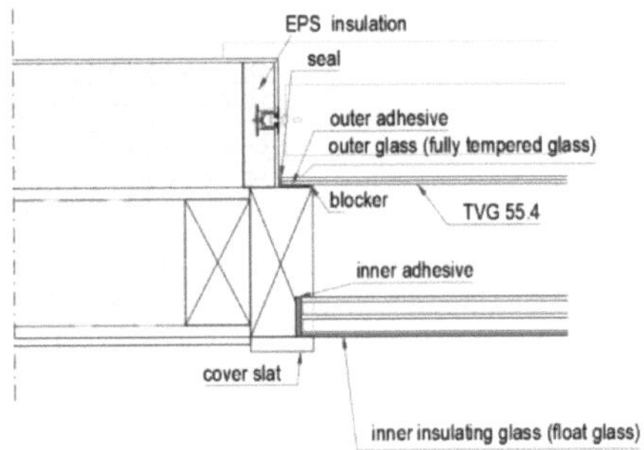

Figure 5. Schematic presentation of a DSF load-bearing structural wall element [19].

3. Mathematical Modelling of Multi-Storey Timber-Framed Buildings

3.1. Mathematical Modelling of Conventional Timber-Framed Wall Elements

To perform the numerical analysis, it was initially necessary to define a suitable mathematical model of the structure. For this purpose, the previously introduced mathematical model with a fictive diagonal for determining the racking stiffness of timber-framed wall elements with conventional OSB or fibre–plaster (FPB) sheathing material [20] was applied and further developed for the timber–glass wall elements stiffness simulation. It is important to point out that in the case of OSB or FPB, sheathing boards are connected to the timber frame with mechanical fasteners, usually located at a constant distance (s_{eff}). Consequently, the effective bending stiffness $(EI)_{eff}$ of such a wall element can be calculated in a semi-analytical way through Equation (1c) using the Gamma method. Following the expressions presented in [20], the fictive diagonal diameter for conventional sheathing boards (OSB or FPB) is further determined in the way that the horizontal displacement of the actual wall element is the same as the horizontal displacement of the simplified model with a fictive diagonal as schematically presented in Figure 6. Finally, the fictive diagonal diameter (d_{fic}) is expressed in the final analytical form, as follows:

$$A_{d,fic} = \frac{k_p \cdot L_d}{E_D \cdot \cos^2 \alpha} \tag{1a}$$

$$k_p = \frac{1}{D_p} = \left(\frac{H^3}{3 \cdot EI_{eff}} + \frac{H}{GA_s} \right)^{-1} \tag{1b}$$

$$(EI)_{eff} = E_b I_b + E_t I_t = E_b \cdot \frac{n_b \cdot t \cdot b^3}{12} + E_t \cdot \left(\frac{2 \cdot a^3 \cdot c}{12} + \frac{d^3 \cdot c}{12} + 2 \cdot \gamma_i \cdot A_t \cdot z^2 \right) \tag{1c}$$

$$d_{fic} = 2 \cdot \sqrt{\frac{A_{d,fic}}{\pi}} \tag{1d}$$

with E_b and E_t being the moduli of elasticity of the board and the timber. E_D is the modulus of elasticity of the diagonal with the fictive cross-section of the diagonal ($A_{d,fic}$) and L_d is the length of the diagonal. The horizontal stiffness of the panel is k_p, D_p is the panel capacity, H is the height of the panel, and GA is the shear stiffness of the panel. The geometrical characteristics in Equation (1c) are schematically presented in Figure 7. However, it is important to point out that the mathematical model developed by [20] can be used only for sheets that are mechanically fastened to the timber frame by staples or nails. The effective stiffness $(EI)_{eff}$ is calculated using the Gamma method following the Eurocode 5 [36] expressions. The stiffness coefficient of fasteners γ_y is defined in accordance with Eurocode 5 [36].

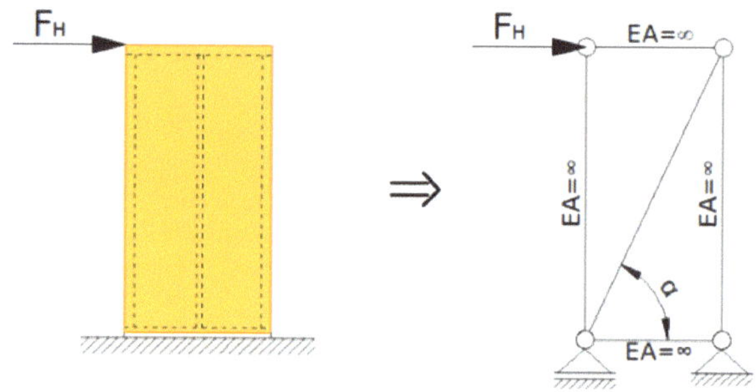

Figure 6. Schematic presentation of the fictive diagonal model.

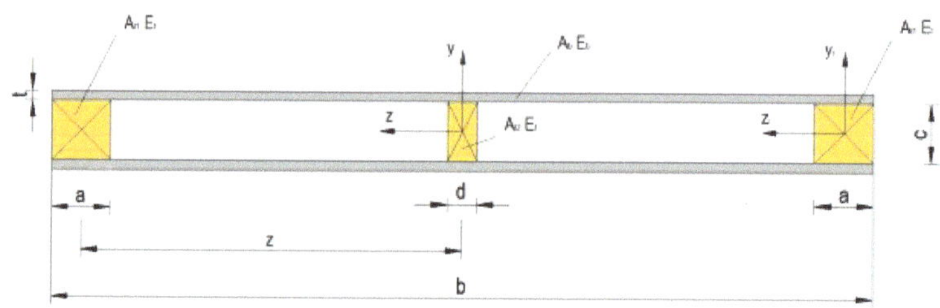

Figure 7. Cross-section of a typical timber-framed wall with a two-sided sheathing board [20].

3.2. Mathematical Modelling of DSF Timber Wall Elements

In case of timber–glass wall elements, where the glass pane is continuously bonded to the timber frame, it is not possible to determine the gamma coefficient and consequently the effective stiffness $(EI)_{eff}$ in Equations (1b) and (1c) directly with the known expressions from the Eurocodes, as it is numerically performed for the conventional sheathing boards (OSB or FPB) in Section 3.1. The diameter of the fictive diagonal (d_{fic}) can be determined using Equation (2):

$$d_{fic} = \sqrt{\frac{4 \cdot F_{cr} \cdot L_d}{w_{cr} \cdot (\cos\alpha)^2 \cdot \pi \cdot E_D}} \qquad (2)$$

in two alternative approaches:

- By using the experimental results from [37] with the measured values for force forming the first crack in the glass pane (F_{cr}) and the corresponding horizontal displacement (w_{cr}) at the top of the wall element. However, this procedure is very expensive and also time-consuming;
- By using the numerical finite element method (FEM) approach, where the flexibility of the bonding line is simulated with elastic spring elements. This approach is briefly described below.

Certain developed mathematical finite element (FE) models with spring elements simulate the flexibility of the bonding line between the glass pane and the timber frame [24]. The model is based on an extensive numerical parametric study performed only for single-skin façade (SSF) elements and presented in [15]. The adhesive bonding of the glazing panes to the timber frame was modelled using elastic linear link elements (springs) as schematically presented in Figure 8 and introduced by Kreuzinger and Niedermaier [38]. In the computational model, the timber frame was modelled with one-dimensional finite elements (beams and studs) and the glazing panels with two-dimensional finite elements of the "composite shell" type, which allows the simulation of multilayer shells. The timber material was considered as an isotropic elastic material (with the modulus of elasticity $E_{0,mean}$) and the elements of the timber frame were modelled as the simple plane stress elements. As glass is a very brittle material, it was therefore modelled as acting linearly elastic in tension and compression; in reality, the adhesive bonding is provided continuously over the whole perimeter, and the stiffness properties of discrete spring elements were defined based on the spacing of the springs (la) in the computational model using Equations (3) and (4) for the bonding line of the inner and outer glass panes (see Figure 5) separately:

$$K_1 = \frac{E_a \cdot A_a}{t_a} = \frac{E_a \cdot (w_a \cdot l_a)}{t_a} \tag{3}$$

$$K_2 = \frac{G_a \cdot A_a}{t_a} = \frac{G_a \cdot (w_a \cdot l_a)}{t_a} \tag{4}$$

where K_1 is stiffness in the direction normal to the connected plane, while K_2 is the shear stiffness in the two perpendicular directions in the connected plane. E_a and G_a are the modulus of elasticity and the shear modulus of the adhesive material, respectively, t_a and w_a designate the thickness and the width of the adhesive layer, respectively, and l_a is the impact length for a single spring element and is equal to the spacing between the springs.

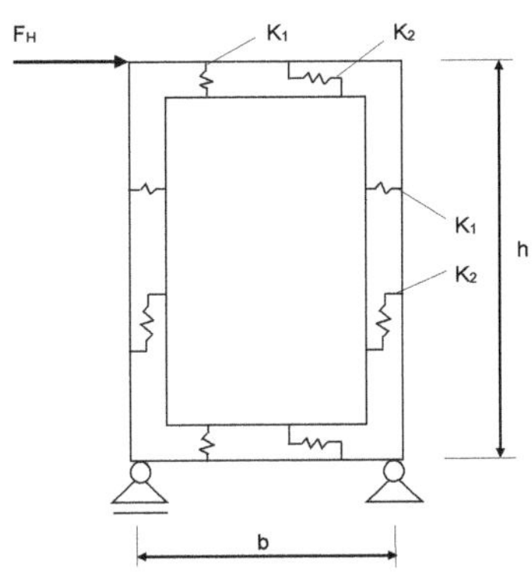

Figure 8. Spring model introduced by Kreuzinger and Niedermaier [38].

With the presented FE model force forming the first crack in the glass pane (F_{cr}), the corresponding horizontal displacement (w_{cr}) at the top of the wall element at this force can be calculated. The diameter of the fictive diagonal (d_{fic}) can be determined now using the expression in Equation (2).

The calculated values of the diameter of the fictive diagonal element (d_{fic}), considering various values for the width (w_a) and the thickness (t_a) of the adhesive in the bonding line, are presented in Tables 1 and 2. The diameter of the fictive diagonal (d_{fic}) was first determined for different widths of the adhesive layer, considering a constant adhesive thickness of t_a = 7 mm (Table 1), and also for different adhesive thicknesses, considering a constant (experimental) width of the adhesive layer of w_a = 28 mm (Table 2). Both tables present the racking stiffness (R) of the DSF wall element calculated with the programme SAP 2000 [21] using the spring mathematical model (Figure 8) and the values for fictive diagonal diameters subsequently obtained by Equation (2).

Table 1. Fictive diagonal diameter and stiffness at different widths of the adhesive layer w_a (E_a = 1.083 MPa).

Width of the Adhesive Layer w_a (mm)	Racking Stiffness R (N/mm)	Diameter of the Fictive Diagonal d_{fic} (mm)
24	792	8.194
28 *	857	8.521
exp.	909	
32	917	8.817
36	976	9.092

* Additional experimental study performed on DSF elements [37].

Table 2. Fictive diagonal diameter and stiffness at different adhesive thicknesses t_a (E_a = 1.083 MPa).

Adhesive Thickness t_a (mm)	Racking Stiffness R (N/mm)	Diameter of the Fictive Diagonal d_{fic} (mm)
3	1563	11.506
5	1080	9.566
7 *	857	8.521
exp.	909	
9	765	8.052

* Additional experimental study performed on DSF elements [37].

A comparison of the numerical and experimental results for the analysed DSF elements is very briefly presented in Figure 9. However, a deep analysis can be found in [19].

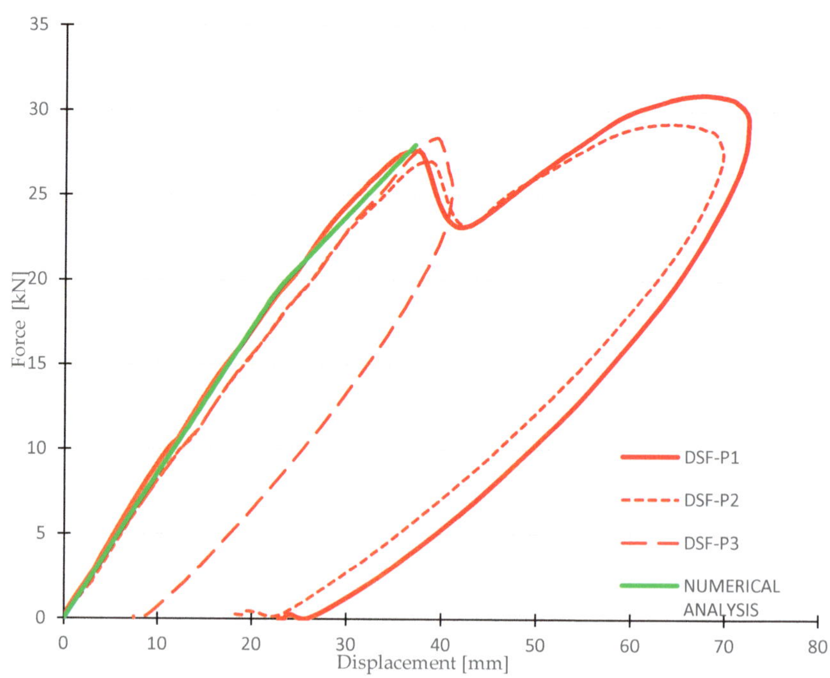

Figure 9. Experimental and numerical force-displacement diagrams for the DSF-P wall element.

The results show that the racking stiffness (R) increases with the increasing width of the adhesive layer (w_a) or the decreasing adhesive thickness (t_a). In the opposite case, stiffness decreases. This was expected, as a higher thickness or lower effective width of the adhesive results in the higher yield strength of the joint between the timber frame and the glass. Similar conclusions were obtained for SSF elements in the parametric numerical study in [16] and will not be further analysed in this paper.

However, in the current form, there is no further developed possible mathematical analytical or semi-analytical correlation between the spring stiffness (K_1 in Equation (3) and K_2 in Equation (4)) and the effective bending stiffness (EI_{eff} in Equation (1c)) to follow the developed expressions from Equations (1a–1d) to analytically determine the needed value of the fictive diagonal parameter (d_{fic}), as it can be very simply analytically performed for conventional FPB or OSB sheathing boards. The DSF modelling presented here is therefore an extension of the previously developed mathematical models of this type by using spring elements (schematically presented in Figure 8) to simulate the bonding line [19]. However, at this point, it is of the utmost importance to point out that using the described FEM procedure with spring elements to determine the racking stiffness of DSF elements is very time consuming. Moreover, the calculation time is too lengthy to be implemented into the whole structural building model for the practical engineering implementation for the seismic analysis of the whole structure.

4. Special Numerical Study on Selected Three-Storey Timber-Framed Building

The seismic resistance of a three-storey prefabricated building in Ljubljana (LJ), with a constant floor plan in each storey as shown in Figure 10, has been analysed. The points ABCD are the corner points on the top storey of the building, where the racking stiffnesses (R) and displacements (U) were calculated. The side view of the building is shown in Figure 11, where the marks 1–8 and A–G are the axes where the load-bearing wall elements are located. The aim of our study is to compare the seismic resistance of a three-storey building, where DSF elements will be considered as load-bearing and non-load-bearing elements, and to assess a possible contribution of the load-bearing DSF elements to the overall seismic resistance of the chosen timber building. Therefore, two alternative structural analyses were performed:

(a) The DSF elements are considered as non-resisting structural wall elements;
(b) The DSF elements are considered as horizontal-load-resisting structural wall elements.

The analyses will be focused primarily on the following two items:

(a) An increase in the overall racking stiffness of the building if the DSF elements are considered as racking resisting;
(b) A decrease in the distortional effect in the first storey if the DSF elements are considered as racking resisting.

The DSF elements are primarily positioned on the south façade (direction X) to increase solar gains through the transparent areas. The seismic analysis of the building was performed using a 3D mathematical model and a modal analysis of the structure by using the commercial finite element model computer programme SAP 2000 Nonlinear v 23.0.0 [21]. Therefore, in alternative (a), only solid wall elements without any openings and DSF wall elements were considered as load-bearing elements, and in the alternative (b), DSF elements with t_a = 7 mm and w_a = 28 mm, which are marked in red on the floor plan (Figure 10), were taken into account in the calculation, using fictive diagonal elements to simulate the racking stiffness of such transparent elements.

Since the length of the wall elements is not ideal (multiples of 1.25 m), we made the assumption for all wall elements that each wall is a multiple of 1.25 m. The principle is that for each wall, the number of full wall elements (1.25 m long) is determined, and the remaining length is considered a full element if the length of the remaining part of the wall is more than half the length of the full wall, i.e., 0.625 m. All wall elements (external and internal) consist of a timber frame and two 2.50 m high sheathing boards. Fibre–plaster boards (FPB) are generally used as sheathing boards. Where the wall elements must be reinforced, OSB sheathing boards were used instead of fibre–plaster boards.

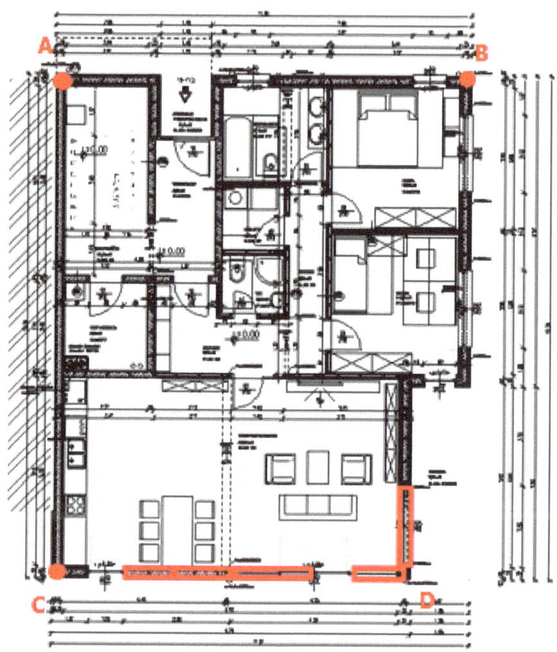

Figure 10. Floor plan of the building and the marked DSF elements (in red).

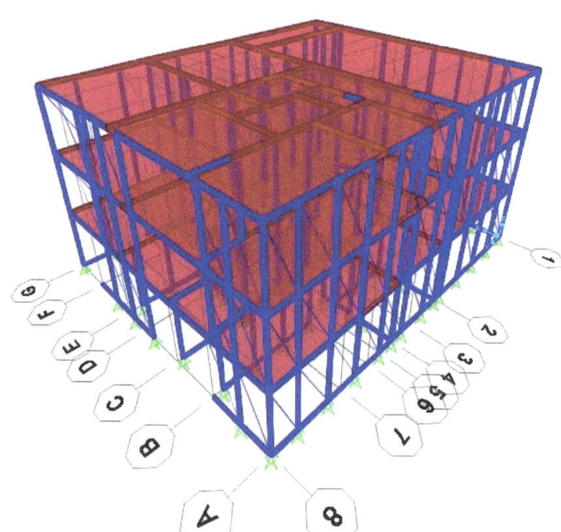

Figure 11. 3D model of the analysed three-storey building.

The structure was modelled using the previously described model with a fictive steel diagonal with a circular cross-section, where the stiffness of the wall element is simulated by the fictive diameter of the diagonal (d_{fic}). Table 3 shows the fictive diagonal diameters and the racking stiffness of each resisting wall element, the conventional sheathing boards (with FPB or OSB sheathing boards for the external and internal wall timber-framed wall elements) and the load-bearing DSF elements. Both values for the DSF element were taken from Tables 1 and 2 by using Equations (2)–(4), while the values for the OSB or FPB timber-framed wall elements were calculated according to Equations (1a)–(1d).

Table 3. Diameter of the fictive diagonals and load-bearing capacities of the wall elements.

Timber-Framed Wall Elements	Racking Stiffness (R) of the Resisting Wall Elements (N/mm)	Diameter of the Fictive Diagonal d_{fic} (mm)
DSF	857	8.52
OSB—external wall	2800	16.90
OSB—internal wall	2482	15.93
FPB—external wall	4192	20.70
FPB—internal wall	3962	20.10

The supports are also simulated as fully rigid. At the same time, the axial stiffness of the frame in the calculation model should be high enough so that the influence of the frame ductility can be eliminated and only the ductility of the diagonals can be considered [20]. The material used for the diagonals was steel with a modulus of elasticity of E = 210 GPa. The surface load subjected to the floor slabs was 2 kN/m^2. A computational model has been carried out for a three-storey timber building with identical floor plan dimensions. Figure 11 shows the 3D model for the three-storey building, while Figure 12 shows the model of the wall in Axis 1 using a load-bearing and non-load-bearing DSF element.

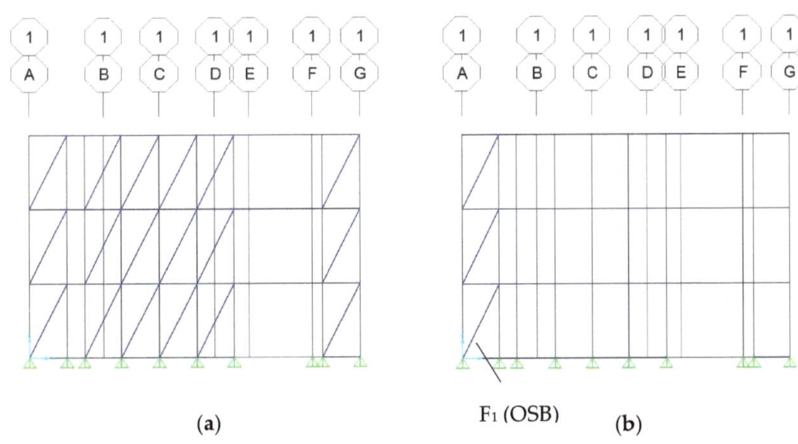

Figure 12. View of Axis 1 of the three-storey wall model using load-bearing (**a**) and non-load-bearing DSF elements (**b**).

5. Discussion of Results

The numerical results for the three-storey prefabricated building have been compared, considering the DSF elements as load-bearing (alternative (b)) and non-load-bearing (alternative (a)). In the case of the load-bearing DSF elements, the thickness of the fictive diagonal was 8.52 mm. The comparison was performed for the location in Ljubljana, where the design ground acceleration ag is 0.25 g. First, the oscillation times of the structure were calculated. Table 4 shows the oscillation times (the first three oscillation modes) of the structure for the two analysed alternatives with resisting and non-resisting DSF elements.

Table 4. Oscillation times of the three-storey building considering load-bearing and non-load-bearing DSF wall elements.

DSF Element	Non-Load-Bearing DSF Elements (a)	Load-Bearing DSF Elements (b)
Oscillation Mode	**Oscillation Times T [s]**	**Oscillation Times T [s]**
1.	0.479	0.438
2.	0.367	0.363
3.	0.308	0.303

As expected, the oscillation times are higher when considering non-load-bearing DSF elements, because the racking stiffness is in this case smaller and the mass is supposed to be unchanged. The calculated horizontal (racking) stiffnesses (R) and displacements of the structure (at the resultant force F_H = 376 kN for non-load-bearing DSF elements and F_H = 386 kN for load-bearing DSF elements) in both global orthogonal directions of the earthquake action (directions X and Y) for the individual selected control points (A–D), considering both load-bearing and non-load-bearing DSF elements, are shown in Table 5.

Table 5. Racking stiffnesses (R) and displacements (U) of the corner points on the top storey of the three-storey building.

DSF Element		Load-Bearing DSF Elements		Non-Load-Bearing DSF Elements	
Earthquake		**Direction X**	**Direction Y**	**Direction X**	**Direction Y**
Location		LJ		LJ	
Point	Displacement (mm)				
A	U_x	10.25	7.96	9.35	8.51
	U_y	1.09	6.37	2.31	6.89
	U_R	10.31	10.20	9.63	10.95
B	U_x	10.25	7.95	9.35	8.51
	U_y	7.87	10.90	11.13	10.44
	U_R	12.92	13.49	14.54	13.47
C	U_x	17.35	6.59	21.64	7.56
	U_y	1.09	6.38	2.32	6.90
	U_R	17.38	9.17	21.76	10.24
D	U_x	17.35	6.59	21.64	7.56
	U_y	7.88	10.91	11.14	10.45
	U_R	19.06	12.75	24.34	12.90
R (N/mm)		15,198	18,859	11,314	18,953

The allowed value of the horizontal displacements for a multi-storey building, according to [22] is H/500, which, in our case, amounts to 15 mm. For Ljubljana, the values of horizontal displacements exceed the prescribed Eurocode limits by about 21% when DSF are considered as resisting and by about 38% when DSF are considered as non-resisting. These values are marked in red in the table. The overall racking stiffness (R) of the whole building also essentially increases if load-bearing DSF elements are used, i.e., by 34% in the X direction, while there is almost no influence for the Y direction because almost all load-bearing DSF elements are placed in the X direction (south face) of the building façade only. Therefore, the use of load-bearing DSF elements is very important in this case.

Table 6 shows acting horizontal forces (F_x and F_y, respectively) due to seismic action in two global perpendicular directions (directions X and Y) and the resulting force (FR) in the external walls of the building façade (axes 1, 8, A, G), which are shown in Figure 11. The horizontal force acting on the corner timber-framed wall element 1 in Axis 1 with the conventional OSB sheathing boards (F_1 marked in Figure 12b) is particularly controlled to assess the influence of the DSF load-bearing elements in decreasing the distortion of the first floor.

Table 6. Horizontal forces in the exterior walls of the building.

DSF Element		Load-Bearing DSF Elements		Non-Load-Bearing DSF Elements	
Earthquake		Direction X	Direction Y	Direction X	Direction Y
Location		LJ		LJ	
Axis	Force (kN)				
1	F_x	52.82	21.67	30.77	11.55
	F_y	126.01	64.38	130.11	58.53
	F_R	136.63	67.93	133.70	59.66
8	F_x	43.78	43.62	39.44	45.94
	F_y	80.13	119.42	117.79	120.68
	F_R	91.31	127.14	124.22	129.13
A	F_x	13.68	76.43	31.71	83.51
	F_y	161.29	138.69	183.43	147.10
	F_R	161.87	158.36	186.15	169.15
G	F_x	28.80	47.96	42.92	47.38
	F_y	130.93	128.20	131.08	125.20
	F_R	134.06	136.88	137.93	133.87
Axis 1: F_1 (kN)		20.20	24.76	24.56	24.68

It is evident from the presented results that the force F_1 decreased notably (by almost 18%) when the load-bearing DSF elements were used for seismic excitation in the X direction. On the other hand, there is practically no influence for seismic excitation in the Y direction, as similar results were observed for the overall racking stiffness of the building in Table 5. Consequently, it can be concluded that the distortion effect on the building can be essentially decreased by using load-bearing DSF elements.

6. Conclusions

Based on the computational analyses performed, the use of the developed DSF elements as load-bearing structural elements to increase the racking load-bearing capacity of the whole structure proved to be very reasonable, since the racking stiffness of the entire building could be increased by up to 35% with the given fixed installation of the DSF elements. It should be noted that the influence of the stiffness of the DSF elements on the overall racking stiffness of the whole building depends on the floor layout of the load-bearing DSF elements on each individual floor. Additionally, as a secondary goal of the study, it is important to point out that by using the DSF elements as racking resistant, the distortion on the first floor essentially decreased because the horizontal action in the checked corner timber-framed wall elements with an OSB sheathing board in Axis 1 was lower by almost 18%.

The development of such racking-resistant timber DSF elements can open many new perspectives in designing contemporary multi-storey timber buildings located in seismic areas with strong winds and with a strong asymmetrical position of the transparent glass areas. On the other hand, this decreases the energy demand for heating and provides better daylight, contributing to better living comfort in the building. However, this opens many structural problems, especially for mid- and high-rise prefabricated timber buildings, which can be solved by the developed lateral resistant DSF elements. The results of the study can also have important socioeconomic effects, as they can significantly contribute to the additional expansion of multi-storey timber buildings throughout Europe, resulting in the better use of forested areas, and thus significantly contributing to the reduction of the environmental impacts of buildings.

However, there are still some problems with the mathematical modelling of resisting DSF elements using a simplified fictive diagonal model, which is only acceptable for a practical and rapid engineering static and dynamical analysis of multi-storey prefabricated timber buildings. In the current stage, the fictive diagonal model can be widely used only for conventional timber-framed wall elements with OSB or FPB boards. However, for load-bearing DSF elements, the effective bending stiffness of the composite wall elements still cannot be calculated in a semi-analytical way through Equation (1c) using the Gamma method. Therefore, the results of experimental tests or the FEM model with spring elements have to be used first to determine the diameter of the fictive diagonal using Equation (2). However, the experimental approach is very expensive and the FEM approach with springs presented in this paper is still very time consuming. Therefore, it is of the utmost importance for further research to develop a semi-analytical approach to determine the fictive diagonal diameter.

Author Contributions: Conceptualization, M.P. and E.K.Š.; methodology, M.P. and E.K.Š.; software, E.K.Š.; validation M.P. and E.K.Š.; formal analysis, M.P. and E.K.Š.; investigation M.P. and E.K.Š.; resources, M.P. and E.K.Š.; data curation, M.P. and E.K.Š.; writing—original draft preparation, M.P. and E.K.Š.; writing—review and editing, M.P. and E.K.Š.; visualization, M.P. and E.K.Š.; supervision, M.P.; project administration, E.K.Š. All authors have read and agreed to the published version of the manuscript.

Funding: Funding for this research was provided by the Slovenian Research Agency and the Ministry of Higher Education, Science and Technology of the Republic of Slovenia, National research program P2-0129 and the investment is co-financed by Republic of Slovenia and the European Union under the European Regional Development Fund (research core funding Development of multifunctional climatically active building wrapper—HOME+).

Institutional Review Board Statement: Not applicable.

Informed Consent Statement: Not applicable.

Data Availability Statement: Data available on request.

Conflicts of Interest: The authors declare no conflict of interest.

References

1. Arguez, A.; Hurley, S.; Inamdar, A.; Mahoney, L.; Sanchez-Lugo, A.; Yang, L. Should We Expect Each Year in the next Decade (2019–28) to Be Ranked among the Top 10 Warmest Years Globally? *Bull. Am. Meteorol. Soc.* **2020**, *101*, E655–E663. [CrossRef]
2. Hamid, A.A.; Farsäter, K.; Wahlström, Å.; Wallentén, P. Literature Review on Renovation of Multifamily Buildings in Temperate Climate Conditions. *Energy Build.* **2018**, *172*, 414–431. [CrossRef]
3. Ascione, F.; Bianco, N.; De Masi, R.F.; Mauro, G.M.; Vanoli, G.P. Energy Retrofit of Educational Buildings: Transient Energy Simulations, Model Calibration and Multi-Objective Optimization towards Nearly Zero-Energy Performance. *Energy Build.* **2017**, *144*, 303–319. [CrossRef]
4. European Commission. *Communication from the Commission to the European Parliament, the European Council, the Council, the European Economic and Social Committee and Committee of the Regions: A Renovation Wave for Europe—Greening Our Buildings, Creating Jobs, Improving Lives*; European Commission: Brussels, Belgium, 2020.
5. Buildings, M.T. Class 2, 3 and 5 Buildings. In *Technical Design Guide issued by FWPA*; WoodSolutions: Canning Vale, WA, USA, 2017.
6. Kuzmanovska, I.; Gasparri, E.; Monne, D.T.; Aitchison, M. Tall Timber Buildings: Emerging Trends and Typologies. In Proceedings of the 2018 World Conference on Timber Engineering, Seoul, Republic of Korea, 20–23 August 2018.
7. Salvadori, V. The Development of a Tall Wood Building. Ph.D. Thesis, Politecnico Milano & Technische Universitat Wien (TU Wien), Vienna, Austria, 2017.
8. Salvadori, V. *Multi-Storey Timber-Based Buildings: An International Survey of Case Studies with Five or More Storeys over the Last Twenty Years*; Technische Universitat Wien (TU Wien): Vienna, Austria, 2021.
9. Krtschil, A.; Orozco, L.; Bechert, S.; Wagner, H.J.; Amtsberg, F.; Chen, T.-Y.; Shah, A.; Menges, A.; Knippers, J. Structural Development of a Novel Punctually Supported Timber Building System for Multi-Storey Construction. *J. Build. Eng.* **2022**, *58*, 104972. [CrossRef]

10. Wurm, J. *Glass Structures: Design and Construction of Self-Supporting Skins*; Walter de Gruyter: Berlin, Germany, 2007; ISBN 3-7643-8317-8.
11. Niedermaier, P. Shear-Strength of Glass Panel Elements in Combination with Timber Frame Constructions. In Proceedings of the 8th International Conference on Architectural and Automotive Glass (Glass Processing Days), Tampere, Finland, 15–18 June 2003; pp. 262–264.
12. Premrov, M.; Serrano, E.; Winter, W.; Fadai, A.; Nicklisch, F.; Dujič, B.; Šušteršič, I.; Brank, B.; Štrukelj, A.; Držečnik, M.; et al. *Workshop Report "WP 6: Testing on Life-Size Specimen Components: Shear Walls, Beams and Columns Including Long-Term Behaviour": WoodWisdom-Net, Research Project, Load Bearing Timber-Glass-Composites, 2012–2014*; Springer: Berlin/Heidelberg, Germany, 2015; p. 151.
13. Neubauer, G.; Schober, P. Weiterentwicklung Und Herstellung von Holz-Glas Verbundkonstruktionen Durch Statisch Wirksames Verkleben von Holz Und Glas Zum Praxiseinsatz Im Holzhausbau. In *Endbericht zum Impulsprojekt V2*; KInd Holz Technologie: Wien, Austria, 2008.
14. Ber, B.; Premrov, M.; Štrukelj, A.; Kuhta, M. Experimental Investigations of Timber–Glass Composite Wall Panels. *Constr. Build. Mater.* **2014**, *66*, 235–246. [CrossRef]
15. Ber, B.; Finžgar, G.; Premrov, M.; Štrukelj, A. On Parameters Affecting the Racking Stiffness of Timber-Glass Walls. *Glass Struct. Eng.* **2019**, *4*, 69–82. [CrossRef]
16. Premrov, M.; Ber, B.; Kozem Šilih, E. Study of Load-Bearing Timber-Wall Elements Using Experimental Testing and Mathematical Modelling. *Adv. Prod. Eng. Manag.* **2021**, *16*, 67–81. [CrossRef]
17. Winter, W.; Hochhauser, W.; Kreher, K. Load Bearing and Stiffening Timber-Glass-Composites (TGC). In Proceedings of the World Conference on Timber Engineering, Trentino, Italy, 20–24 June 2010.
18. Premrov, M.; Žegarac Leskovar, V.; Kozem Šilih, E.; Ber, B. Development of Double Glazing Façade Prefabricated Wooden Wall Element. Patent SI 26095 A, 29 April 2022.
19. Šilih, E.K.; Premrov, M. Numerical Study of Racking Resistance of Timber-Made Double-Skin Façade Elements. *Adv. Prod. Eng. Manag.* **2022**, *17*, 231–242. [CrossRef]
20. Pintarič, K.; Premrov, M. Mathematical Modelling of Timber-Framed Walls Using Fictive Diagonal Elements. *Appl. Math. Model.* **2013**, *37*, 8051–8059. [CrossRef]
21. Computer & Structures, Inc. *CSI. SAP2000 (V23.0.0). Structural Analyis and Design 2021*; CSI: Berkeley, CA, USA, 2021.
22. Code, P. *Eurocode 8: Design of Structures for Earthquake Resistance-Part 1: General Rules, Seismic Actions and Rules for Buildings*; European Committee for Standardization: Brussels, Belgium, 2005.
23. European Committee for Standardization. *For Eurocode 1: Actions on Structures—Part 1–4: General Actions—Wind Actions*; European Committee for Standardization: Brussels, Belgium, 2005.
24. Hochhauser, W. *Ein Beitrag Zur Berechnung Und Bemessung von Geklebten Und Geklotzten Holz-Glas-Verbundscheiben*; Vienna University of Technology: Vienna, Austria, 2011.
25. Blyberg, L. *Timber/Glass Adhesive Bonds for Structural Applications*; Institutionen för teknik, Linnéuniversitetet: Växjö, Sweden, 2011.
26. Šilih, E.K.; Premrov, M. Analysis of Timber-Framed Wall Elements with Openings. *Constr. Build. Mater.* **2010**, *24*, 1656–1663. [CrossRef]
27. Alibaba, H.Z.; Ozdeniz, M.B. Energy Performance and Thermal Comfort of Double-Skin and Single-Skin Facades in Warm-Climate Offices. *J. Asian Archit. Build. Eng.* **2016**, *15*, 635–642. [CrossRef]
28. Pomponi, F.; D'Amico, B. Holistic Study of a Timber Double Skin Façade: Whole Life Carbon Emissions and Structural Optimisation. *Build. Environ.* **2017**, *124*, 42–56. [CrossRef]
29. Ghaffarianhoseini, A.; Ghaffarianhoseini, A.; Berardi, U.; Tookey, J.; Li, D.H.W.; Kariminia, S. Exploring the Advantages and Challenges of Double-Skin Façades (DSFs). *Renew. Sustain. Energy Rev.* **2016**, *60*, 1052–1065. [CrossRef]
30. Chan, A.L.S.; Chow, T.T.; Fong, K.F.; Lin, Z. Investigation on Energy Performance of Double Skin Façade in Hong Kong. *Energy Build.* **2009**, *41*, 1135–1142. [CrossRef]
31. Musa, B.T.; Alibaba, H.Z. Evaluating the Use of Double Skin Facade Systems for Sustainable Development. *J. Med. Appl. Biosci.* **2016**, *9*, 2017.
32. Xu, L.; Ojima, T. Field Experiments on Natural Energy Utilization in a Residential House with a Double Skin Façade System. *Build. Environ.* **2007**, *42*, 2014–2023. [CrossRef]
33. Saelens, D.; Roels, S.; Hens, H. Strategies to Improve the Energy Performance of Multiple-Skin Facades. *Build. Environ.* **2008**, *43*, 638–650. [CrossRef]
34. Huckemann, V.; Leão, É.B.; Leao, M. Acoustic Comfort in Office Buildings with Double Skin Glass Façades. *Bauphysik* **2009**, *31*, 305–312. [CrossRef]
35. Saroglou, T.; Theodosiou, T.; Givoni, B.; Meir, I.A. A Study of Different Envelope Scenarios towards Low Carbon High-Rise Buildings in the Mediterranean Climate-Can DSF Be Part of the Solution? *Renew. Sustain. Energy Rev.* **2019**, *113*, 109237. [CrossRef]
36. *Code Eurocode 5: Design of Timber Structures, Part 1-1 General Rules and Rules for Buildings*; European Committee for Standardization: Brussels, Belgium, 2005.

37. Premrov, M.; Žegarac Leskovar, V.; Ber, B.; Kozem Šilih, E.; Vogrinec, K.; Lešnik Nedelko, M. *Development of Multifunctional Air Conditioning Active Supporting Wrapper of Buildings—Home+: Final Research Project Report: Working Package A1.2 Load Capacity and Rigidity Survey DSF Wall Elements (UM FGPA, TRL 3-4): Project Duration (TRL 3-4): 1. 3. 2019-28. 2. 2021*; University of Maribor, Faculty of Civil Engineering, Transport Engineering and Architecture: Maribor, Slovenia, 2021.
38. Kreuzinger, H.; Niedermaier, P. *Glas Als Schubfeld. Tagungsband Ingenieurholzbau, Karlsruher Tage*; KIT Scientific Publishing: Karlsruhe, Germany, 2005.

sustainability

Article

Optimal Dimensions of Post-Tensioned Concrete Cylindrical Walls Using Harmony Search and Ensemble Learning with SHAP

Gebrail Bekdaş [1,*], Celal Cakiroglu [2], Sanghun Kim [3] and Zong Woo Geem [4,*]

1 Department of Civil Engineering, Istanbul University-Cerrahpasa, 34320 Istanbul, Turkey
2 Department of Civil Engineering, Turkish-German University, 34820 Istanbul, Turkey; cakiroglu@tau.edu.tr
3 Department of Civil and Environmental Engineering, Temple University, Philadelphia, PA 19122, USA; sanghun.kim@temple.edu
4 Department of Smart City & Energy, Gachon University, Seongnam 13120, Republic of Korea
* Correspondence: bekdas@iuc.edu.tr (G.B.); geem@gachon.ac.kr (Z.W.G.)

Abstract: The optimal design of prestressed concrete cylindrical walls is greatly beneficial for economic and environmental impact. However, the lack of the available big enough datasets for the training of robust machine learning models is one of the factors that prevents wide adoption of machine learning techniques in structural design. The current study demonstrates the application of the well-established harmony search methodology to create a large database of optimal design configurations. The unit costs of concrete and steel used in the construction, the specific weight of the stored fluid, and the height of the cylindrical wall are the input variables whereas the optimum thicknesses of the wall with and without post-tensioning are the output variables. Based on this database, some of the most efficient ensemble learning techniques like the Extreme Gradient Boosting (XGBoost), Light Gradient Boosting Machine (LightGBM), Categorical Gradient Boosting (CatBoost) and Random Forest algorithms have been trained. An R^2 score greater than 0.98 could be achieved by all of the ensemble learning models. Furthermore, the impacts of different input features on the predictions of different machine learning models have been analyzed using the SHapley Additive exPlanations (SHAP) methodology. The height of the cylindrical wall was found to have the greatest impact on the optimal wall thickness, followed by the specific weight of the stored fluid. Also, with the help of individual conditional expectation (ICE) plots the variations of predictive model outputs with respect to each input feature have been visualized. By using the genetic programming methodology, predictive equations have been obtained for the optimal wall thickness.

Keywords: optimization; machine learning; XGBoost; SHAP; prestressed concrete; post-tensioning; genetic programming

Citation: Bekdaş, G.; Cakiroglu, C.; Kim, S.; Geem, Z.W. Optimal Dimensions of Post-Tensioned Concrete Cylindrical Walls Using Harmony Search and Ensemble Learning with SHAP. *Sustainability* **2023**, *15*, 7890. https://doi.org/10.3390/su15107890

Academic Editors: Uroš Klanšek and Tomaž Žula

Received: 10 April 2023
Revised: 30 April 2023
Accepted: 9 May 2023
Published: 11 May 2023

1. Introduction

Prestressing has well-known benefits in terms of increasing the load-carrying capacity and durability of concrete structures like liquid storage tanks, silos, and nuclear facilities [1]. It significantly reduces concrete cracks at the tension side by pre-compressing those parts of the concrete structures under tensile strain before the service loads are applied thereby countering the effects of tension caused by bending. Post-tensioning is a type of prestressing where the tendons are stressed after the concrete has been poured and hardened. Post-tensioning is a particularly favorable technique when the serviceability requirements on a large-scale structure cannot be met using conventional reinforcement techniques while keeping the dimensions of the structure in an economically feasible range [2].

Previous studies of the reinforced concrete structures optimization with and without prestressing have demonstrated the applicability of metaheuristic optimization techniques like the harmony search (HS) methodology to reduce the structural cost and environmental impact [2–6]. Metaheuristic optimization techniques have been applied to a variety of problems in civil and structural engineering including the dimensioning of trusses [7–13], retaining walls [14–22], laminated composite plates [23–30], and steel plate girders [31]. On the other hand, post-tensioning in structures is an area where metaheuristic optimization techniques found relatively few applications. The research in the area of structural post-tensioning can be grouped into studies related to beams [32], slabs [33–35], experimental studies [33,34,36], and numerical studies [32,34–36]. Elbelbisi et al. [32] performed a parametric study of post-tensioned fiber reinforced polymer (FRP) systems. The flexural behavior of beams externally post-tensioned with FRP tendons was investigated using finite element analysis. The application of external post-tensioning significantly increased the load-carrying capacity of the beams. Elsheshtawy et al. [33] investigated the effect of prestressing force and layout of strands on the punching shear strength of slab-column connections and found that the banded layout of the post-tensioning strands was most favorable to increase the punching shear strength of the post-tensioned flat slabs. Attia et al. [34] examined the load-carrying capacity of two-way flat slabs under post-tensioning with external FRP laminates using finite element analysis. A parametric study was carried out and it was shown that post-tensioning with FRP laminates increase the ductility. Furthermore, strengthening near the supports was shown to be more effective than in the middle of the slab. Tahmasebinia et al. [35] investigated post-tensioned concrete flat slabs under dynamic loading. According to finite element analysis, increasing the slab thickness and damping ratio is favorable towards better vibration serviceability. Vavrus and Kralovanec [36] investigated the application of steel fiber reinforced concrete in the anchorage zones of post-tensioning tendons to increase the load-carrying capacity. Based on numerical analysis, increasing the fiber content near the anchorage plate is favorable for the increase of the load carrying capacity. In some of the more recent studies in the area of post-tensioned structural members, bearing capacity of anchorage zones [37], precast post-tensioned girders [38], and post-tensioned self-compacting concrete beams with recycled coarse aggregate [39] have been experimentally and numerically investigated. Lei et al. [37] proposed a new formula for the bearing capacity of anchorage zones using the stress field approach. Joyklad et al. [38] carried out an experimental program with full-scale pre-cast post-tensioned (PCPT) girders accompanied with numerical simulations. Two PCPT girders of 29.95 m length, 1.8 m depth and 0.69 m width were manufactured. The upward deflection response of these girders, post-tensioning tendon strains, and reinforcement strains were measured with displacement transducers and strain gauges for 120 days. Furthermore, the thermal response of fresh concrete was measured using thermocouple wires. All measured parameters were observed to increase during post-tensioning. Yu et al. [39] carried out four-point bending tests on post-tensioned beams with and without recycled coarse aggregate. The specimens made with recycled coarse aggregate and self-compacting concrete were demonstrated to have similar flexural capacity as the specimens with natural coarse aggregate. To the best of the authors' knowledge the only research studies involving the application of metaheuristic optimization techniques to post-tensioned reinforced concrete cylindrical containers was carried out by Bekdaş et al. [40]. It was found that the addition of multiple layers of post-tensioning cables reduces the amount of carbon emission associated with the manufacturing process. Harmony search (HS), teaching–learning based optimization (TLBO), and flower pollination algorithm (FPA) were utilized for minimizing the carbon footprint.

The current study aims to develop predictive models using ensemble machine learning techniques and genetic programming in order to predict the optimal value of wall thickness for a post-tensioned reinforced concrete cylindrical wall. In order to train these predictive models a large database of optimal design configurations has been generated using the harmony search algorithm. The statistical analysis of this data set as well as the harmony search methodology have been presented in Section 2 which also contains the details of the genetic programming and ensemble learning algorithms. Using the genetic programming algorithm, closed-form equations have been proposed for the prediction of the optimal wall thickness for cylindrical walls with and without post-tensioning. The performances of the ensemble learning and genetic programming algorithms in predicting the optimal wall thickness have been presented in Section 3. In the same section also, the input feature impact analysis has been carried out using SHAP and individual conditional expectation plots. The major novelty of the current paper is the proposal of a new methodology to overcome the data scarcity for the training of reliable predictive models using machine learning algorithms. Since reliable machine learning models necessitate large data sets which are not readily available in case of post-tensioned cylindrical walls, the current proposed technique can make a valuable contribution to this area.

2. Machine Learning and Optimization Methods

In this section, the procedures of building data sets of optimum post-tensioned cylindrical walls and developing machine learning (ML) models based on these data sets are explained. The database for the training of the ML models is generated using the harmony search optimization algorithm. In this process the serviceability requirements on the cylindrical wall such as the limitation on the crack size according to ACI 318 code [41] are considered as the constraints of the optimization. The structural analysis of the cylindrical walls has been carried out using the superposition method (SPM), which is based on St. Venant's principle according to which stresses sufficiently far from an applied load are not significantly altered if this load is changed to a statically equivalent load combination [42].

$$M_A + \frac{P_{0A}}{4\beta} + \frac{P_{0B}}{4\beta}C_{\beta l} + \frac{M_{0A}}{2} + \frac{M_{0B}}{2}D_{\beta l} = 0, \quad (1)$$

$$Q_A - \frac{P_{0A}}{2} + \frac{P_{0B}}{2}D_{\beta l} - \frac{\beta M_{0A}}{2} + \frac{\beta M_{0B}}{2}A_{\beta l} = 0, \quad (2)$$

$$M_B + \frac{P_{0A}}{4\beta}C_{\beta l} + \frac{P_{0B}}{4\beta} + \frac{M_{0A}}{2}D_{\beta l} + \frac{M_{0B}}{2} = 0, \quad (3)$$

$$Q_B - \frac{P_{0A}}{2}D_{\beta l} + \frac{P_{0B}}{2} - \frac{\beta M_{0A}}{2}A_{\beta l} - \frac{\beta M_{0B}}{2} = 0, \quad (4)$$

$$A_{\beta l} = e^{-\beta l}(\cos\beta l + \sin\beta l),\ B_{\beta l} = e^{-\beta l}\sin\beta l \quad (5)$$

$$C_{\beta l} = e^{-\beta l}(\cos\beta l - \sin\beta l),\ D_{\beta l} = e^{-\beta l}\cos\beta l \quad (6)$$

The process of SPM has been visually depicted in Figure 1 where the beam in the top left shows a section of length L from an infinite wall, q is a distributed load representing the forces acting on the wall, M_A, M_B, Q_A, Q_B are the moments and shear forces at the positions A and B, respectively and P_0, M_0 at the bottom left portion of Figure 1 are the end conditioning forces and moments, respectively which should be statically equivalent and in opposite direction to M_A, M_B, Q_A, Q_B such that these moments and shear forces at the positions A,B are cancelled and the infinite wall is equivalent to a finite wall with free ends. The end-conditioning forces can be found by solving Equations (1) to (4). Further details of the SPM can be found in Hetenyi [43] and Bekdaş [2].

The post-tensioning forces are denoted with P_1 to P_n in the right side of Figure 1. The optimal distances of these forces from the ground as well as their optimal magnitudes were determined using the harmony search technique. The wall thickness, radius of the cylindrical wall and the height of the wall are denoted with t, r, and H, respectively. Using the methodology described in the previous paragraph, data sets of 1925 samples were generated for cylindrical walls with different numbers of post-tensioning loads. The data samples were generated by considering the total cost of the resulting structures and minimizing the objective function given in Equation (7).

$$f(\mathbf{x}) = C_c V_c + C_s W_s + C_{pt} W_{pt} + C_{fw} A_{fw} \quad (7)$$

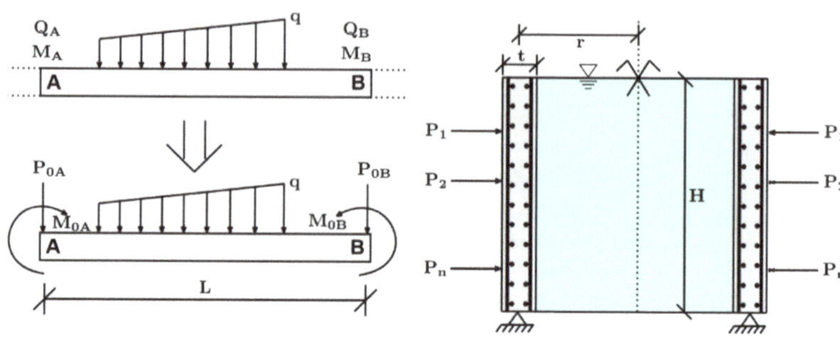

Figure 1. SPM for cylindrical walls.

In Equation (7), $f(\mathbf{x})$ denotes the function that outputs the total cost of the resulting structure, **x** denotes a vector containing the values for the wall dimensions and unit costs, and C_c, C_s, C_{pt}, C_{fw} are the cost of concrete per unit volume, cost of steel rebars per unit weight, cost of post-tensioning per unit weight and cost of formwork per unit area, respectively. The details of the objective function can be found in [2]. The database consisting of the optimal design configurations has been taken for the prediction of the optimal wall thickness. The height of the wall (H), the unit cost of concrete (C_c), specific weight of the liquid (γ), and unit cost of steel (C_s), have been used as the input features defining the load cases. The following section shows the details of each data set used in the development of the predictive models.

2.1. *Analysis of the Data Set*

The data sets used in the prediction of the optimal wall thickness for the case with and without post-tensioning consist of 1908 and 1925 data samples, respectively. The data sets were generated using the harmony search algorithm to find the optimal wall thickness value corresponding to certain combinations of wall height, fluid specific weight and unit costs of concrete and steel. These data sets are split into a training set and a test set in 70% to 30% ratio. The ensemble learning models have been trained on the training set using the 10-fold cross-validation process. Afterward, the performances of these models have been measured on the test set which consists of data samples yet unseen by the models. In the 10-fold cross-validation process the training set is split into 10 disjoint subsets. The models are trained using 9 of these subsets while the tenth subset is being used for model validation. After 10 passes of the training set the best performing model is being selected which is used for performance evaluation on the test set.

The correlation plots in Figures 2 and 3 show the Pearson correlation coefficients between different features in the upper right triangular parts. It can be seen that the wall height is the most correlated input feature with the wall thickness with correlation coefficients of 0.89 and 0.91 followed by the fluid specific weight (γ) with correlation coefficients of 0.34 and 0.37. The strength of the correlation is also denoted by stars in Figures 2 and 3. A Pearson correlation coefficient close to 1 would indicate a highly linear relationship between to variables. The computation formula for the Pearson correlation coefficient is given in Equation (8) where x_i, y_i denote two data series of equal length, n is the length of these data series, and r_{xy} is the Pearson correlation coefficient between these two data series.

$$r_{xy} = \frac{n\sum_{i=1}^{n} x_i y_i - \sum_{i=1}^{n} x_i \sum_{i=1}^{n} y_i}{\sqrt{n\sum_{i=1}^{n} x_i^2 - \left(\sum_{i=1}^{n} x_i\right)^2}\sqrt{n\sum_{i=1}^{n} y_i^2 - \left(\sum_{i=1}^{n} y_i\right)^2}} \tag{8}$$

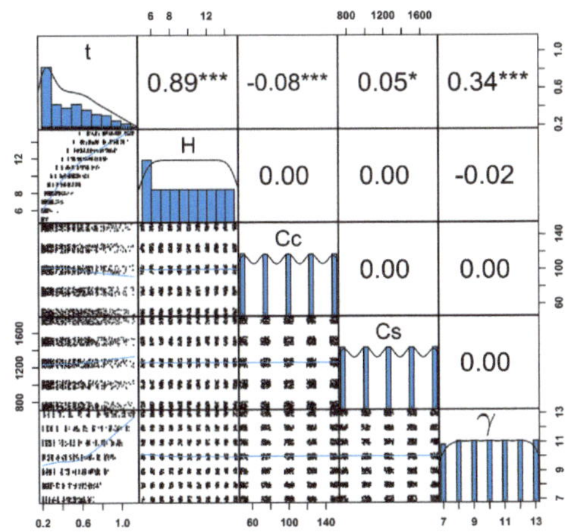

Figure 2. Correlation plot of the data set with one layer of post-tensioning cables.

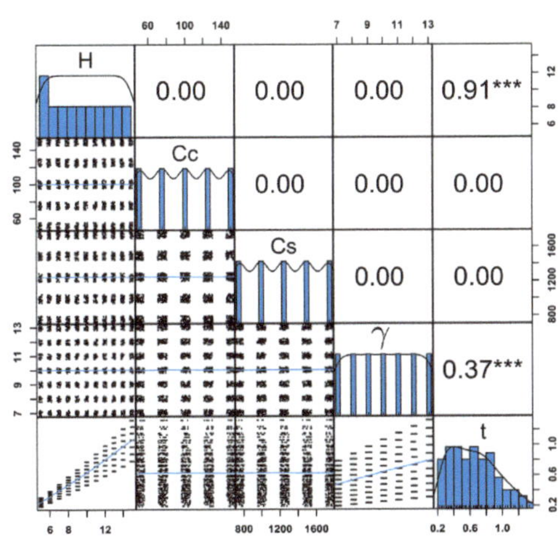

Figure 3. Correlation plot of the data set without post-tensioning cables.

In addition to the Pearson coefficient, Figures 2 and 3 also display the distributions of different features. The diagonals of each plot contain frequency distributions of the features whereas the lower left triangular area contains bivariate scatter plots with regression lines. For each feature denoted on a diagonal tile, the value range of this feature is shown in both the horizontal and the vertical axis.

2.2. Harmony Search Algorithm

The harmony search (HS) algorithm was developed by Geem et al. [44] and has been used for solving numerous optimization problems in broad areas including the numerical solution of differential equations [45], project scheduling [46], weapon target assignment [47], law enforcement [48], internet of things [49], robotics [50], soil stability analysis [51], and structural design [52–54]. The HS method is based on the evolutionary improvement of an initially randomly generated population of optimal solution candidates denoted with x in Equations (10) and (11). The size of this population is denoted with HMS in Equation (9) which stands for harmony memory size. As the members of the population go through the HS iterations given in Equations (9) to (12), new and better-performing population members replace the members of the previous generations.

$$k = \mathrm{int}(\mathrm{rand}\cdot \mathrm{HMS}),\ \mathrm{rand} \in (0,\ 1) \tag{9}$$

$$x_{i,new} = x_{i,min} + \mathrm{rand}\cdot(x_{i,max} - x_{i,min}),\ \text{if HMCR} > \mathrm{rand} \tag{10}$$

$$x_{i,new} = x_{i,k} + \mathrm{rand}\cdot \mathrm{PAR}\cdot(x_{i,max} - x_{i,min}),\ \text{if HMCR} \leq \mathrm{rand} \tag{11}$$

$$\mathrm{HMCR} = 0.5\left(1 - \frac{i}{\max(i)}\right), \mathrm{PAR} = 0.05\left(1 - \frac{i}{\max(i)}\right) \tag{12}$$

2.3. *Ensemble Learning Algorithms*

Ensemble learning methods are based on the idea of combining the predictions of multiple predictive models to obtain a strong learning algorithm. In this study XGBoost, Random Forest, LightGBM, and Catboost ensemble learning algorithms have been applied. The ensemble learning models have been trained using the scikit-learn library available for the Python programming language. The XGBoost algorithm iteratively generates decision trees while each newly generated tree corrects the errors of the previous trees. In these iterations the XGBoost algorithm aims at minimizing the objective function given in Equation (13) where M is the size of the training set, N is the number of decision trees, l is a loss function and Ω is a penalty function. The XGBoost algorithm is summarized in Equation (14) where g denotes the strong learner model which is a linear combination of N weak learners denoted with g_k and $\hat{y}$ is the model prediction [55].

$$L = \sum_{i=1}^{M} l(y_i, \hat{y}_i) + \sum_{k=1}^{N} \Omega(f_k) \tag{13}$$

$$g(\mathbf{x}) = \sum_{k=1}^{N} g_k(\mathbf{x}) = \hat{y} \tag{14}$$

The LightGBM, Random Forest, and CatBoost algorithms function on similar principals as the XGBoost algorithm by iteratively adding weak learner trees to the model and combining their output to build a strong learner model. The distinguishing feature of the LightGBM algorithm is its computational speed. The LightGBM algorithm implements techniques such as histogram-based algorithm, and Gradient-based One-Side Sampling (GOSS) to achieve improved performance [56]. On the other hand, the CatBoost algorithm stands out by its ability to work with categorical features more efficiently. The CatBoost algorithm incorporates ordered boosting, greedy method, and L2-regularization to improve model performance [57].

2.4. *Genetic Programming*

Genetic programming (GP) is a population based evolutionary algorithm. The algorithm generates a population of programs where each program is represented by a tree structure. Each node in these programs represents an operation (such as addition or multiplication) or a numerical value. Figure 4 shows an example of representing a program as a tree structure where nodes containing numerical values, unary operators and binary operators are shown in different colors. The fitness of each program is evaluated by running on a test set, and the programs that perform the best are selected for producing the next generation of programs through crossover and mutation [58–60]. The process of producing new programs from the existing ones using crossover and mutation operations is schematically explained in Figure 5. In the crossover part of Figure 5 the identical parts of the programs are shown inside rectangles of the same color whereas in the mutation part the mutated operator is shown inside a red rectangle.

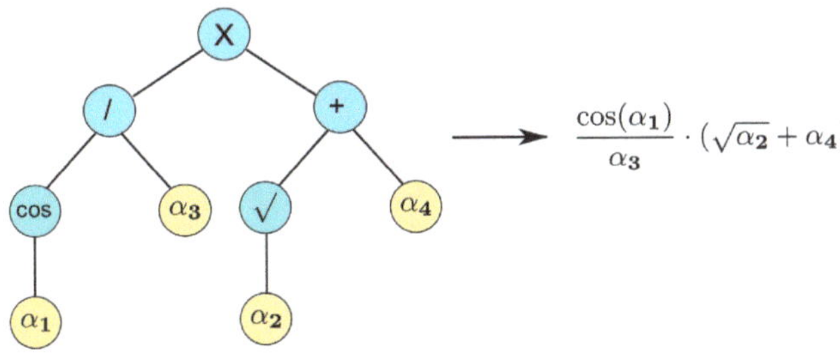

Figure 4. Tree representation of a program consisting of 5 operators and 4 parameters.

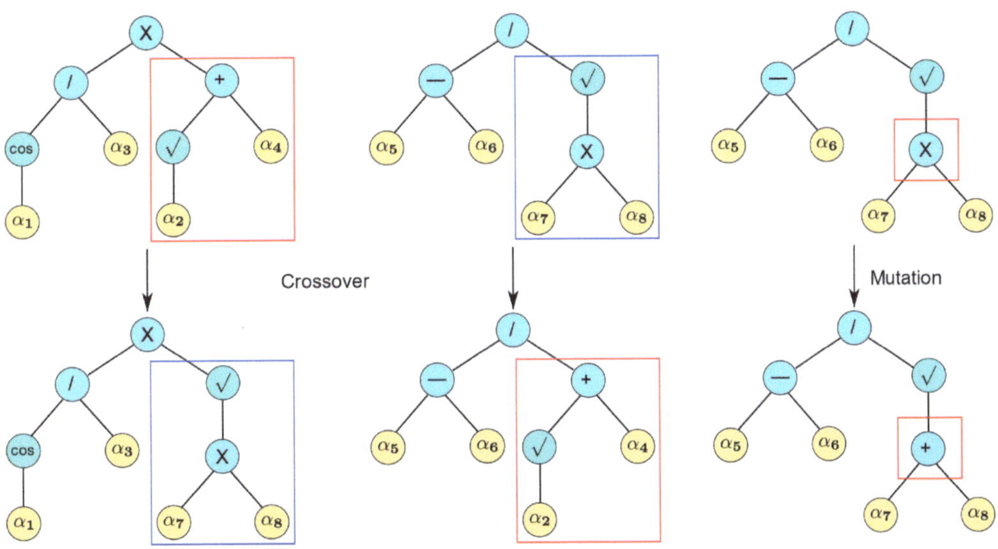

Figure 5. Visualizations of pragram updates by crossover and mutation.

3. Results

This section gives a detailed presentation of the predictive performances of the machine learning models. The predicted and actual optimal wall thickness values are plotted against each other and the percentage deviations of the predicted values from the actual thickness values have been shown. The prediction accuracies of ML models have been quantified using root mean squared error, mean absolute error, and the coefficient of determination (R^2). ML models have been trained on datasets of wall geometries with and without post-tensioning cables. Figure 6 shows the prediction of the optimal wall thickness as a function of the concrete and steel unit costs, liquid specific weight, and wall height for a cylindrical wall without post-tensioning. The wall radius has been fixed at a constant value in all the databases in this study since the liquid pressure on the wall is not affected by this quantity. It can be observed from Figure 6 that the predicted and actual optimal values for the thickness perfectly overlap.

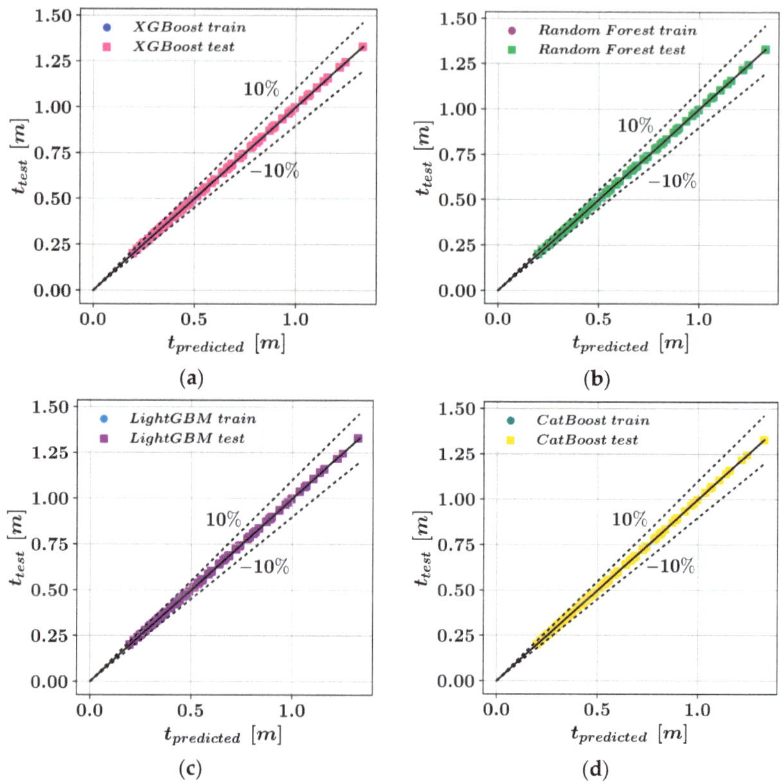

Figure 6. Comparison of the predicted and actual wall thickness values without post-tensioning cables using (**a**) XGBoost, (**b**) Random Forest, (**c**) LightGBM, (**d**) CatBoost.

The performances of four different ML models have been quantified and listed in Table 1. It can be observed that all of the ML models were able to predict the optimal wall thickness values with near-perfect accuracy since the R^2 scores are close to 1 for all ML models. Table 1 also lists the performance values for the genetic programming model (GP). Clearly, in terms of both accuracy and computational efficiency the GP model performed worse than the ensemble learning models. However, the GP algorithm has been included in this study, because this algorithm delivers closed-form equations that could be easily incorporated into practical engineering applications for the prediction of the optimal wall thickness.

Table 1. Model accuracies in predicting the optimal wall thickness without post-tensioning.

Algorithm	R^2		MAE		RMSE		Duration [s]
	Train	Test	Train	Test	Train	Test	
XGBoost	0.9999	0.9999	0.0001	0.0002	0.0003	0.0003	4.81
Random Forest	0.9999	0.9999	10^{-5}	3×10^{-5}	9×10^{-5}	0.0002	3.87
LightGBM	0.9999	0.9999	0.0005	0.0006	0.0012	0.0014	4.52
CatBoost	0.9999	0.9999	0.0002	0.0003	0.0002	0.0004	32.82
GP	0.9584	0.9573	0.0465	0.0460	0.0573	0.0570	359

The predicted and actual optimal wall thickness values in the case of a single layer of post-tensioning have been plotted in Figure 7. It can be observed that the introduction of post-tensioning into the models caused a slight reduction in the accuracy of the ML models. However, Table 2 shows that all ensemble models were capable of predicting the optimal wall thickness with an R^2 score greater than 0.98 on the test set.

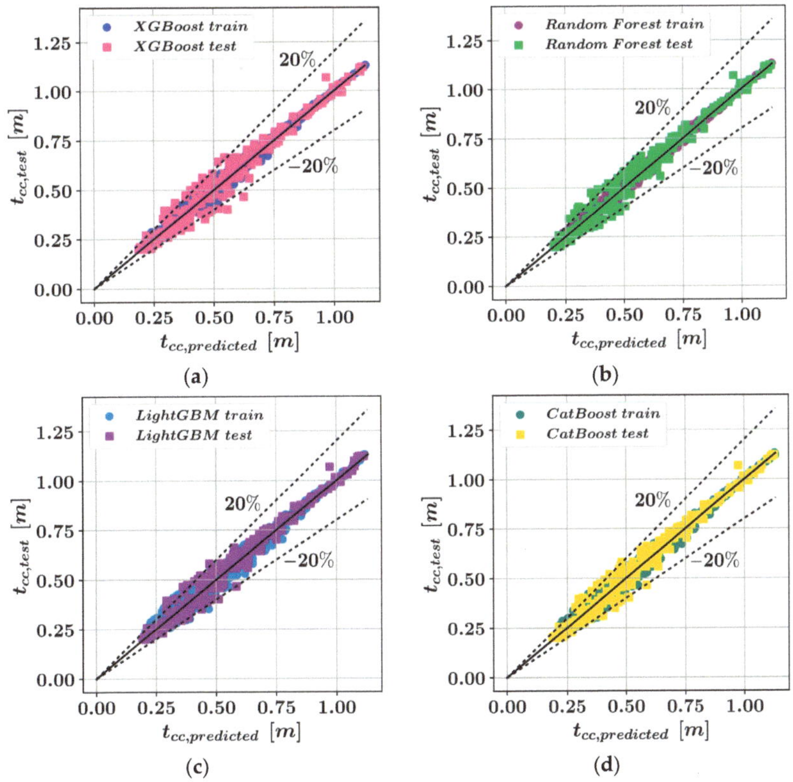

Figure 7. Comparison of the predicted and actual optimal wall thicknesses with post-tensioning for (**a**) XGBoost, (**b**) Random forest, (**c**) LightGBM, (**d**) CatBoost.

Table 2. Model accuracies in predicting the optimal wall thickness with post-tensioning.

Algorithm	R^2		MAE		RMSE		Duration [s]
	Train	Test	Train	Test	Train	Test	
XGBoost	0.9967	0.9825	0.0087	0.021	0.0139	0.0346	5.66
Random Forest	0.9973	0.9850	0.0077	0.019	0.0126	0.0320	5.14
LightGBM	0.9908	0.9864	0.0153	0.019	0.0234	0.0305	4.39
CatBoost	0.9931	0.9863	0.0129	0.018	0.0203	0.0306	25.15
GP	0.9588	0.9581	0.0422	0.045	0.0501	0.0531	150

To have a clear visualization of the model performance, the percentage errors on the training and test sets have been plotted for the CatBoost model in Figure 8. Figure 8a shows the overlap between the actual and predicted optimal wall thickness values. According to Figure 8b, the error percentages fluctuate in the ±10% and ±30% ranges for the training and test sets respectively. Figure 8c,d show the distributions of the error percentages for the training and test sets respectively. Figure 8d, shows that most of the error percentage fluctuations on the test set are accumulated in the ±20% range whereas smaller error percentages are observed in Figure 8c in the ±10% range. The entire data set was split into a training set and a test set in 70% to 30% ratio. In terms of mean absolute error (MAE), the CatBoost model performed best on the test set, whereas the LightGBM model performed best in terms of RMSE and R^2. It should be noted that all of the four ensemble learning models tested in this study demonstrated similar performances and the differences in performance are deemed negligible. On the other hand, a significant difference in terms of the computational speed could be observed for the CatBoost model. It was observed that the duration of training and testing the models is significantly longer in the case of CatBoost. The LightGBM model was observed to be the most efficient model in terms of computational speed, although the XGBoost and Random Forest models also have similar performances to LightGBM in terms of computational speed. Finally, the closed-form equation obtained from the genetic programming (GP) algorithm performed less accurately than the ensemble learning models.

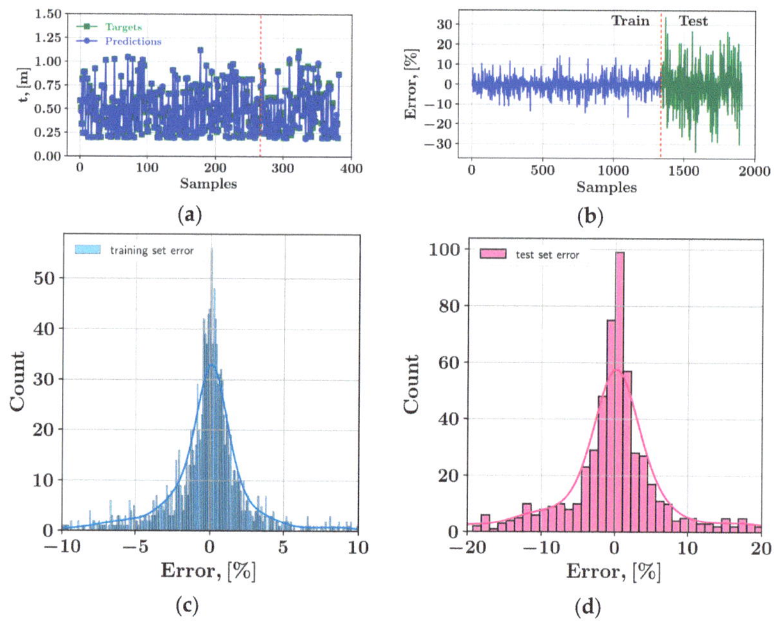

Figure 8. Error percentages of the CatBoost model.

Figure 9 shows the first of the decision trees that constitute the XGBoost model developed for the case with one layer of post-tensioning cables. The XGBoost model consists of a total of 100 iteratively added decision trees such that each new decision tree corrects the errors of the trees before itself. The final prediction of the XGBoost model is the sum of all the predictions made by the decision trees. The decision tree in Figure 9 consists of a root node which is split according to the level of wall height, 41 internal nodes, and 43 leaf nodes. The internal nodes are split at different levels of the fluid specific weight, and the unit costs of steel and concrete. The green colored leaf nodes contain the possible contributions of the decision tree to the final model prediction.

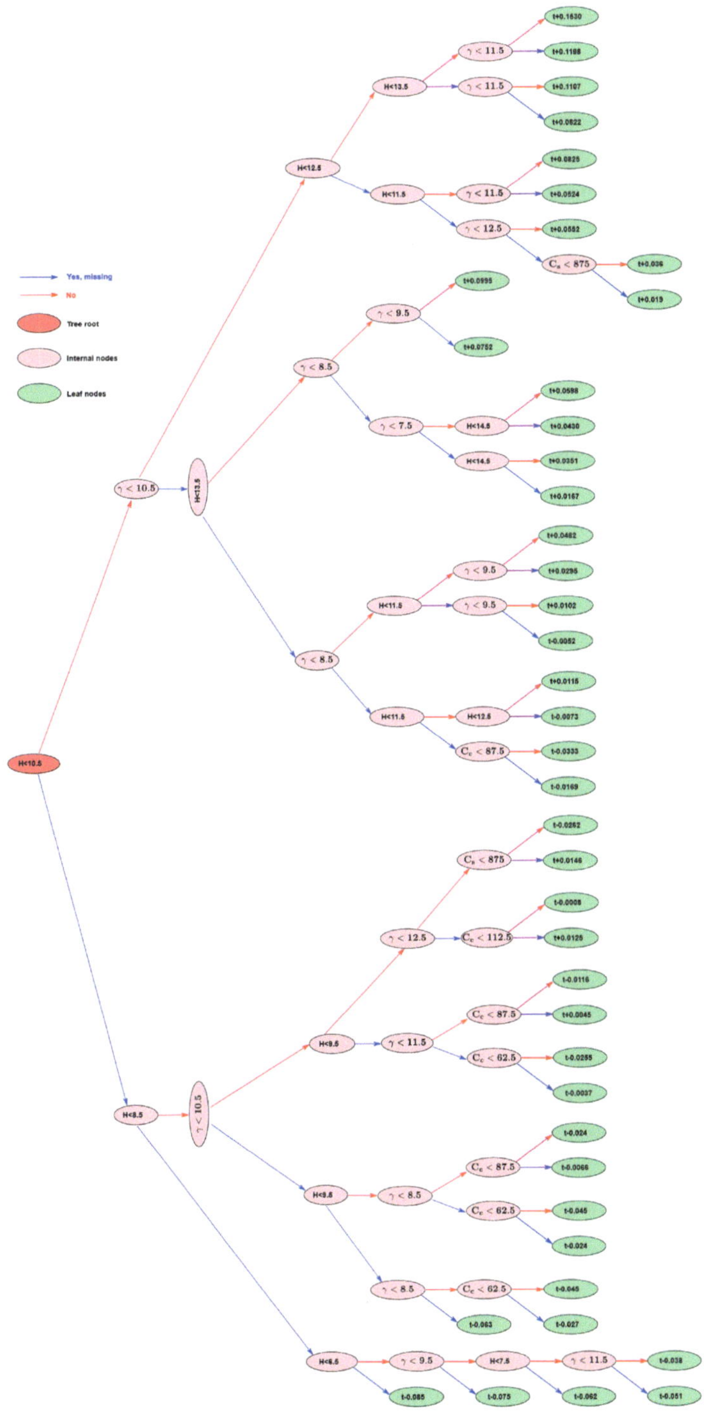

Figure 9. First tree of the XGBoost model.

3.1. Interpretation of the Ensemble Learning Models Using SHAP Approach

The SHAP summary plots in Figures 10 and 11 visualize the impact of each input feature on the CatBoost model predictions. The load cases with and without post-tensioning are visualized in Figures 10 and 11, respectively. In the SHAP summary plots each dot represents one of the data points in the data set. The SHAP value corresponding to an input feature in a data point is represented on the horizontal axis, whereas the value of the input feature is color coded as shown with a color bar on the right-hand side of the plot. The higher values of an input feature are displayed with shades of red, while blue colors represent the lower values of an input feature. The input features are sorted by the magnitude of their impact, in decreasing order from top to bottom. The SHAP value measures the contribution of a feature to the model prediction such that positive SHAP values indicate an increasing effect of an input feature on the model output and negative SHAP values indicate a decreasing effect on the model output. The computation of the SHAP values can be described as in Equation (15) where F is the set of all input features and S is a subset of F where the feature with the index i has been withheld. In Equation (15), $\mathbf{x}$ and ϕ_i represent a vector of input feature values and the corresponding SHAP value respectively [61].

$$\phi_i = \sum_{S \subseteq F \setminus \{i\}} \frac{|S|!(|F| - |S| - 1)!}{|F|!} \left[f_{S \cup \{i\}} \left(\mathbf{x}_{S \cup \{i\}} \right) - f_S(\mathbf{x}_S) \right] \tag{15}$$

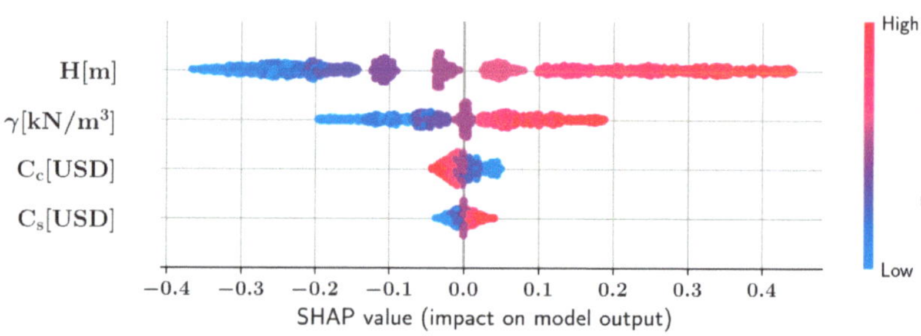

Figure 10. SHAP summary plot with one layer of post-tensioning.

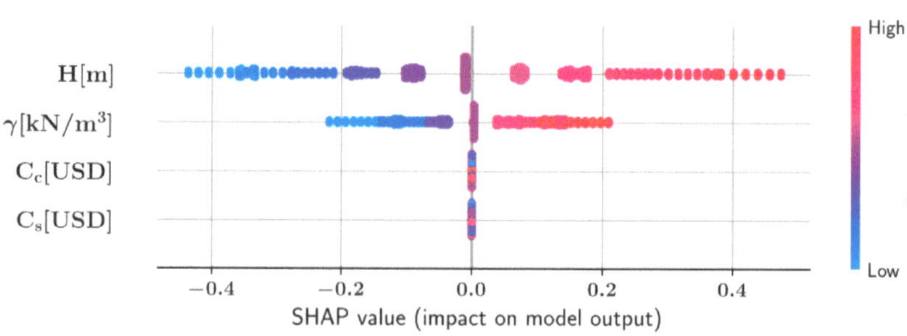

Figure 11. SHAP summary plot without post-tensioning.

According to Figures 10 and 11 the height of the cylindrical wall has the greatest impact on the predicted optimal wall thickness for both load cases. The fluid specific weight is the second most impactful input feature whereas the concrete and steel unit costs have a relatively minor impact on the model predictions. Particularly, in the load case without post-tensioning, the impacts of the concrete and steel unit costs are significantly less than the fluid specific weight and the wall height.

Figures 12 and 13 show feature dependence plots where the y-axis represents the SHAP value for the feature of interest, and the x-axis represents the actual value of the feature for each point in the dataset for the cases with and without post-tensioning respectively. The color of each dot on the feature dependence plots represents the value of a second feature that is most correlated with the main feature represented in the horizontal axis. The feature dependence plots convey information about how the predictions of the model are affected as the value of an input feature varies. A positive relationship between a feature's value and its corresponding SHAP value indicates that an increase in that feature's value will result in an increase in the model predictions. In both Figures 12 and 13, there is a positive relationship between the height of the wall and the corresponding SHAP values which indicates that increased wall height leads to increased predictions of the wall thickness. The coloring of the dots in Figures 12 and 13 indicate that for any given value of the wall height, an increase in the fluid specific weight is associated with a greater increasing impact on the model output wall heights greater than 10 m. On the other hand, for wall heights less than 10 m, in both load cases with and without post-tensioning, the fluid specific weight has the opposite effect on the predicted wall thickness. A similar relationship is also observed between the fluid specific weight and its SHAP value. For values greater than 10 kN/m^3 the addition of this variable into the models has an increasing impact on the model output whereas specific weight values less than 10 kN/m^3 are associated with a decrease in the predicted wall thickness values. Furthermore, for any given value of the fluid specific weight, increased wall height also increases the impact of γ on the model output as the coloring of the data points shows.

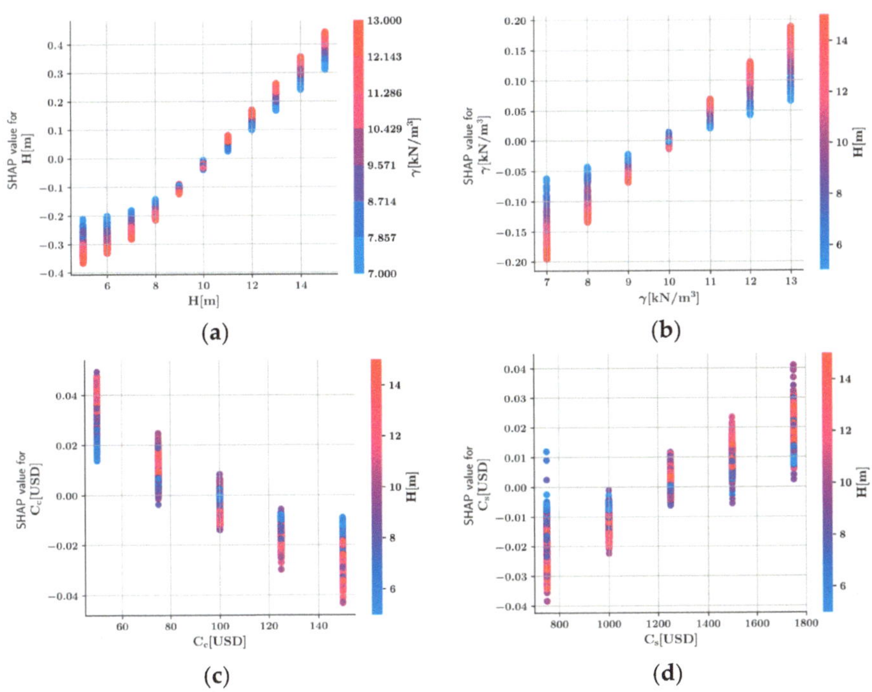

Figure 12. Feature dependence plots (CatBoost) with post-tensioning for (**a**) H (**b**) γ (**c**) Cc (**d**) Cs.

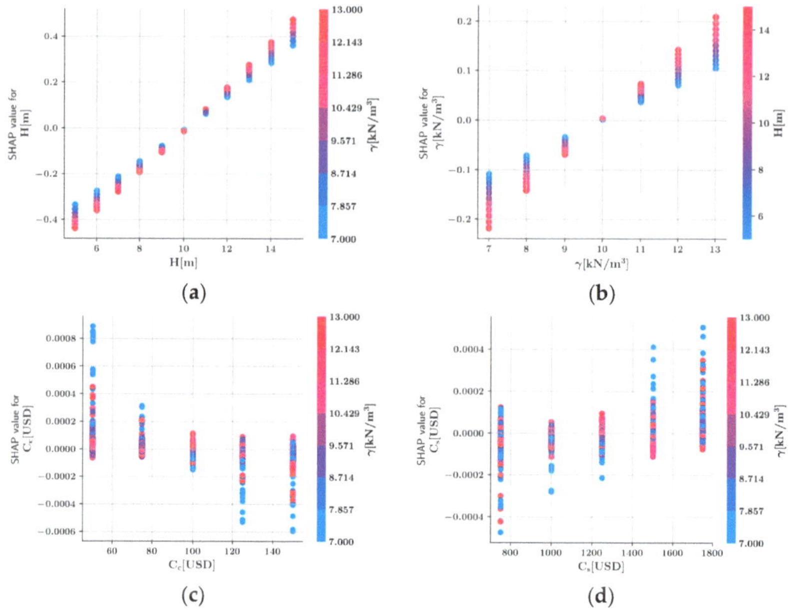

Figure 13. Feature dependence plots (CatBoost) without post-tensioning for (**a**) H (**b**) γ (**c**) Cc (**d**) Cs.

Individual conditional expectation (ICE) plots are helpful for gaining a better understanding of how a single feature affects the predictions made by a machine learning model. The ICE plots in Figures 14 and 15 show the variation of the model predictions concerning a single input feature while the values of the remaining input features are kept constant. Each line in the ICE plot represents one of the data points in the data set and the average value of the predictions for all the data points is plotted with a thick blue line. Figures 14a and 15a clearly show that the predicted optimal wall thickness values increase with the wall height while the range of optimal wall thickness values becomes wider as the wall height increases. A less steep increase in the optimal wall thickness values can be observed with respect to the γ values in Figures 14b and 15b. Also, for any given γ value, the optimal wall thicknesses take values on a much wider range compared to the wall height. Finally, the ICE plots of Cc and Cs show that these features have significantly less impact on the model output compared to H and γ.

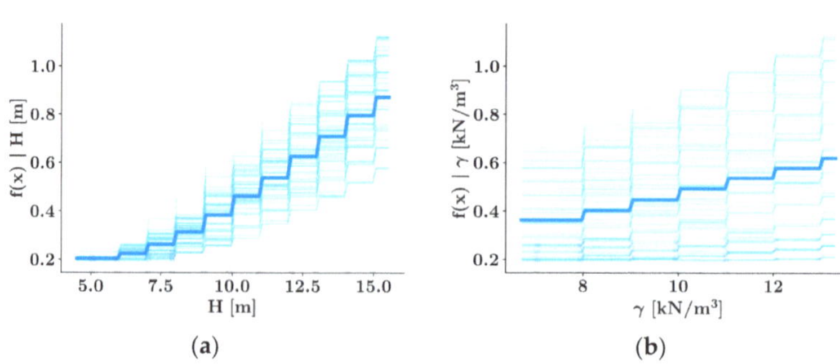

Figure 14. *Cont.*

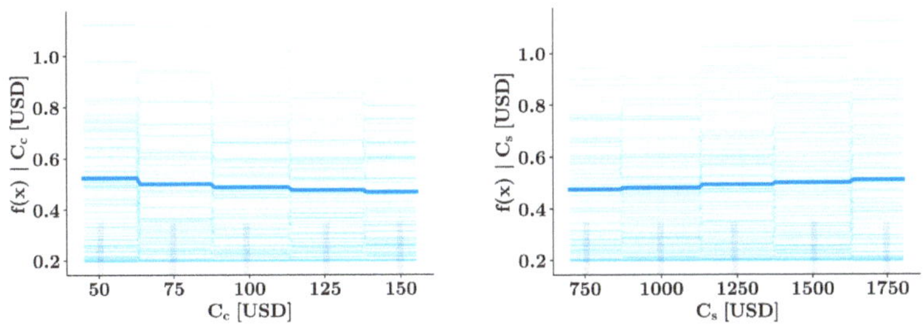

(c) (d)

Figure 14. ICE plots (XGBoost) with post-tensioning for (**a**) H (**b**) γ (**c**) Cc (**d**) Cs.

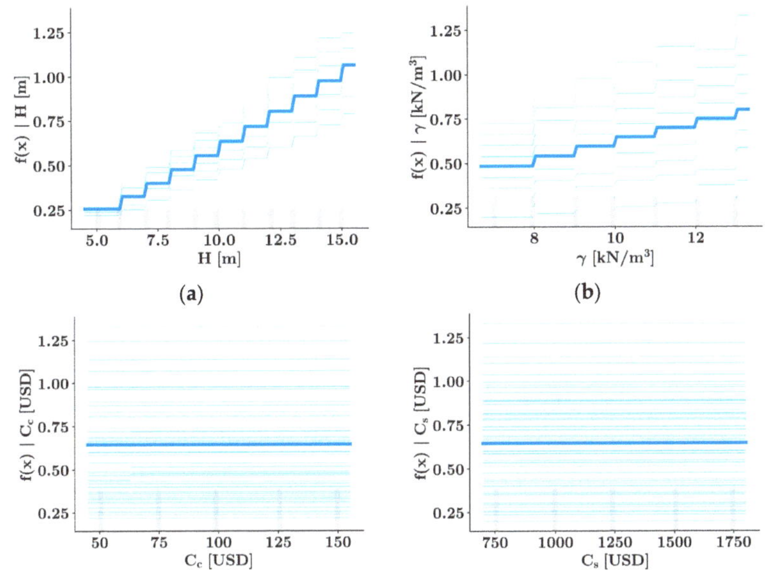

(c) (d)

Figure 15. ICE plots (XGBoost) without post-tensioning for (**a**) H (**b**) γ (**c**) Cc (**d**) Cs.

3.2. Genetic Programming

Equations (16) and (17) have been obtained by the genetic programming algorithm for the prediction of the optimal wall thickness as a function of the wall height and liquid specific weight for the case of no post-tensioning and one post-tensioning cable, respectively. The tree representations of Equations (16) and (17) are shown in Figure 16. The tree population size was 5000 and mean absolute error (MAE) was used as the error metric. MAE values of 0.041 and 0.046 could be achieved in less than 50 iterations in cases with and without post-tensioning, respectively.

$$t(H,\gamma) = 0.095 \cdot H \cdot \log(0.878 \cdot \log(\gamma)) \quad (16)$$

$$t(H,\gamma) = \tan\left(\tan\left(0.093 \cdot H \cdot \tan\left(\log\left(\sqrt{\log(\gamma)}\right)\right)\right)\right) \quad (17)$$

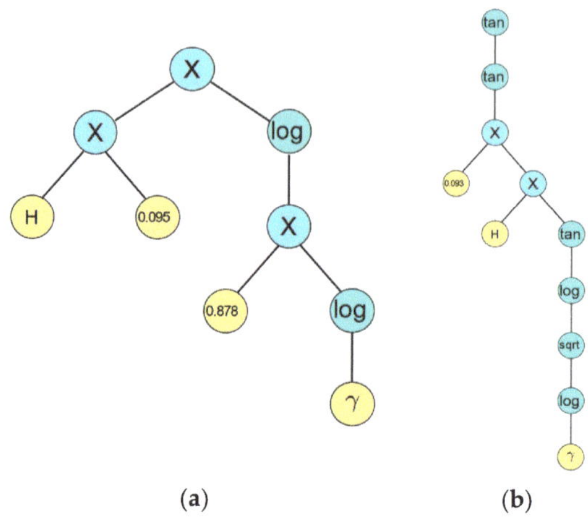

Figure 16. Tree representations of (**a**) Equation (16), (**b**) Equation (17).

Figures 17 and 18 show the development of the equation length and accuracy throughout the generations. It can be seen from Figure 17 that after 50 iterations an MAE value of 0.046 could be achieved with a predictive equation that consists of 9 components in the load case without post-tensioning. These 9 components consist of the log function which appears 2 times, the multiplication operator which appears 3 times, the coefficients 0.095 and 0.878, and the variables H and γ. Figure 18 shows that after 50 iterations an MAE value of 0.041 is achieved with a predictive equation consisting of 11 components. These components are the tangent function which appears 3 times, the log function which appears 2 times, the square root function which appears 1 time, the multiplication operator which appears 2 times, the coefficient 0.093, and the variables H and γ. The larger MAE value associated with the case without post-tensioning is due to the larger wall thicknesses necessary in the absence of post-tensioning. A complete list of the hyperparameters modified in obtaining Equations (16) and (17) is given in Table 3. For the remaining hyperparameters, the default values have been used.

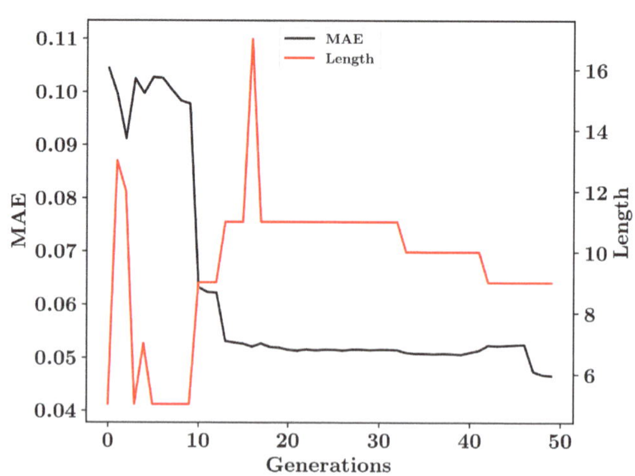

Figure 17. Development of the genetic programs without post-tensioning cables.

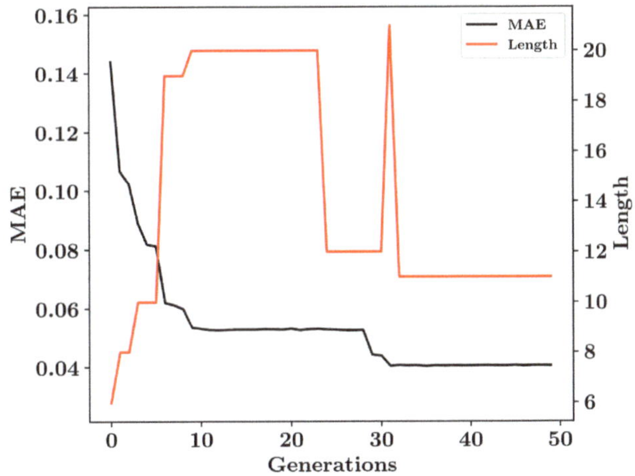

Figure 18. Development of the genetic programs for a single level of post-tensioning cables.

Table 3. Hyperparameters for the Genetic Programmin (GP) Model of the wall thickness.

Model	Parameter Name	Value
GP	population_size	5000
	p_crossover	0.7
	p_subtree_mutation	0.1
	p_hoist_mutation	0.05
	p_point_mutation	0.1
	tournament_size	150
	function_set	('add', 'sin', 'cos', 'tan', 'log', 'sub', 'mul', 'div', 'sqrt')

4. Discussion

The current study demonstrates the application of the harmony search optimization algorithm in generating large data sets for training state-of-the-art machine learning models. The problem of predicting the optimal wall thickness of a cylindrical post-tensioned reinforced concrete wall of a liquid container has been investigated. In addition to ensemble learning models such as XGBoost and CatBoost also, the genetic programming methodology has been utilized to obtain closed-form predictive equations. The ensemble learning models have been trained with the height of the wall, specific weight of the liquid, and unit costs of steel and concrete as the input features. It was shown that the inclusion of post-tensioning cables leads to a decrease in the optimal wall thickness. The output of the ensemble learning models has been further analyzed using the SHAP approach and individual conditional expectation plots. The SHAP analysis showed that the wall height is the most impactful input feature affecting the model predictions of the optimal wall thickness followed by the liquid specific weight. An increase in the wall height and liquid specific weight was shown to also increase the optimal wall thickness. Furthermore, the genetic programming methodology provided predictive equations as functions of wall height and liquid specific weight for the load cases with and without post-tensioning. All predictive models were able to provide highly accurate predictions of the optimal wall thickness, with of the ensemble learning models achieving R^2 scores greater than 0.98 on the test sets. The LightGBM model delivered the fastest and most accurate predictions in the case of walls with post-tensioning whereas the CatBoost model was the slowest.

5. Conclusions

The application of optimization techniques to generate big data sets is a novel approach to overcome the lack of experimental data points for the training of ML models related to structural engineering. Furthermore, the availability of closed-form equations for the prediction of optimal design configurations has significant practical benefits. The current study demonstrated the application of the genetic programming methodology in obtaining closed-form equations for the prediction of the optimal wall thickness for a liquid containing post-tensioned cylindrical wall. The data set necessary for this procedure was generated using the harmony search optimization technique. The obtained predictive equations could be incorporated into the engineering design process to facilitate structural optimization. However, further research needs to be done with larger data sets for the performance validation of the presented equations. On the other hand, higher R^2 scores could be obtained from the ensemble learning algorithms. A limitation of the current study is the range of design variables included in the data set generation process. Future research in this area can incorporate additional input features into the data set such as the mechanical properties of steel and concrete or the positions of post-tensioning cables. Data sets can also be further enhanced with the results of numerical studies.

Author Contributions: Writing—original draft preparation, G.B. and C.C.; Conceptualization, G.B. and C.C.; data curation, C.C.; visualization, C.C.; writing—review & editing, G.B., S.K. and Z.W.G.; supervision G.B. and Z.W.G. All authors have read and agreed to the published version of the manuscript.

Funding: This research received no external funding.

Data Availability Statement: Data will be made available upon reasonable request.

Conflicts of Interest: The authors declare no conflict of interest.

Nomenclature

C_c	Unit cost of concrete	C_c	Unit cost of steel
C_{pt}	Unit cost of post-tensioning	C_{fw}	Unit cost of formwork
F	Set of all input features	FPA	Flower pollination algorithm
FRP	Fiber reinforced polymer	HMCR	Harmony memory consideration rate
HMS	Harmony memory size	i	Index of a design variable
γ	Specific weight of the liquid	GOSS	Gradient-based One-Side Sampling
GP	Genetic programming	H	Height of the wall
HS	Harmony search	ICE	Individual conditional expectation
k	Index of a population member	M	Size of the training set
MAE	Mean absolute error	ML	Machine learning
N	Number of decision trees	Ω	Penalty function
$P_1 \ldots P_n$	Post-tensioning forces	PAR	Pitch adjustment rate
ϕ_i	SHAP value of the i-th feature	PCPT	Pre-cast post-tensioned
r	Radius of the cylindrical wall	R^2	Coefficient of determination
RMSE	Root mean squared error	r_{xy}	Pearson correlation coefficient
SHAP	SHapley Additive exPlanations	t	Wall thickness
TLBO	Teaching learning based optimization	SPM	Superposition method
V_c	Volume of concrete	W_{pt}	Weight of post-tensioning cables
W_s	Weight of steel	$\hat{y}$	Model prediction
A_{fw}	Area of formwork		

References

1. *Post-Tensioning Manual*, 6th ed.; Post-Tensioning Institute: Phoenix, AZ, USA, 2006; ISBN 0977875202.
2. Bekdaş, G. Harmony Search Algorithm Approach for Optimum Design of Post-Tensioned Axially Symmetric Cylindrical Reinforced Concrete Walls. *J. Optim. Theory Appl.* **2014**, *164*, 342–358. [CrossRef]
3. Bekdas, G. Optimum design of axially symmetric cylindrical reinforced concrete walls. *Struct. Eng. Mech.* **2014**, *51*, 361–375. [CrossRef]

4. Bekdaş, G. New improved metaheuristic approaches for optimum design of posttensioned axially symmetric cylindrical reinforced concrete walls. *Struct. Des. Tall Spéc. Build.* **2018**, *27*, e1461. [CrossRef]
5. Bekdaş, G.; Nigdeli, S.M. Optimum Reduction of Flexural Effect of Axially Symmetric Cylindrical Walls with Post-tensioning Forces. *KSCE J. Civ. Eng.* **2017**, *22*, 2425–2432. [CrossRef]
6. Bekdaş, G.; Cakiroglu, C.; Islam, K.; Kim, S.; Geem, Z.W. Optimum Design of Cylindrical Walls Using Ensemble Learning Methods. *Appl. Sci.* **2022**, *12*, 2165. [CrossRef]
7. Kaveh, A.; Zolghadr, A. Meta-heuristic methods for optimization of truss structures with vibration frequency constraints. *Acta Mech.* **2018**, *229*, 3971–3992. [CrossRef]
8. Renkavieski, C.; Parpinelli, R.S. Meta-heuristic algorithms to truss optimization: Literature mapping and application. *Expert Syst. Appl.* **2021**, *182*, 115197. [CrossRef]
9. Mortazavi, A.; Togan, V. Metaheuristic Algorithms for Optimal Design of Truss Structures. In *Advances in Structural Engineering—Optimization. Studies in Systems, Decision and Control*; Nigdeli, S.M., Bekdaş, G., Kayabekir, A.E., Yucel, M., Eds.; Springer: Cham, Switzerland, 2021; Volume 326. [CrossRef]
10. Mortazavi, A. The Performance Comparison of Three Metaheuristic Algorithms on the Size, Layout and Topology Optimization of Truss Structures. *Mugla J. Sci. Technol.* **2019**, *5*, 28–41. [CrossRef]
11. Bekdaş, G.; Yucel, M.; Nigdeli, S.M. Evaluation of Metaheuristic-Based Methods for Optimization of Truss Structures via Various Algorithms and Lèvy Flight Modification. *Buildings* **2021**, *11*, 49. [CrossRef]
12. Grzywinski, M.; Selejdak, J.; Dede, T. Shape and size optimization of trusses with dynamic constraints using a metaheuristic algorithm. *Steel Compos. Struct.* **2019**, *33*, 747–753. [CrossRef]
13. Yousefpoor, H.; Kaveh, A. Chaos Embedded Meta-heuristic Algorithms for Optimal Design of Truss Structures. *Sci. Iran.* **2022**, *29*, 2868–2885. [CrossRef]
14. Uray, E.; Tan, Ö.; Çarbaş, S.; Erkan, H. Metaheuristics-based Pre-Design Guide for Cantilever Retaining Walls. *Tek. Dergi* **2021**, *32*, 10967–10993. [CrossRef]
15. Sharma, S.; Saha, A.K.; Lohar, G. Optimization of weight and cost of cantilever retaining wall by a hybrid metaheuristic algorithm. *Eng. Comput.* **2021**, *38*, 2897–2923. [CrossRef]
16. Aydoğdu, I. Comparison of metaheuristics on multi objective (Cost-C02) optimization of RC cantilever retaining walls. *Pamukkale Univ. J. Eng. Sci.* **2017**, *23*, 221–231. [CrossRef]
17. Bekdaş, G.; Cakiroglu, C.; Kim, S.; Geem, Z.W. Optimal Dimensioning of Retaining Walls Using Explainable Ensemble Learning Algorithms. *Materials* **2022**, *15*, 4993. [CrossRef]
18. Kaveh, A.; Hamedani, K.B.; Bakhshpoori, T. Optimal Design of Reinforced Concrete Cantilever Retaining Walls Utilizing Eleven Meta-Heuristic Algorithms: A Comparative Study. *Period. Polytech. Civ. Eng.* **2020**, *64*, 156–158. [CrossRef]
19. Kalemci, E.N.; Ikizler, S.B.; Dede, T.; Angın, Z. Design of reinforced concrete cantilever retaining wall using Grey wolf optimization algorithm. *Structures* **2019**, *23*, 245–253. [CrossRef]
20. Aral, S.; Yılmaz, N.; Bekdaş, G.; Nigdeli, S.M. Jaya Optimization for the Design of Cantilever Retaining Walls with Toe Projection Restriction. In Proceedings of the 6th International Conference on Harmony Search, Soft Computing and Applications (ICHSA 2020), Istanbul, Turkey, 22–24 April 2020; Nigdeli, S.M., Kim, J.H., Bekdaş, G., Yadav, A., Eds.; Advances in Intelligent Systems and Computing. Springer: Singapore, 2021; Volume 1275. [CrossRef]
21. Kalemci, E.N. Rao-3 algorithm for the weight optimization of reinforced concrete cantilever retaining wall. *Geomech. Eng.* **2020**, *20*, 527–536. [CrossRef]
22. Cakiroglu, C.; Islam, K.; Bekdaş, G.; Nehdi, M.L. Data-driven ensemble learning approach for optimal design of cantilever soldier pile retaining walls. *Structures* **2023**, *51*, 1268–1280. [CrossRef]
23. Jalili, S.; Khani, R.; Maheri, A.; Hosseinzadeh, Y. Performance assessment of meta-heuristics for composite layup optimisation. *Neural Comput. Appl.* **2021**, *34*, 2031–2054. [CrossRef]
24. Nicholas, P.E.; Dharmaraja, C.; Sofia, A.S.; Vasudevan, D. Optimization of laminated composite plates subjected to nonuniform thermal loads. *Polym. Polym. Compos.* **2019**, *27*, 314–322. [CrossRef]
25. Cakiroglu, C.; Islam, K.; Bekdaş, G.; Kim, S.; Geem, Z.W. Metaheuristic Optimization of Laminated Composite Plates with Cut-Outs. *Coatings* **2021**, *11*, 1235. [CrossRef]
26. Nicholas, P.E.; Padmanaban, K.; Vasudevan, D. Buckling optimization of laminated composite plate with elliptical cutout using ANN and GA. *Struct. Eng. Mech.* **2014**, *52*, 815–827. [CrossRef]
27. Javidrad, F.; Nazari, M.; Javidrad, H.R. An Innovative Optimized Design for Laminated Composites in terms of a Proposed Bi-Objective Technique. *J. Soft Comput. Civ. Eng.* **2020**, *4*, 1–28. [CrossRef]
28. Topal, U.; Vo-Duy, T.; Dede, T.; Nazarimofrad, E. Buckling load optimization of laminated plates resting on Pasternak foundation using TLBO. *Struct. Eng. Mech. Int. J.* **2018**, *67*, 617–628. [CrossRef]
29. Jafari, M.; Chaleshtari, M.H.B.; Khoramishad, H.; Altenbach, H. Minimization of thermal stress in perforated composite plate using metaheuristic algorithms WOA, SCA and GA. *Compos. Struct.* **2023**, *304*, 116403. [CrossRef]
30. Cakiroglu, C.; Bekdaş, G.; Geem, Z.W. Harmony Search Optimisation of Dispersed Laminated Composite Plates. *Materials* **2020**, *13*, 2862. [CrossRef]
31. Cakiroglu, C.; Bekdaş, G.; Kim, S.; Geem, Z.W. Optimisation of Shear and Lateral–Torsional Buckling of Steel Plate Girders Using Meta-Heuristic Algorithms. *Appl. Sci.* **2020**, *10*, 3639. [CrossRef]

32. Elbelbisi, A.H.; El-Sisi, A.A.; Hassan, H.A.; Salim, H.A.; Shabaan, H.F. Parametric Study on Steel–Concrete Composite Beams Strengthened with Post-Tensioned CFRP Tendons. *Sustainability* **2022**, *14*, 15792. [CrossRef]
33. Elsheshtawy, S.S.; Shoeib, A.K.; Hassanin, A.; Ors, D.M. Influence of the Distribution and Level of Post-Tensioning Force on the Punching Shear of Flat Slabs. *Designs* **2022**, *7*, 1. [CrossRef]
34. Attia, M.M.; Khalil, A.H.H.; Mohamed, G.N.; Samaan, M.F.; Katunský, D. Nonlinear Behavior of Bonded and Unbonded Two-Way Post-Tensioned Slabs Pre-Strengthened with CFRP Laminates. *Buildings* **2022**, *13*, 35. [CrossRef]
35. Tahmasebinia, F.; Hu, Z.; Wei, Q.; Ma, W. Numerically Evaluation of Dynamic Behavior of Post-Tensioned Concrete Flat Slabs under Free Vibration. *Sustainability* **2023**, *15*, 845. [CrossRef]
36. Vavrus, M.; Kralovanec, J. Study of Application of Fiber Reinforced Concrete in Anchorage Zone. *Buildings* **2023**, *13*, 524. [CrossRef]
37. Lei, H.; Liu, Z.; Alsomiri, M. Bearing capacity of concentric anchorage zones in post-tensioned members: A stress field solution. *Structures* **2023**, *50*, 1368–1375. [CrossRef]
38. Joyklad, P.; Ali, N.; Chaiyasarn, K.; Suparp, S.; Hussain, Q. Time-dependent behavior of full-scale precast post-tensioned (PCPT) girders: Experimental and finite element analysis. *Case Stud. Constr. Mater.* **2022**, *17*, e01310. [CrossRef]
39. Yu, F.; Wang, M.; Yao, D.; Liu, Y. Experimental research on flexural behavior of post-tensioned self-compacting concrete beams with recycled coarse aggregate. *Constr. Build. Mater.* **2023**, *377*, 131098. [CrossRef]
40. Bekdaş, G.; Yucel, M.; Nigdeli, S.M. Generation of eco-friendly design for post-tensioned axially symmetric reinforced concrete cylindrical walls by minimizing of CO_2 emission. *Struct. Des. Tall Spéc. Build.* **2022**, *31*, e1948. [CrossRef]
41. *ACI 318M–05*; Building Code Requirements for Structural Concrete and Commentary. American Concrete Institute: Farmington Hills, MI, USA, 2005.
42. Love, A. A Treatise on the Mathematical Theory of Elasticity. 1, 1892. ⟨hal-01307751⟩. Available online: https://hal.science/hal-01307751 (accessed on 23 February 2023).
43. Hetenyi, M. *Beams on Elastic Foundation*; The University of Michigan Press: Ann Arbor, MI, USA, 1946.
44. Geem, Z.W.; Kim, J.H.; Loganathan, G.V. A new heuristic optimization algorithm: Harmony search. *Simulation* **2001**, *76*, 60–68. [CrossRef]
45. Yadav, N.; Ngo, T.T.; Kim, J.H. An algorithm for numerical solution of differential equations using harmony search and neural networks. *J. Appl. Anal. Comput.* **2022**, *12*, 1277–1293. [CrossRef]
46. Geem, Z.W. Multiobjective Optimization of Time-Cost Trade-Off Using Harmony Search. *J. Constr. Eng. Manag.* **2010**, *136*, 711–716. [CrossRef]
47. Chang, Y.-Z.; Li, Z.-W.; Kou, Y.-X.; Sun, Q.-P.; Yang, H.-Y.; Zhao, Z.-Y. A New Approach to Weapon-Target Assignment in Cooperative Air Combat. *Math. Probl. Eng.* **2017**, *2017*, 2936279. [CrossRef]
48. Shih, P.-C.; Chiu, C.-Y.; Chou, C.-H. Using Dynamic Adjusting NGHS-ANN for Predicting the Recidivism Rate of Commuted Prisoners. *Mathematics* **2019**, *7*, 1187. [CrossRef]
49. Xu, C.; Wang, X. Transient content caching and updating with modified harmony search for Internet of Things. *Digit. Commun. Netw.* **2018**, *5*, 24–33. [CrossRef]
50. Gonzalez, P.; Mora, A.; Garrido, S.; Barber, R.; Moreno, L. Multi-LiDAR Mapping for Scene Segmentation in Indoor Environments for Mobile Robots. *Sensors* **2022**, *22*, 3690. [CrossRef]
51. Cheng, Y.; Li, L.; Lansivaara, T.; Chi, S.; Sun, Y. An improved harmony search minimization algorithm using different slip surface generation methods for slope stability analysis. *Eng. Optim.* **2008**, *40*, 95–115. [CrossRef]
52. Lee, K.S.; Geem, Z.W. A new structural optimization method based on the harmony search algorithm. *Comput. Struct.* **2004**, *82*, 781–798. [CrossRef]
53. Lee, K.S.; Geem, Z.W.; Lee, S.-H.; Bae, K.-W. The harmony search heuristic algorithm for discrete structural optimization. *Eng. Optim.* **2005**, *37*, 663–684. [CrossRef]
54. García-Segura, T.; Yepes, V.; Alcalá, J. Computer-support tool to optimize bridges automatically. *Int. J. Comput. Methods Exp. Meas.* **2017**, *5*, 171–178. [CrossRef]
55. Aladsani, M.A.; Burton, H.; Abdullah, S.A.; Wallace, J.W. Explainable Machine Learning Model for Predicting Drift Capacity of Reinforced Concrete Walls. *ACI Struct. J.* **2022**, *119*, 191–204. [CrossRef]
56. Ke, G.; Meng, Q.; Finley, T.; Wang, T.; Chen, W.; Ma, W.; Ye, Q.; Liu, T.Y. Lightgbm: A highly efficient gradient boosting decision tree. *Adv. Neural Inf. Process. Syst.* **2017**, *30*, 3146–3154.
57. Dorogush, A.V.; Ershov, V.; Gulin, A. CatBoost: Gradient boosting with categorical features support. *arXiv* **2018**, arXiv:1810.11363.
58. Gandomi, A.H.; Roke, D.A. Assessment of artificial neural network and genetic programming as predictive tools. *Adv. Eng. Softw.* **2015**, *88*, 63–72. [CrossRef]
59. Madár, J.; Abonyi, J.; Szeifert, F. Genetic Programming for the Identification of Nonlinear Input−Output Models. *Ind. Eng. Chem. Res.* **2005**, *44*, 3178–3186. [CrossRef]

60. Gondia, A.; Ezzeldin, M.; El-Dakhakhni, W. Mechanics-Guided Genetic Programming Expression for Shear-Strength Prediction of Squat Reinforced Concrete Walls with Boundary Elements. *J. Struct. Eng.* **2020**, *146*, 04020223. [CrossRef]
61. Mangalathu, S.; Hwang, S.-H.; Jeon, J.-S. Failure mode and effects analysis of RC members based on machine-learning-based SHapley Additive exPlanations (SHAP) approach. *Eng. Struct.* **2020**, *219*, 110927. [CrossRef]

sustainability

MDPI

Article

Sustainable Design of Circular Reinforced Concrete Column Sections via Multi-Objective Optimization

Primož Jelušič * and Tomaž Žula

Faculty of Civil Engineering, Transportation Engineering and Architecture, University of Maribor, Smetanova 17, 2000 Maribor, Slovenia; tomaz.zula@um.si

* Correspondence: primoz.jelusic@um.si

Abstract: An optimization model for reinforced concrete circular columns based on the Eurocodes is presented. With the developed optimization model, which takes into account the exact distribution of the steel reinforcement, which is not the case when designing with conventional column design charts, an optimal design for the reinforced concrete cross section is determined. The optimization model uses discrete variables, which makes the results more suitable for actual construction practice and fully exploits the structural capacity of the structure. A parametric study of the applied axial load and bending moment was performed for material cost and CO_2 emissions. The results based on a single objective function show that the optimal design of the reinforced concrete column cross section obtained for the material cost objective function contains a larger cross-sectional area of concrete and a smaller area of steel compared with the optimization results when CO_2 emissions are determined as the objective function. However, the optimal solution in the case where the material cost was assigned as the objective function has much more reserve in axial load capacity than in the optimal design where CO_2 was chosen as the objective function. In addition, the multi-objective optimization was performed to find a set of solutions that provide the best trade-offs between the material cost and CO_2 emission objectives.

Keywords: reinforced concrete columns; circular cross section; cost; CO_2 emissions; multi-objective optimization; genetic algorithm

Citation: Jelušič, P.; Žula, T. Sustainable Design of Circular Reinforced Concrete Column Sections via Multi-Objective Optimization. *Sustainability* **2023**, *15*, 11689. https://doi.org/10.3390/su151511689

Academic Editor: Syed Minhaj Saleem Kazmi

Received: 5 June 2023
Revised: 27 July 2023
Accepted: 27 July 2023
Published: 28 July 2023

1. Introduction

Column design charts are commonly used to determine the steel reinforcement required for a given axial loading and bending moment (Figure 1a). Each pair of axial load (N) and bending moment (M) corresponds to a specific neutral axis position and can be plotted as a point on an N-M graph. Connecting those points to a curve forms an envelope of column resistance against the axial load and bending moment. A stress state (caused by axial and bending) outside the curve is not possible since the column resistance is overreached. Alternatively, the area of steel reinforcement required is determined by non-dimensional design charts such as those presented in Figure 1b. Steel reinforcement configuration affects the column axial and bending resistances. As steel reinforcement increases the cross section, the overall load-carrying capacity of the member increases, making it more resistant to cracking and failure under applied loads. With more steel, the cross section also becomes more ductile, allowing it to deform more before failure. This results in a more gradual and favorable load–moment interaction diagram, as the capacity of the cross section is utilized to a greater extent. In the design of reinforced concrete, a balanced steel reinforcement ratio is often sought. This ratio is the point at which the concrete and steel simultaneously reach their respective limits so that both materials can be used as efficiently as possible.

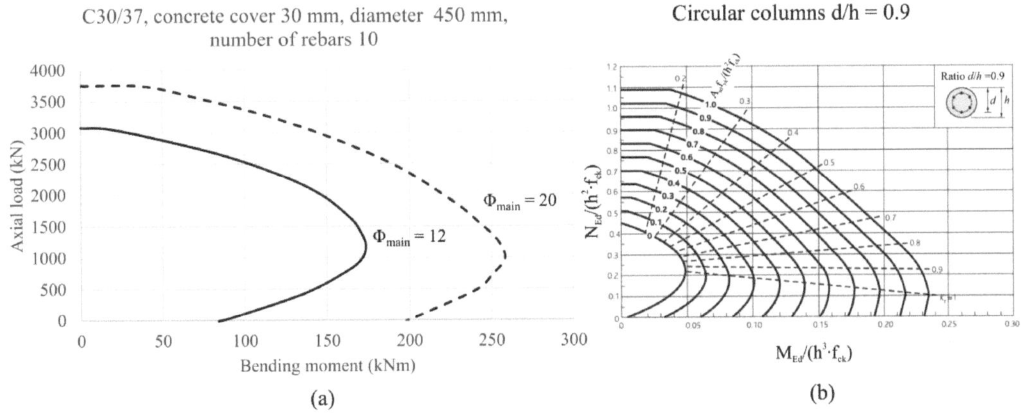

Figure 1. Typical column design charts. (**a**) Axial and bending resistance, (**b**) non-dimensional design charts.

Research on the optimization of reinforced concrete columns shows the evolution of optimization methods, from linear programming to nonlinear programming [1] to genetic algorithms [2] and particle swarm optimization [3]. Recent research has also introduced multi-objective optimization methods that can simultaneously consider different objectives and constraints, which enables cost-effective, environmentally friendly, and safe design of reinforced concrete columns. The main objective of the optimization models is to find the optimal cross-sectional dimensions and arrangement of steel reinforcement that can support the applied loads with minimum cost, minimum CO_2 emissions, and maximum safety.

In the field of optimization and sustainability within structural engineering, researchers have introduced various optimization techniques and objectives for different types of structural elements. One such study, conducted by Zaforteza et al. [4], focused on the application of the simulated annealing algorithm (SA) in optimizing reinforced concrete frames. They considered two objective functions: the embedded CO_2 emissions and the economic cost. Another study, by Camp and Huq [5], utilized a hybrid algorithm called big bang–big crunch (BB-BC) for the optimal design of reinforced concrete frames. Their objective was to minimize either the total cost or the CO_2 emissions associated with the structures. Trinh et al. [6] employed a branch-and-reduce deterministic algorithm to optimize the design of flat plate buildings based on carbon footprint. Alonso and Berdasco [7] proposed a method to assess the carbon footprint of sawn timber products. Yeo and Gabbai [8] introduced a sustainable design approach for rectangular beams, aiming to minimize both the embodied energy and cost. Zhang and Zhang [9] presented a study where a multi-objective genetic algorithm was adopted for the sustainable design of reinforced concrete members, considering both the embodied emissions and costs. Jayasinghe et al. [10] minimized the embodied carbon in three different optimization approaches, namely theoretical optimum shape finding, feasible optimum shape finding, and optimizing prismatic beams. Sahebi and Dehestani [11] considered the objectives of cost and CO_2 footprint in optimizing the sustainable design of reinforced beams.

The work of Ahmed et al. [12] and Tayem and Najmi [13] concerned with the optimal design of circular reinforced concrete columns using a nonlinear optimization approach to minimize the material cost considering constraints on axial load capacity, bending capacity, and maximum steel ratio. The hybrid optimization algorithm was also used to minimize the total material cost and for predictive modeling of circular reinforced concrete columns [14]. Camp and Assadollah [15] also presented a hybrid optimization algorithm for the CO_2 and cost optimization of reinforced concrete foundations. The work of Zhao et al. [16] aimed to optimize the design of reinforced concrete columns strengthened with square steel tubes and sandwiched concrete. Jelušič and Žlender [17] optimized the reinforced cross sections of the geothermal energy piles using design column charts fitted with approximation functions. Payá-Zaforteza et al. [18] proposed a multi-objective optimization approach to optimize the cost and sustainability of reinforced concrete building frames considering the constraints on structural performance, environmental impact, and economic feasibility. Hong et al. [19] investigated an artificial-neural-network-based Lagrangian optimization approach for a multi-objective optimization model in which both the cost and performance of the reinforced concrete circular columns were optimized.

In this paper, a genetic algorithm is used to solve a multi-objective optimization problem, since it offers several advantages over other optimization techniques [20–22]. Genetic algorithms preserve diversity within the population, which helps to explore a wide range of solutions and avoid premature convergence to suboptimal solutions. This is important in multi-objective optimization, where there may be multiple solutions that are equally good but differ in terms of the tradeoffs between objectives. In addition, genetic algorithms do not require differentiable functions, which is advantageous for complex nonlinear functions that are common in multi-objective optimization. Finally, the genetic algorithm can be easily parallelized to perform multiple evaluations simultaneously, allowing faster convergence to optimal solutions. This is useful in multi-objective optimization, where the evaluation of solutions can be computationally expensive. However, using genetic algorithms for multi-objective optimization also has drawbacks, such as slow convergence speed, non-deterministic results, and scalability issues. Tuning parameters or selecting appropriate values such as population size, mutation rate, crossover probability, etc., can also be challenging, and incorrect selection can affect the performance of the genetic algorithm.

The main objective of this paper is to present an optimization model based on mixed-integer nonlinear programming solved by a genetic algorithm. The optimization model was developed, and optimal designs were determined using MATLAB [23]. The development optimization model is used to determine the difference in the design of reinforced concrete circular columns in terms of the environmental and economic aspects.

Parametric optimization is performed separately for different combinations of applied axial load and bending moments and for the two objective functions of material cost and CO_2 emissions, which are generated during the production of the reinforced concrete. Furthermore, a multi-objective optimization was executed with the aim of identifying a range of solutions that offer optimal balances between material cost and CO_2 emission objectives.

2. Optimization Model: Reinforced Circular Concrete Section (RCCS)

The optimization model, named the reinforced circular concrete section (RCCS), was developed to minimize material cost, minimize CO_2 consumption, or minimize both material cost and CO_2 consumption simultaneously through multi-objective optimization. Therefore, the optimization model includes input data, two objective functions, and constraints derived from the structural analysis of a reinforced circular concrete section. The structural analysis of the reinforced concrete section considers the relationship between the axial force and the bending moment in the different positions of the neutral axis.

The input data represent the design and economic data (constants) for the optimization. The design data comprise the design value of the applied axial load N_{Ed} (kN), the design value of the applied bending moment M_{Ed} (kNm), the length of the column section L (m), the characteristic value of the compressive strength of the concrete f_{ck} (MPa), the characteristic value of the tensile strength of the steel f_{yk} (MPa), the modulus of elasticity of steel E_s (MPa), the steel density ρ_{steel} (kg/m^3), the diameter of shear reinforcement Φ_{link} (mm), a coefficient taking account of sustained compression α_{cc}, the safety factor for concrete γ_c, safety factor for steel γ_s, and the concrete cover c_{con}. The input data also included all the defined values of the necessary material cost and CO_2 consumption coefficients included in the objective functions, as well as all the other coefficients included in the objective function and the structural analysis of a circular reinforced concrete section (see Table 1) and the discrete alternatives of the circular reinforced concrete column section (see Table 2). It should be noted that the construction costs and the CO_2 emissions that arise during the construction of reinforced concrete are not included in the objective functions. The circular reinforced concrete section has the following design variables: column diameter (Φ), location of neutral axis (c_x), steel reinforcement area ($A_{s,main}$), which is determined by the number of rebars (n), and reinforcement diameter (Φ_{main}), see Figure 2.

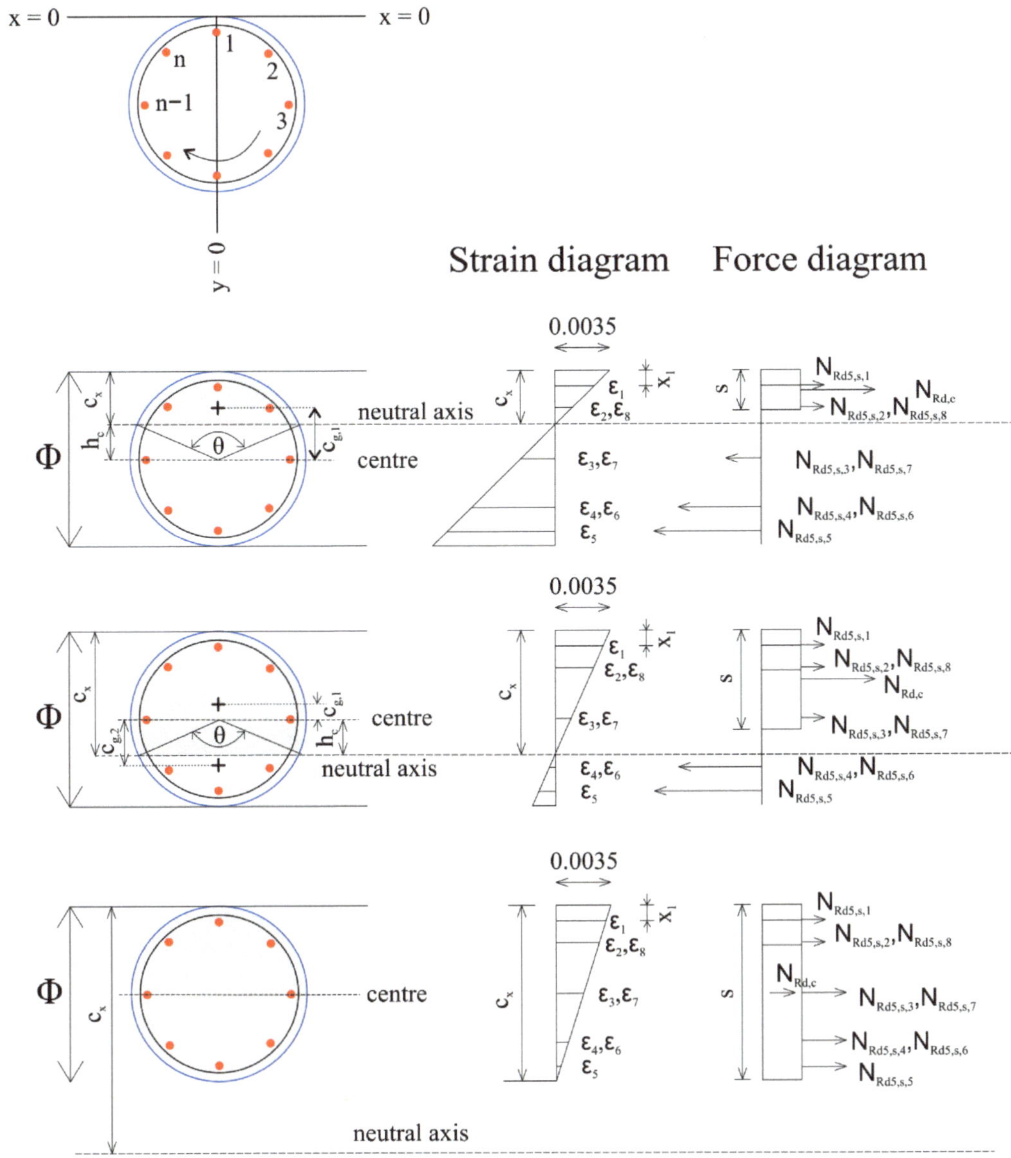

Figure 2. Strain and force diagrams under different positions of the neural axis.

2.1. Objective Functions of the Optimization Model RCCS

Material cost optimization refers to the process of reducing the cost of producing a product or service by selecting the least expensive materials or minimizing the amount of materials used. This process is primarily focused on reducing the financial cost of producing a product. Optimizing the amount of CO_2 generated by the use of materials, on the other hand, refers to the process of reducing the greenhouse gas emissions generated in the production of a product or service. This process is primarily aimed at reducing the environmental impact of producing a product. Although these two concepts may overlap to some degree, they are ultimately different. For example, it is possible to reduce the material cost of a product by selecting a cheaper material, but this may result in a higher amount of CO_2 emissions during production. Conversely, it is possible to reduce the amount of CO_2 emissions by using more environmentally friendly materials, but this may result in higher material costs. Ultimately, companies and investors need to find a balance between optimizing the cost of materials and optimizing the amount of CO_2 generated by the use of materials in order to achieve their economic and environmental goals. Therefore, the two objective functions are determined in the optimization model RCCS. The material cost objective function is defined with Equation (1):

$$min:\ COST = c_{con} \cdot \left(\pi \cdot \Phi^2/4\right) \cdot L + c_{steel} \cdot \rho_{steel} \cdot \left(\pi \cdot {\Phi_{main}}^2/4\right) \cdot L \cdot n \quad (1)$$

whereas the objective function for the quantity of CO_2 emissions during the production of reinforced concrete sections is defined by Equation (2):

$$min:\ CO_2 = CO_{2,con} \cdot \left(\pi \cdot \Phi^2/4\right) \cdot L + CO_{2,steel} \cdot \rho_{steel} \cdot \left(\pi \cdot {\Phi_{main}}^2/4\right) \cdot L \cdot n \quad (2)$$

In Equation (1), c_{con} (€/m^3) represents the unit price of concrete and c_{steel} (€/kg) is the unit price of the steel reinforcement, whereas in Equation (2), the $CO_{2,con}$ (kgCO_2/m^3) represents the unit emissions of CO_2 generated by the use of concrete and the $CO_{2,steel}$ (kgCO_2/kg) is the unit emissions of CO_2 generated by the use of steel reinforcement.

2.2. Structural Analysis of a Reinforced Circular Concrete Section and Design Constraints

Both objective functions aim to minimize the amount of concrete and steel reinforcement; however, the circular reinforced concrete cross section must be able to resist the applied loads with a sufficient amount of material. The resistance of a circular reinforced concrete section is calculated based on the location of the neutral axis. The neutral axis is located at a distance cx below the compression face, where the cross section experiences neither compression nor tension, resulting in zero strain at that level. Five main conditions are defined in accordance with the Eurocode 2 [24] standard in the form of five inequality constraints:

Condition 1: the design bending moment M_{Ed} in the circular reinforced concrete cross section needs to be limited to under the design bending resistance M_{Rd}, see Equation (3).

$$M_{Ed} \leq M_{Rd} \quad (3)$$

Condition 2: the design axial load N_{Ed} applied to the circular reinforced concrete cross section needs to be limited to under the design axial resistance or cross section N_{Rd}, see Equation (4).

$$N_{Ed} \leq N_{Rd} \quad (4)$$

Condition 3: the minimum area of the main reinforcement must be provided, see Equation (5).

$$A_{s,total} \geq A_{s,min} \quad (5)$$

Condition 4: the distance between the neutral axis c_x and the edge of the cross section must always be positive, see Equation (6).

$$c_x \geq 0 \quad (6)$$

Condition 5: the maximum area of the main reinforcement must not be exceeded, see Equation (5).

$$A_{s,total} \leq A_{s,max} \quad (7)$$

The calculation procedure to determine the design bending resistance M_{Rd} and the design axial resistance N_{Rd} of the reinforced concrete circular cross section is presented in Equations (8)–(50). First, the design compressive strength of the concrete f_{cd} (see Equation (8)) and the design tensile strength of the steel f_{yd} (see Equation (9)) are calculated:

$$f_{cd} = \alpha_{cc} \cdot f_{ck} / \gamma_c, \quad (8)$$

$$f_{yd} = f_{yk} / \gamma_s \quad (9)$$

In the proposed model, each rebar is assigned a unique identification number (ID) along with its exact position in the coordinate system. The diameter of circle that joins the centroid of the rebars can be calculated with Equation (10), and the circumference of this internal circle is calculated by Equation (11):

$$\Phi_{in} = \Phi - 2 \cdot c_{con} - 2 \cdot \Phi_{link} - \Phi_{main}, \quad (10)$$

$$P_{in} = \pi \cdot \Phi_{in} \quad (11)$$

It should be noted that c_{con} represents the depth of concrete cover. The center to center spacing between rebars that are evenly placed is calculated by using Equation (12):

$$s_{main} = \pi \cdot \Phi_{in} / n \quad (12)$$

The clear spacing between rebars s_{clear} is therefore calculated by Equation (13); the angle between each adjacent rebar θ_r is calculated by Equation (14):

$$s_{clear} = s - \Phi_{main}, \quad (13)$$

$$\theta_r = s_{main} / r \quad (14)$$

where

$$r = \Phi_{in} / 2, \quad (15)$$

$$s = 0.8 \cdot c_x \quad (16)$$

The steel reinforcement of area A_s of a single rebar is calculated as:

$$A_s = \pi \cdot {\Phi_{main}}^2 / 4 \quad (17)$$

For each rebar (where i is from 1 to n), the position relative to the center of the circular cross section can be defined with:

$$x_{nc,i} = r \cdot \cos \theta_{cumm,i}, \quad (18)$$

$$y_{nc,i} = r \cdot \sin \theta_{cumm,i} \quad (19)$$

where the cumulative angle from the vertical in a clockwise direction is determined as:

$$\theta_{cumm,i} = (i-1) \cdot \theta_r \tag{20}$$

The location according to selected coordinate system (see, Figures 1 and 2) is therefore defined with:

$$x_{n,i} = \Phi_{main}/2 - x_{nc,i}, \tag{21}$$

$$y_{n,i} = y_{nc,i}, \tag{22}$$

To calculate the strain in each rebar ε_i at various locations of neutral axis, Equations (23)–(31) are used:

$$xi_i = c_x \cdot (i-(i-1)), \tag{23}$$

$$T_i = xi_i - x_{n,i}, \tag{24}$$

$$U_i = x_{n,i} - xi_i, \tag{25}$$

$$A_i = 0.0035 \cdot (T_i/xi_i), \tag{26}$$

$$B_i = 0.0035 \cdot (U_i/xi_i), \tag{27}$$

$$C_i = f_{yd}/E_s, \tag{28}$$

$$D_i = \min(B_i; C_i), \tag{29}$$

$$zero_i = 0, \tag{30}$$

$$\varepsilon_i = \begin{cases} A_i;\ T_i \geq zero_i \text{ and s} \leq \Phi \\ D_i;\ T_i < zero_i \\ 0.00175 \end{cases} \tag{31}$$

The resistance force developed in rebars ($N_{Rd5,s,i}$) under compression ($N_{Rd3,s,i}$) or under tension ($N_{Rd4,s,i}$) is determined by Equations (32)–(36):

$$N_{Rd1,s,i} = \min\left(E_s \cdot \varepsilon_i \cdot A_s; f_{yd} \cdot A_s\right), \tag{32}$$

$$N_{Rd2,s,i} = \min\left((E_s \cdot \varepsilon_i - (\alpha_{cc}/\gamma_c) \cdot f_{ck}) \cdot A_s; \left(f_{yd} - (\alpha_{cc}/\gamma_c) \cdot f_{ck}\right) \cdot A_s\right), \tag{33}$$

$$N_{Rd3,s,i} = \begin{cases} N_{Rd1,s,i};\ \text{s} \leq x_{n,i} \\ N_{Rd2,s,i} \end{cases}, \tag{34}$$

$$N_{Rd4,s,i} = -\min\left(E_s \cdot \varepsilon_i \cdot A_s; f_{yd} \cdot A_s\right), \tag{35}$$

$$N_{Rd5,s,i} = \begin{cases} N_{Rd3,s,i};\ T_i > zero_i \\ N_{Rd4,s,i} \end{cases} \tag{36}$$

The design bending resistance of concrete and reinforcing bars is calculated by multiplying the developed resistance force by the lever arm from the center of the column cross section, as determined in Figure 2. Since the cross section is symmetrical, the center point is located at the center of the column cross section. The lever arm for concrete varies with the position of the neutral axis, but the lever arm for each rebar can be determined by Equation (37):

$$l_{n,i} = \Phi/2 - x_{n,i} \tag{37}$$

Once the resistance force and the lever arm for each rebar is determined, the bending moment resistance ($M_{Rd,s,i}$) for each rebar is calculated by Equation (38):

$$M_{Rd,s,i} = N_{Rd5,s,i} \cdot l_{n,i} \tag{38}$$

Finally, the total axial load resistance ($N_{Rd,s,total}$) and the total bending moment resistance ($M_{Rd,s,total}$) due to all reinforcing bars are calculated using Equation (39) and Equation (40), respectively:

$$N_{Rd,s,total} = \sum\nolimits_{i=1}^{n} N_{Rd5,s,i}, \tag{39}$$

$$M_{Rd,s,total} = \sum\nolimits_{i=1}^{n} M_{Rd,s,i} \tag{40}$$

The distance from neutral axis and the center of the section h_c is determined as:

$$h_c = \begin{cases} \Phi/2 - s\,;\ s \leq \Phi/2 \\ s - \Phi/2;\ s \leq \Phi \text{ and } s > \Phi/2 \\ 0 \end{cases} \tag{41}$$

To calculate the bending moment resistance due to the concrete resisting force, the lever arm for the concrete force ($c_{g,1}$ or $c_{g,2}$) and the concrete area under compression (A_c) are calculated using Equations (42)–(47), see Figure 2:

$$A_c = \begin{cases} \frac{\theta}{2\cdot\pi}\cdot\frac{\pi\cdot\Phi^2}{4} - \frac{1}{2}\cdot\left(\frac{\Phi}{2}\right)^2\cdot\sin\theta;\ s \leq \Phi/2 \\ \frac{(2\cdot\pi-\theta)}{2\cdot\pi}\cdot\frac{\pi\cdot\Phi^2}{4} + \frac{1}{2}\cdot\left(\frac{\Phi}{2}\right)^2\cdot\sin\theta;\ s \leq \Phi \text{ and } s > \Phi/2 \\ \frac{\pi\cdot\Phi^2}{4};\ s > \Phi \end{cases} \tag{42}$$

where the inner angle θ is calculated as:

$$\theta = 2\cdot\cos^{-1}\left(\frac{h_c}{(\Phi/2)}\right), \tag{43}$$

$$A_{c1} = \frac{1}{2}\cdot\frac{\pi\cdot\Phi^2}{4}, \tag{44}$$

$$A_{c2} = \frac{\theta}{2\cdot\pi}\cdot\frac{\pi\cdot\Phi^2}{4} - \frac{1}{2}\cdot\left(\frac{\Phi}{2}\right)^2\cdot\sin\theta, \tag{45}$$

$$c_{g,1} = \begin{cases} \frac{4\cdot\left(\frac{\Phi}{2}\right)\cdot\sin^3\left(\frac{\theta}{2}\right)}{3\cdot(\theta-\sin\theta)};\ s \leq \Phi/2 \\ \frac{4\cdot\left(\frac{\Phi}{2}\right)\cdot\sin^3\left(\frac{\pi}{2}\right)}{3\cdot(\pi-\sin\pi)} \end{cases}, \tag{46}$$

$$c_{g,2} = \begin{cases} \frac{4\cdot\left(\frac{\Phi}{2}\right)\cdot\sin^3\left(\frac{\theta}{2}\right)}{3\cdot(\theta-\sin\theta)};\ s < \Phi/2 \\ 0 \end{cases} \tag{47}$$

Therefore, the bending moment resistance provided by concrete for any position of the neutral axis ($M_{Rd,c}$) is determined by Equation (48), whereas the axial resistance of concrete ($N_{Rd,c}$) is determined by Equation (49):

$$M_{Rd,c} = \begin{cases} A_c\cdot f_{cd}\cdot c_{g,1};\ s \leq \Phi/2 \\ A_{c1}\cdot f_{cd}\cdot c_{g,1} - \left(A_{c1}\cdot f_{cd}\cdot c_{g,1} - A_{c2}\cdot f_{cd}\cdot c_{g,2}\right);\ s \leq \Phi \text{ and } s > \Phi/2, \\ 0 \end{cases} \tag{48}$$

$$N_{Rd,c} = f_{cd} \cdot A_c \tag{49}$$

Summing the total axial resistance of all reinforcing bars ($N_{Rd,s,total}$) and the axial resistance of the concrete ($N_{Rd,c}$) gives the design axial resistance or cross section (N_{Rd}) according to Equation (50), whereas summing the total bending moment resistance of all reinforcing bars ($M_{Rd,s,total}$) and the bending moment resistance of the concrete ($M_{Rd,c}$) gives the design moment resistance or cross section (M_{Rd}) according to Equation (51).

$$N_{Rd} = N_{Rd,c} + N_{Rd,s,total}, \tag{50}$$

$$M_{Rd} = M_{Rd,c} + M_{Rd,s,total} \tag{51}$$

Once the design axial and bending moment resistance or cross section is available, conditions 1 and 2 can be verified. However, for the verification of condition 3, the required minimum area of steel reinforcement ($A_{s,min}$) and the total area of steel reinforcement ($A_{s,total}$) must be calculated according to Equation (52) and Equation (53), respectively:

$$A_{s,min} = \max\left(\frac{0.1 \cdot N_{Ed}}{f_{yd}}; 0.002 \cdot \frac{\pi \cdot \Phi^2}{4}\right), \tag{52}$$

$$A_{s,total} = A_s \cdot n \tag{53}$$

The upper limit of the steel reinforcement area ($A_{s,max}$) is determined by Equation (54).

$$A_{s,max} = 0.04 \cdot \left(\frac{\pi \cdot \Phi^2}{4}\right) \tag{54}$$

The structural analysis of a circular reinforced concrete cross section also includes design (in)equality constraints that ensure that the dimensions of the circular reinforced concrete cross section do not lie outside the specified limits. In addition, the discrete/standardized values for the dimensions are used in the RCCS optimization model (see Table 2). However, the diameter of the concrete cross section Φ (m) is limited by Equation (55), the number of reinforcing bars n (-) varies between the lower and upper limits, see Equation (56), and, finally, the diameter of the reinforcing bars Φ_{main} (mm) is limited by Equation (57).

$$\Phi^{LO} \leq \Phi \leq \Phi^{UP}, \tag{55}$$

$$\mathrm{n}^{LO} \leq \mathrm{n} \leq \mathrm{n}^{UP}, \tag{56}$$

$$\Phi^{LO}_{main} \leq \Phi_{main} \leq \Phi^{LO}_{main} \tag{57}$$

Table 1. The input data involved in objective functions and structural analysis of a reinforced circular concrete section.

Symbol	Description	Value
c_{con}	unit price of concrete C30/37	115 EUR/m^3
c_{steel}	unit price of the steel reinforcement S500	1.45 EUR/kg
$CO_{2,con}$ *	unit emissions of CO_2 for concrete	308.2 kgCO_2/m^3
$CO_{2,steel}$ *	unit emissions of CO_2 for steel reinforcement	0.87 kgCO_2/kg

Table 1. *Cont.*

Symbol	Description	Value
ρ_{steel}	steel density	7850 kg/m^3
L	length of the column section	1 m
f_{ck}	the compressive strength of the concrete	30 MPa
f_{yk}	tensile strength of the steel	500 MPa
E_s	modulus of elasticity of steel	200,000 MPa
γ_c	safety factor for concrete	1.5
γ_s	safety factor for steel	1.15
α_{cc}	coefficient for sustained compression	0.85
Φ_{link}	diameter of shear reinforcement	6 mm
c_{con}	concrete cover	30 mm

* The carbon footprint emission factors used in the study are taken from the literature [25].

Table 2. Discrete alternatives for the dimensions of the circular reinforced concrete cross section.

Variable	Discrete Alternatives
Φ (mm)	400, 450, 500, 550, 600, 650, 700, 750, 800, 850, 900, 950, 1000
n (-)	6, 8, 10, 12, 14, 16, 18, 20, 22
Φ_{main} (mm)	12, 14, 16, 18, 20, 22, 24, 26, 28

2.3. Genetic Algorithm

In MATLAB [23], the genetic algorithm (GA) provides built-in functions for implementing genetic algorithms. To use GAs for mixed-integer design problems, the following steps were performed: First, the objective functions (*COST* and CO_2 emissions) and constraints (four conditions) were defined. The design variables were also defined. One variable is continuous (neutral axis location), whereas the other three variables are discrete (number of rebars, diameter of steel reinforcement, and diameter of column). The fitness function was determined to evaluate the objective(s) of the optimization problem. This function was included in the design variables and returns a scalar fitness value. Constraints that must be satisfied are also defined, such as upper and lower bounds for the design variables. The parameters of GA, such as population size, maximum number of generations, tolerance threshold for the fitness function, mutation rate, and crossover proportion, are selected to improve the optimization results. Once the best solution is found, the integer variables are mapped to a discrete set. To map integer variables to a discrete set, the MATLAB toolbox provides built-in options for integer constraints. These methods were used to constrain integer variables to a finite set of values. The results of the optimization, including the fitness value and the values of the design variables, are evaluated separately for each objective function in the multiparametric optimization section and simultaneously for both objective functions in the multi-objective optimization section, whereas the optimal solution is presented as a set of Pareto-optimal solutions. In the multi-objective optimization problem presented, the main objective is to minimize the first objective function subject to the condition that the value of the second objective function does not exceed the threshold. This threshold represents the maximum allowable value for the second objective function that the decision maker is willing to accept given the importance of the first objective and the tradeoffs between the two objectives. The threshold values are implicitly determined by the fitness scaling function and the selection criteria used in the genetic algorithm. The fitness scaling function maps the objective values of the solutions in the population to a common scale that allows the algorithm to compare solutions with different objective function values. The selection criteria are used to select the solutions that will be used to generate the next generation of solutions. These criteria consider both the objective values of the solutions and their dominance relationship with other solutions in the population. The main options of the genetic algorithm were set as follows: a population size of 500, a maximum number of generations of 200, a number of elites of 20, and a function tolerance for the fitness function of 1×10^{-8}. In the multi-objective optimization, a large population size of 1000 was used to increase the diversity of the population and improve the chance of finding better solutions; a maximum of 50 was used to limit the number of generations without progress and 100 generations was used to set the maximum number of generations for the execution of the GA. By setting these parameters correctly, it is possible to optimize the GA for multi-objective optimization and obtain high-quality solutions. The obtained solutions satisfy the optimization criteria, and the genetic algorithm has effectively explored the search space, efficiently evaluated potential solutions, and found solutions that are close to the best possible solution.

3. Multiparametric Optimization

The previously defined optimization model RCCS was used to obtain the optimal designs for different combinations of applied axial load N_{Ed} and bending moments M_{Ed} and for the two objective functions material cost and CO_2 emissions separately. The multiparametric optimization was therefore performed for 30 combinations between the following different design parameters:

- Three different axial loads: 1000 kN, 3000 kN, and 5000 kN;
- Five different bending moments: 100 kNm, 300 kNm, 500 kNm, 700 kNm, and 1000 kNm;
- Two objective functions: material cost and quantity of CO_2 emissions.

The results of the 30 individual optimizations performed are shown in Tables 3 and 4 and in Figure 3. Table 3 shows the optimal design variables and associated material costs, as well as the amount of CO_2 emissions generated for the case where the material cost is set as the objective function. Table 4 also shows optimal solutions for various design parameters, but the results correspond to the case where the objective function was the amount of CO_2 emissions caused by the production of the reinforced concrete member. Figure 3 directly compares the material costs where the optimization function was the material costs and where the optimization function was the CO_2 emissions. Note that the dotted curves represent the values for the CO_2 emissions. It can be concluded that when the material *COST* was used as the objective function, a different design for the concrete cross section was obtained than when CO_2 emissions were chosen as the objective function. The optimal design of the reinforced concrete cross section obtained for the material cost objective function contains a larger cross-sectional area of concrete and a smaller area of steel compared with the optimization results when CO_2 emissions are determined as the objective function. In general, exploitation of condition 1 (bending moment resistance) was the top priority in both optimization models, regardless of whether material cost or CO_2 emissions were chosen as the objective function. However, the optimal solution in the case where the material cost was assigned as the objective function has much more reserve in axial load capacity than in the optimal design where CO_2 was chosen as the objective function.

Table 3. Optimal design for the case where the material cost has been assigned as an objective function.

N_{Ed} (kN)	M_{Ed} (kNm)	c_x (mm)	n (-)	Φ_{main} (mm)	Φ (mm)	N_{Rd} (kN)	M_{Rd} (kNm)	$A_{s,min}$ (cm^2)	$A_{s,total}$ (cm^2)	*COST* (€/m)	CO_2 (kg/m)
1000	100	274.13	6	12	400	1304.0	107.6	2.51	6.79	22.18	43.36
1000	300	345.24	6	12	600	2196.8	342.7	5.65	6.79	40.24	91.78
1000	500	420.77	8	12	700	3180.4	545.8	7.70	9.05	54.56	124.79
1000	700	412.14	6	18	750	3230.4	701.9	8.84	15.27	68.18	146.59

Table 3. *Cont.*

N_{Ed} (kN)	M_{Ed} (kNm)	c_x (mm)	n (-)	Φ_{main} (mm)	Φ (mm)	N_{Rd} (kN)	M_{Rd} (kNm)	$A_{s,min}$ (cm^2)	$A_{s,total}$ (cm^2)	$COST$ (€/m)	CO_2 (kg/m)
1000	1000	518.16	14	12	850	4795.6	1000.8	11.35	15.83	83.28	185.70
3000	100	484.11	8	12	500	3021.5	122.0	6.90	9.05	32.88	66.69
3000	300	433.11	8	12	600	3021.3	333.7	6.90	9.05	42.81	93.32
3000	500	457.15	8	12	700	3560.1	538.6	7.70	9.05	54.56	124.79
3000	700	470.55	6	18	750	3927.1	701.8	8.84	15.27	68.18	146.59
3000	1000	487.00	14	12	850	4385.9	1002.8	11.35	15.83	83.28	185.70
5000	100	670.25	6	18	600	5002.9	102.8	11.50	15.27	49.89	97.57
5000	300	670.67	6	16	700	5699.5	335.1	11.50	12.06	57.99	126.85
5000	500	623.14	6	16	750	5571.6	558.7	11.50	12.06	64.54	144.40
5000	700	559.91	6	16	800	5111.5	787.9	11.50	12.06	71.54	163.16
5000	1000	539.41	16	12	850	5091.3	1011.3	11.50	18.10	85.85	187.25

Table 4. Optimal design for the case where CO_2 emissions were assigned as the objective function.

N_{Ed} (kN)	M_{Ed} (kNm)	c_x (mm)	n (-)	Φ_{main} (mm)	Φ (mm)	N_{Rd} (kN)	M_{Rd} (kNm)	$A_{s,min}$ (cm^2)	$A_{s,total}$ (cm^2)	COST (€/m)	CO_2 (kg/m)
1000	100	301.79	6	12	400	1483.3	100.9	2.51	6.79	22.18	43.36
1000	300	231.21	8	26	450	1080.1	300.0	3.18	42.47	66.64	78.02
1000	500	241.57	18	22	500	1064.8	502.3	3.93	68.42	100.46	107.25
1000	700	276.92	18	22	600	1387.2	703.7	5.65	68.42	110.40	133.87
1000	1000	353.53	18	26	650	2591.7	1000.5	6.64	95.57	146.94	167.54
3000	100	455.84	20	12	450	3001.1	100.2	6.90	22.62	44.04	64.47
3000	300	421.93	12	16	550	3067.8	301.2	6.90	24.13	54.79	89.70
3000	500	391.52	20	16	600	3015.0	501.2	6.90	40.21	78.29	114.60
3000	700	386.66	10	26	650	3015.2	700.6	6.90	53.09	98.59	138.53
3000	1000	387.00	14	26	700	3024.4	1002.7	7.70	74.33	128.86	169.37
5000	100	671.73	6	18	600	5011.1	100.9	11.50	15.27	49.89	97.57
5000	300	615.70	16	12	650	5063.0	300.4	11.50	18.10	58.76	114.63
5000	500	575.86	18	12	700	5001.6	501.3	11.50	20.36	67.43	132.51
5000	700	551.17	12	16	750	5003.8	702.2	11.50	24.13	78.27	152.64
5000	1000	524.82	20	16	800	5008.2	1000.8	11.50	40.21	103.58	182.38

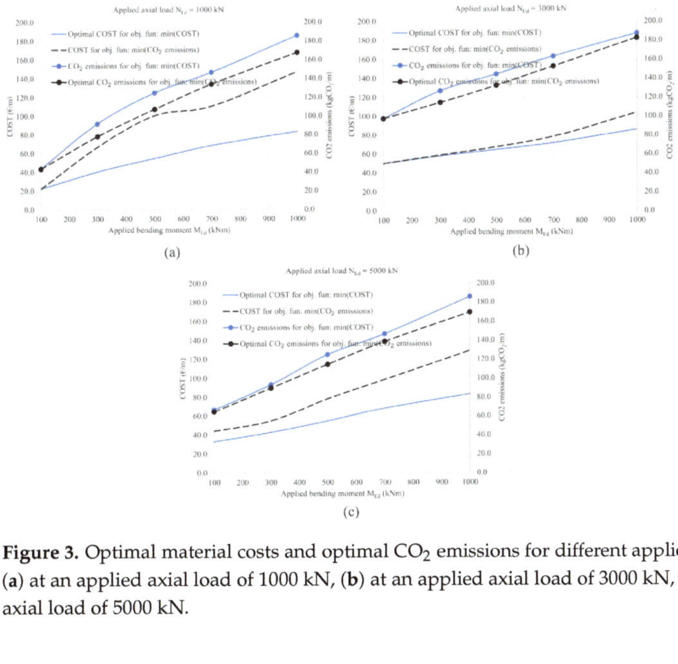

Figure 3. Optimal material costs and optimal CO_2 emissions for different applied bending moments (**a**) at an applied axial load of 1000 kN, (**b**) at an applied axial load of 3000 kN, and (**c**) at an applied axial load of 5000 kN.

The performance of the genetic algorithm is illustrated in the case where CO_2 emissions were assigned as the objective function and the reinforced concrete section was subjected to an axial load of N_{Ed} = 1000 kN and bending moments of M_{Ed} = 1000 kNm. The progress of the genetic algorithm, in terms of the best score value and mean score value, is plotted out in Figure 4a. The genetic algorithm stopped when the average relative change in the best fitness function value over stall generations was less than or equal to the function tolerance (see, Figure 4b). The maximum number of iterations for the genetic algorithm to perform was assigned to 200. Figure 4c shows a histogram of the parents and a population size of 500 individuals. The score at each generation is plotted in Figure 4d.

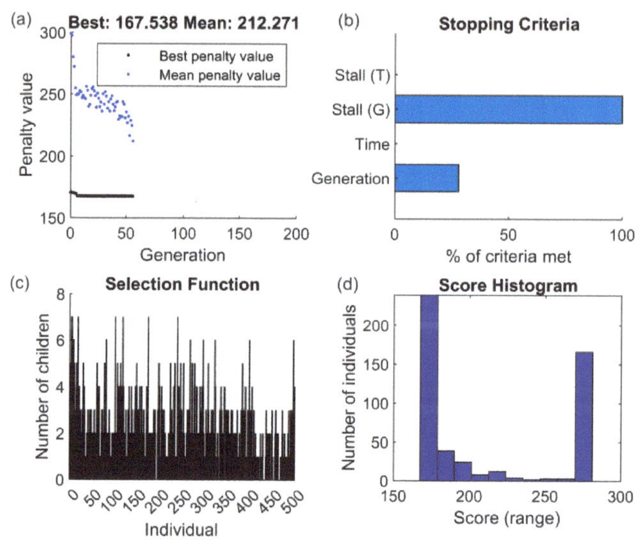

Figure 4. Performance of the genetic algorithm (objective function: CO_2 emissions, N_{Ed} = 1000 kN and M_{Ed} = 1000 kNm): (**a**) progress of the genetic algorithm; (**b**) stopping criteria; (**c**) population size; (**d**) score histogram.

4. Multi-Objective Optimization

In the above optimizations, where a single objective function (material cost or CO_2 emissions) was used, the optimization problem is relatively simple where the objective is to find the optimal value of the chosen objective function. In a multi-objective optimization, there is no unique optimal solution because the optimization problem involves tradeoffs between the objectives of material cost and CO_2 emissions. The objective of the multi-objective function is to find a set of solutions that provide the best trade-offs between the different objectives. In multi-objective optimization, the optimal solution is represented as a set of Pareto-optimal solutions, where no solution can be improved in one objective without making it worse in at least one other objective. The Pareto-optimal solutions represent the best possible tradeoffs among the different objectives, and the goal is to find the set of solutions that is most desirable given the decision maker's preferences. The main difference between the Pareto front in optimization with discrete variables and optimization with continuous variables lies in the nature of the design variables. Discrete variables can only take a limited number of values, whereas continuous variables can take any value within a certain range. In this optimization model, RCCS, discrete variables are used so that the Pareto front consists of a set of discrete points, each point representing a combination of the design variables that gives an optimal solution. In contrast, when continuous variables are used, the Pareto front is a continuous curve representing an infinite number of optimal solutions. Moreover, optimization with discrete variables results in a Pareto front that is less smooth and more jagged than the Pareto front obtained with continuous variables. The reason for this is that the set of feasible solutions is limited by the discrete nature of the design variables. Figures 5–7 show the Pareto front for axial loads of 1000 kN, 3000 kN, and 5000 kN, respectively. From all figures, it can be seen that not many solutions were found, which is due to the use of discrete variables. It can also be seen that the curves are flat, which means that a small reduction in CO_2 emissions is accompanied by a high increase in material costs, especially for an axial load of 1000 kN.

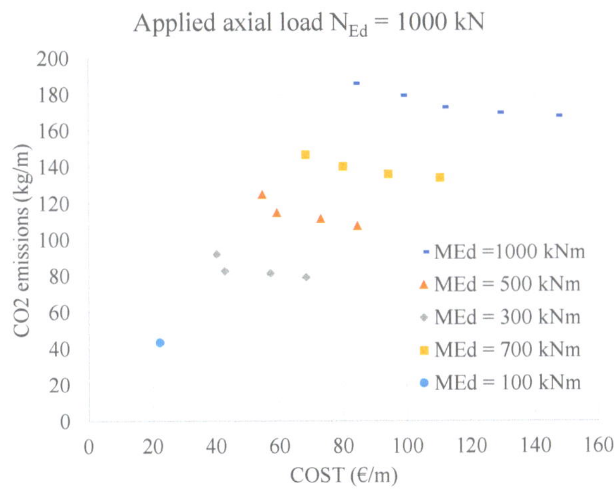

Figure 5. Pareto front for a reinforced concrete circular cross section loaded with an axial load of 1000 kN for different applied bending moments.

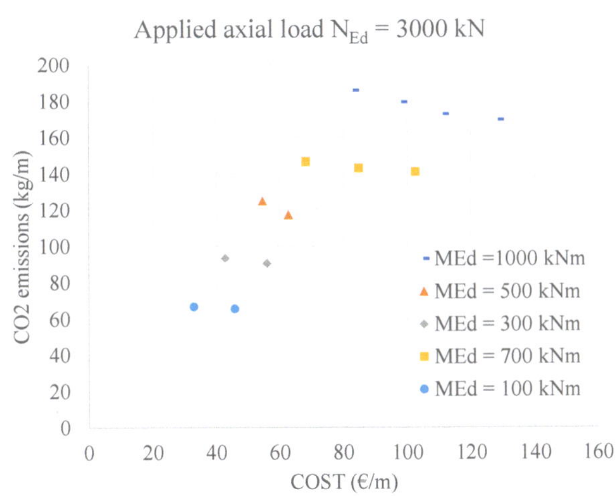

Figure 6. Pareto front for a reinforced concrete circular cross section loaded with an axial load of 3000 kN for different applied bending moments.

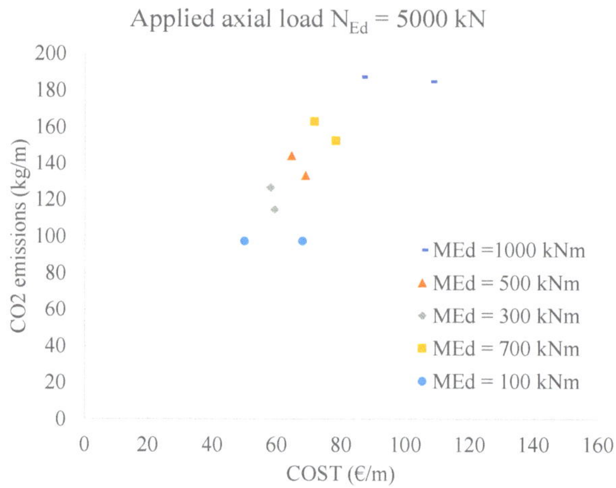

Figure 7. Pareto front for a reinforced concrete circular cross section loaded with an axial load of 5000 kN for different applied bending moments.

Since the above Pareto notation only shows the tradeoff between CO_2 emissions and material costs and not the optimal solutions of the design variables, the parallel coordinate representation is provided. A parallel coordinate representation is a visualization technique that can be used to display multidimensional data in a compact and informative manner. In a parallel coordinate plot, the data points are shown as connected lines and the parallel axes represent the different variables (applied bending moment, number of rebars, values of optimal CO_2 emissions, and material costs) and the optimal dimensions (rebar and column diameters). The data points are grouped based on the applied bending moment and plotted in different colors. Figure 8 shows that five different combinations of reinforcement number, rebar diameter, and section diameter for an axial load of 1000 kN and CO_2 emissions for a bending moment of 1000 kNm result in five different optimal costs and CO_2 emissions, whereas only one optimal solution was obtained for an applied bending moment of 100 kNm. Figures 9 and 10 show that for an applied bending moment of 100 kNm, the choice of different optimal designs achieves only a small reduction in CO_2 emissions but causes a significant increase in material costs. The parallel plot can be read as follows, as in Figure 9, see blue lines: for an applied axial load of N_{Ed} = 3000 kN and a bending moment of M_{Ed} = 100 kNm, two optimal solutions are determined. The first option includes the following values for the design variables: number of rebars n = 12, rebar diameter Φ_{main} = 16 mm, cross-sectional diameter Φ = 450 mm, CO_2 emissions = 65.5 $kgCO_2/m$, and production COST = 45.8 EUR/m. The second option includes the following values for the design variables: number of rebars n = 8, rebar diameter Φ_{main} = 12 mm, cross-sectional diameter Φ = 500 mm, CO_2 emissions = 66.7 $kgCO_2/m$, and production COST = 32.9 EUR/m. It can be seen that a small increase in CO_2 emissions (by 1.8%) leads to a significant reduction in production costs (by 28.2%). The general observation is also that when designing the reinforced concrete cross section in multi-objective optimization, not many optimal solutions are found for the designer to choose from due to the discrete set of variables. In this case, the multi-objective optimization of material cost and CO_2 emissions no longer causes difficulties for the designer due to the large number of different optimal solutions.

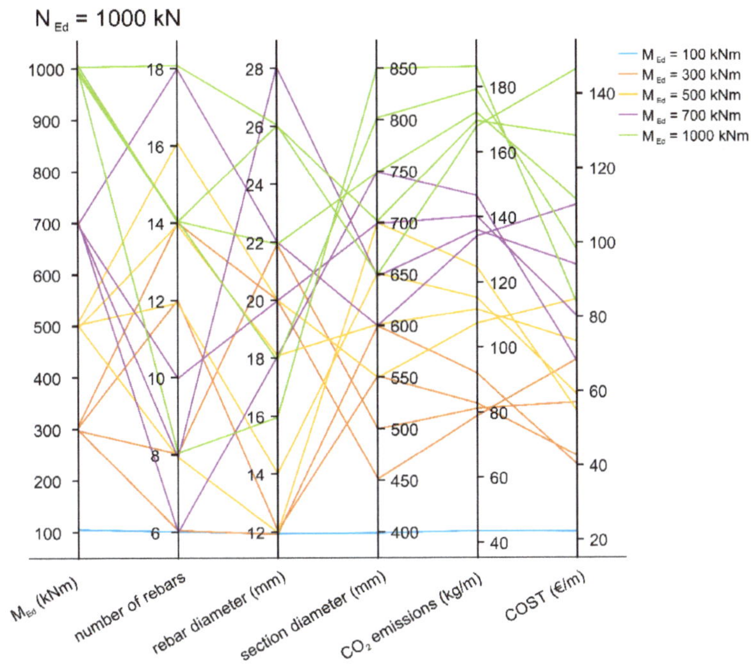

Figure 8. Parallel coordinate representation of the optimum design variables of a reinforced concrete circular cross section loaded with an axial load of 1000 kN for different applied bending moments.

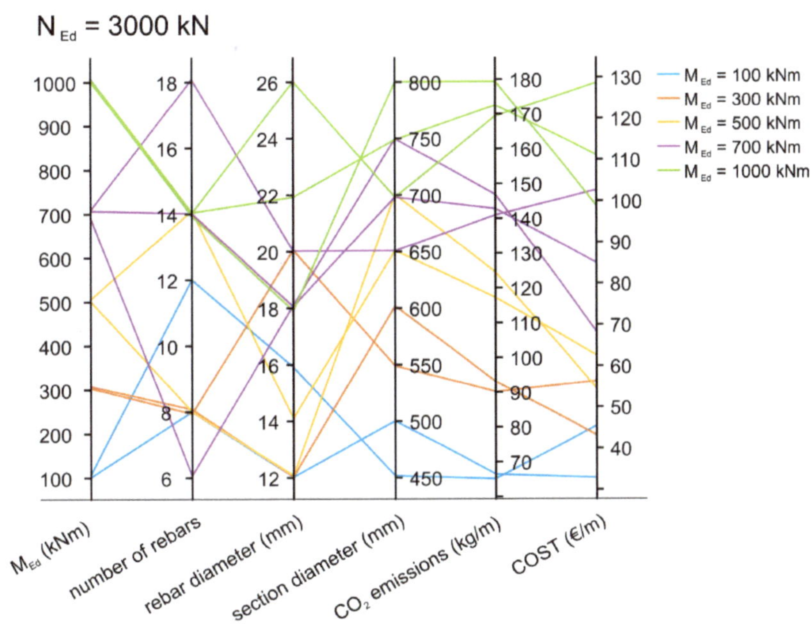

Figure 9. Parallel coordinate representation of the optimum design variables of a reinforced concrete circular cross section loaded with an axial load of 3000 kN for different applied bending moments.

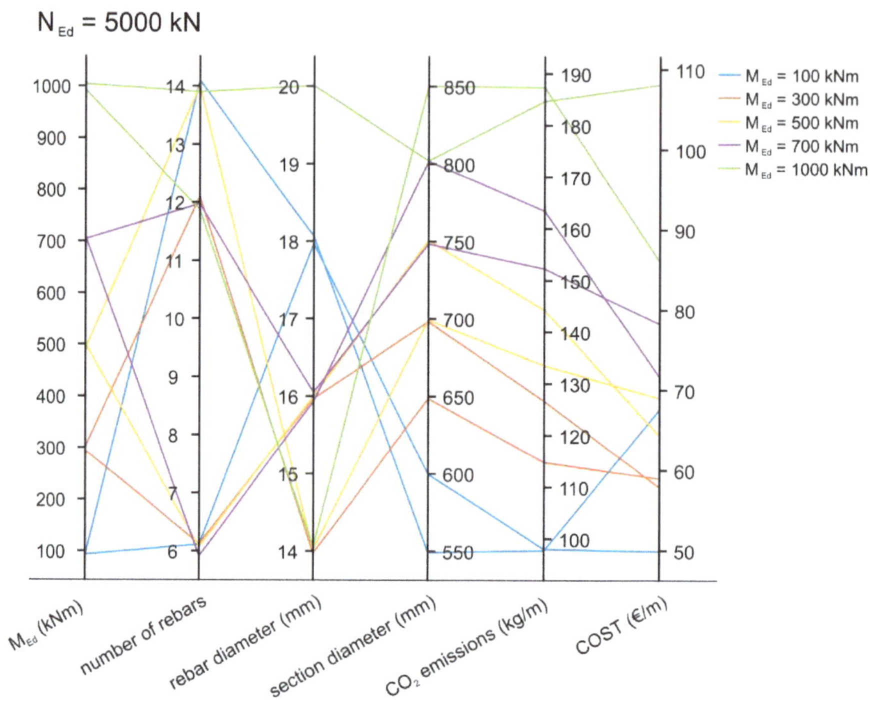

Figure 10. Parallel coordinate representation of the optimum design variables of a reinforced concrete circular cross-section loaded with an axial load of 5000 kN for different applied bending moments.

5. Summery and Conclusions

An optimization model for reinforced concrete circular columns based on the Eurcode 2 standard is presented. Discrete variables are used for the practical design of the cross section, which is particularly important for the number of reinforcing bars included in the cross section. The optimization model enables a comparison of the optimal solutions for the objective function material costs and CO_2 emissions. The genetic algorithm was used to solve the optimization problem, and the entire model was created in MATLAB software (R2021a). The parametric study of applied axial load and bending moment was performed for material cost and CO_2 emissions. The results based on a single objective function show that the optimal design of the reinforced concrete column cross section obtained for the material cost objective function contains a larger cross-sectional area of concrete and a smaller area of steel compared with the optimization results when CO_2 emissions are determined as the objective function. The utilization of bending moment resistance was the top priority for both optimal solutions, regardless of whether material cost or CO_2 emissions were chosen as the objective function. However, the optimal solution in which material cost was assigned as the objective function has much more reserve in axial load carrying capacity than the optimal design in which CO_2 was selected as the objective function. Furthermore, the multi-objective optimization was performed to find a set of solutions that provide the best trade-offs between the material cost and CO_2 emission objectives. The general observation that emerges from the multi-objective optimization is that when designing the reinforced concrete cross section, due to the discrete set of variables, there are not many optimal solutions that the designer can choose from. However, material costs are much more sensitive to the choice of optimal design than CO_2 emissions. The mixed-integer nonlinear optimization model RCCS was developed in a general form that can provide an optimal solution for various design parameters including different concrete strength properties. Based on numerical analysis, the following conclusions can be stated:

- The optimal design of the reinforced concrete cross section, considering the material cost as the objective function, results in a larger cross-sectional area of concrete and a smaller area of steel compared with the optimization results when CO_2 emissions are considered as the objective function;
- The optimal solution obtained with material cost as the objective function exhibits a significantly higher reserve in axial load capacity than the optimal design when CO_2 emissions are selected as the objective function;
- Analyzing the Pareto front reveals that a marginal decrease in CO_2 emissions is accompanied by a substantial increase in material costs;
- In addition, the model can be integrated into the design of structural elements such as columns and piles.

Author Contributions: Conceptualization, P.J. and T.Ž.; methodology, P.J. and T.Ž.; software, P.J. and T.Ž.; validation, P.J. and T.Ž.; writing—original draft preparation, P.J. and T.Ž.; writing—review and editing, P.J. and T.Ž. All authors have read and agreed to the published version of the manuscript.

Funding: This research was funded by the Slovenian Research Agency (grant number P2-0268 and P2-0129) and the GEOLAB project (grant number 101006512).

Institutional Review Board Statement: Not applicable.

Informed Consent Statement: Not applicable.

Data Availability Statement: The data presented in this study are available on request from the corresponding author.

Conflicts of Interest: The authors declare no conflict of interest.

References

1. Kanagasundaram, S.; Karihaloo, B.L. Minimum-cost design of reinforced concrete structures. *Comput. Struct.* **1991**, *41*, 1357–1364. [CrossRef]
2. Menezes, I.S.; Tinoco, V.N.V.; Christoforo, A.L.; Bomfim Junior, F.C.; Narques, T.V.N. Optimization of reinforced concrete columns via genetic algorithm. *Acta Sci. Technol.* **2022**, *45*, e61562. [CrossRef]
3. de Oliveira, L.C.; de Almeida, F.S.; Gomes, H.M. Optimization of RC polygonal cross-sections under compression and biaxial bending with QPSO. *Comput. Concr.* **2022**, *30*, 127–141. [CrossRef]
4. Paya-Zaforteza, I.; Yepes, V.; Hospitaler, A.; González-Vidosa, F. CO2-optimization of reinforced concrete frames by simulated annealing. *Eng. Struct.* **2009**, *31*, 1501–1508. [CrossRef]
5. Camp, C.V.; Huq, F. CO_2 and cost optimization of reinforced concrete frames using a big bang-big crunch algorithm. *Eng. Struct.* **2013**, *48*, 363–372. [CrossRef]
6. Trinh, H.T.M.K.; Chowdhury, S.; Nguyen, M.T.; Liu, T. Optimising flat plate buildings based on carbon footprint using Branch-and-Reduce deterministic algorithm. *J. Clean. Prod.* **2021**, *320*, 128780. [CrossRef]
7. Martínez-Alonso, C.; Berdasco, L. Carbon footprint of sawn timber products of Castanea sativa Mill. in the north of Spain. *J. Clean. Prod.* **2015**, *102*, 127–135. [CrossRef]
8. Yeo, D.; Gabbai, R.D. Sustainable design of reinforced concrete structures through embodied energy optimization. *Energy Build.* **2011**, *43*, 2028–2033. [CrossRef]
9. Zhang, X.; Zhang, X. Sustainable design of reinforced concrete structural members using embodied carbon emission and cost optimization. *J. Build. Eng.* **2021**, *44*, 102940. [CrossRef]
10. Jayasinghe, A.; Orr, J.; Ibell, T.; Boshoff, W.P. Minimising embodied carbon in reinforced concrete beams. *Eng. Struct.* **2021**, *242*, 112590. [CrossRef]
11. Sahebi, M.; Dehestani, M. Sustainability assessment of reinforced concrete beams under corrosion in life-span utilizing design optimization. *J. Build. Eng.* **2023**, *65*, 105737. [CrossRef]
12. Ahmed, M.A.; Amin, A.H.; Khalil, A.B.I. Optimal circular reinforced concrete columns. *J. Eng. Appl. Sci.* **1997**, *44*, 69–84.
13. Tayem, A.; Najmi, A. Design of Round Reinforced-Concrete Columns. *J. Struct. Eng.* **1996**, *122*, 1062–1071. [CrossRef]
14. Bekdaş, G.; Cakiroglu, C.; Kim, S.; Geem, Z.W. Optimization and Predictive Modeling of Reinforced Concrete Circular Columns. *Materials* **2022**, *15*, 6624. [CrossRef]
15. Camp, C.V.; Assadollahi, A. CO_2 and cost optimization of reinforced concrete footings using a hybrid big bang-big crunch algorithm. *Struct. Multidiscip. Optim.* **2013**, *48*, 411–426. [CrossRef]

16. Zhao, X.; Lu, Y.; Liang, H.; Wang, Y.; Yan, Y. Optimal design of reinforced concrete columns strengthened with square steel tubes and sandwiched concrete. *Eng. Struct.* **2021**, *244*, 112723. [CrossRef]
17. Jelušič, P.; Žlender, B. Determining optimal designs for conventional and geothermal energy piles. *Renew. Energy* **2020**, *147*, 2633–2642. [CrossRef]
18. Payá-Zaforteza, I.; Yepes, V.; González-Vidosa, F.; Hospitaler, A. Cost versus sustainability of reinforced concrete building frames by multiobjective optimization. In Proceedings of the Life-Cycle Civil Engineering—1st International Symposium on Life-Cycle Civil Engineering, IALCCE '08, Commo, Italy, 10–14 June 2008; pp. 953–958.
19. Hong, W.-K.; Le, T.-A.; Nguyen, M.C.; Pham, T.D. ANN-based Lagrange optimization for RC circular columns having multiobjective functions. *J. Asian Archit. Build. Eng.* **2023**, *22*, 961–976. [CrossRef]
20. Sun, L.; Zhang, B.; Wang, P.; Gan, Z.; Han, P.; Wang, Y. Multi-Objective Parametric Optimization Design for Mirrors Combined with Non-Dominated Sorting Genetic Algorithm. *Appl. Sci.* **2023**, *13*, 3346. [CrossRef]
21. Khouri Chalouhi, E.; Zelmanovitz Ciulla, G.; García-Brioles Bueno, J.; Pacoste, C.; Karoumi, R. Environmental and economical optimization of reinforced concrete overhang bridge slabs. *Struct. Multidiscip. Optim.* **2023**, *66*, 66. [CrossRef]
22. Di, L.; Sun, Z.; Zhi, F.; Wan, T.; Yang, Q. Assessment of an Optimal Design Method for a High-Energy Ultrasonic Igniter Based on Multi-Objective Robustness Optimization. *Sustainability* **2023**, *15*, 1841. [CrossRef]
23. The MathWorks, Inc., R. How the Genetic Algorithm Works. 2021. Available online: https://se.mathworks.com/help/gads/how-the-genetic-algorithmworks.html (accessed on 5 May 2023).
24. *EN1992-1-1 2004*; CEN Eurocode 2 Design of Concrete Structures–Part 1-1: General Rules and Rules for Buildings. EU: Brussels, Belgium, 2004.
25. The Model of the Eco-costs/Value Ratio (EVR). Available online: www.ecocostsvalue.com/eco-costs/eco-costs-emissions/ (accessed on 5 May 2023).

sustainability

MDPI

Article

Potential of Using Waste Materials in Flexible Pavement Structures Identified by Optimization Design Approach

Primož Jelušič [1], Süleyman Gücek [2], Bojan Žlender [1,*], Cahit Gürer [2], Rok Varga [1], Tamara Bračko [1], Murat V. Tacıroğlu [3], Burak E. Korkmaz [2], Şule Yarcı [2] and Borut Macuh [1]

1 Faculty of Civil Engineering, Transportation Engineering and Architecture, University of Maribor, Smetanova 17, 2000 Maribor, Slovenia; primoz.jelusic@um.si (P.J.); rok.varga@um.si (R.V.); tamara.bracko@um.si (T.B.); borut.macuh@um.si (B.M.)

2 Faculty of Engineering, Afyon Kocatepe University, Afyonkarahisar 03200, Turkey; sgucek@aku.edu.tr (S.G.); cgurer@aku.edu.tr (C.G.); eniskorkmaz@aku.edu.tr (B.E.K.); syarci@aku.edu.tr (Ş.Y.)

3 Faculty of Engineering, Department of Civil Engineering, Mersin University, Mersin 33343, Turkey; mtaciroglu@mersin.edu.tr

* Correspondence: bojan.zlender@um.si

Abstract: This paper presents the design of geosynthetic reinforced flexible pavements and their modification by incorporating waste materials into bonded and unbonded layers of the pavement structure. The optimal design of flexible pavements was achieved by minimizing the construction cost of the pavement. The incorporation of waste materials into the pavement structure affects the material properties. Therefore, along with the traffic load, the effects of the material properties of the asphalt concrete, base layer, sub-base layer, and subgrade were analyzed in terms of pavement structure costs and CO_2 emissions of materials used in pavement construction. In addition, a comparison was made between pavements with and without geosynthetic reinforcement in terms of design, optimum construction cost, and CO_2 emissions. The use of geosynthetics is even more effective in pavement structures that contain waste materials in an unbound layer, both in terms of cost and CO_2 emissions. The minimum value of the California Bearing Ratio of the subgrade was determined at which the use of geosynthetic reinforcement for pavement structure with and without the inclusion of waste materials is economically and sustainably justified. The use of geosynthetics could result in a 15% reduction in pavement structure cost and a 9% reduction in CO_2 emissions due to the reduced thickness of unbound layers. In addition, reducing the CBR of the unbound layer from 100% to 30% due to the inclusion of waste materials implies a cost increase of up to 13%. While the present study is based on an empirical pavement design method in which pavement thickness is limited by the pavement thickness index, the same minimum thicknesses are obtained in the optimization process regardless of whether the objective function is the minimum construction cost or minimum CO_2 emissions.

Keywords: pavement design; waste materials; optimization; minimum construction cost; CO_2 emissions; geosynthetics

Citation: Jelušič, P.; Gücek, S.; Žlender, B.; Gürer, C.; Varga, R.; Bračko, T.; Tacıroğlu, M.V.; Korkmaz, B.E.; Yarcı, Ş.; Macuh, B. Potential of Using Waste Materials in Flexible Pavement Structures Identified by Optimization Design Approach. *Sustainability* **2023**, *15*, 13141. https://doi.org/10.3390/su151713141

Academic Editor: Marinella Giunta

Received: 27 July 2023
Revised: 17 August 2023
Accepted: 17 August 2023
Published: 31 August 2023

1. Introduction

The construction of buildings, roads, railroads, power grids, and other infrastructure often generates large amounts of clean and contaminated waste, most of which ends up in landfills. Therefore, measures that focus on sustainable management are important. Smarter methods are needed to reduce waste and ensure that its reuse does not pose a risk to public health or the environment. An analysis of current waste reuse practices has identified the main barriers (legal, organizational, logistical, and material quality) to effective reuse. The (re)use of waste, including excavated contaminated or uncontaminated soils, offers the following benefits [1]:

- Reduction in costs associated with disposal.
- Preservation of landfill capacity.
- Conservation of mined natural resources.
- Reduction in environmental and ecological impacts.

Waste management plays a critical role in road construction and encompasses a number of important aspects, as pointed out by Hale et al. [1]. First and foremost, a legal framework ensures that construction projects comply with environmental regulations and waste disposal laws to minimize potential negative impacts on the ecosystem. The organizational aspect involves effective planning and coordination among stakeholders to promote efficient waste sorting, collection, and disposal processes. Logistical and economic considerations optimize resource allocation, reduce costs, and minimize the environmental footprint associated with waste transportation and disposal. In addition, maintaining material quality is critical to ensure that waste materials incorporated into road construction meet required standards and improve the overall infrastructure durability and performance. Convergence of these aspects is essential to ensure that waste is not simply disposed of, but reused through appropriate management practices, contributing to sustainable road construction practices.

In the design of road pavements, many of the technical conditions set out in standards and technical specifications must be met to ensure the quality of the material, which is difficult to achieve when using waste. Waste may be contaminated depending on its origin, so its level of contamination must be well characterized and evaluated as a first step for reuse. In addition, these wastes may have quality differences and therefore need to be characterized in order to assess whether the waste in question can be reused for the proposed purpose (Figure 1). For this reason, several research works have been carried out to precisely determine the performance of the material containing the different wastes. Huang et al. [2] studied the different types of waste materials that can be used in road pavements. The study included waste glass, steel slag, tires, and plastics, and investigated the wet and dry methods, in which the waste material is mixed with bitumen, and the dry method, in which the waste material replaces the fine aggregates in asphalt. The study concluded that waste materials could prove to be valuable substitutes in the construction process. Similar studies were conducted and confirmed the statement that waste materials can be a valuable substitute for virgin material [3–5]. The properties of subgrade when plastic waste is added were also analyzed. Abukhettala and Fall [6] concluded that the amount of plastic waste added and the change in the CBR value of the subgrade is not linearly correlated, but that depending on the type and form of plastic waste added, there is a critical value beyond which the CBR value decreases. This study also suggests that this is related to the reinforcing capacity of the plastic waste, as a smaller amount of plastic is more easily distributed in the subgrade and therefore has a greater chance of increasing the CBR value of the subgrade. Subgrade in particular clay was also reinforced using fly ash. Studies showed a significant improvement in the CBR value of the subgrade when 15% fly ash was incorporated into the subgrade layer, reducing the required thickness and lowering the cost of the asphalt pavement structure [7,8]. To determine the foundation properties of bitumen containing plastic waste, models have been developed using various machine learning algorithms [9]. Machine learning algorithms have also been applied to predict the Marshall Stability of asphalt concrete based on bitumen content, plastic content, bitumen grade, and plastic size [10]. Lee and Le [11] quantitatively investigated the effects of adding waste plastic aggregates and magnesium-based additives in asphalt mixtures and evaluated various properties such as the deformation strength, indirect tensile strength, rut depth, and dynamic stability.

Figure 1. Typical types of waste: (**a**) Lignin, (**b**) car tire waste, (**c**) glass, and (**d**) nylon bags.

Because waste materials have been shown to be useful and enhance performance or be on par with virgin materials, various studies have been conducted on the economic and environmental impacts that using waste material has over virgin materials. The cost of asphalt pavement is mainly caused by the asphalt itself. Approximately 60% of the asphalt cost is bitumen, so the main objective is to reduce the bitumen content to significantly affect the cost savings [12]. Therefore, studies have been conducted on the usage of waste plastic in bitumen to save costs. Vasudevan et al. [13] concluded that using waste plastic is an ecofriendly disposal of plastic and that the performance of these roads is on par with roads constructed with virgin materials while the costs were also reduced. The same conclusions were also documented by several authors for the usage of crumb rubber to also reduce the pavement costs for the same or even better performance [14,15].

Furthermore, following the costs, the environmental impact of using waste materials has also been documented by several authors. To estimate carbon emissions from recycled construction waste and other materials, a grey model that can be used in various situations to estimate carbon emissions from recycling activities and carbon emissions from alternative materials can benefit the environment. Wang et al. [16] proved the environmental benefits of recycling and the use of waste materials in construction. Mechanical and environmental performance was also studied and showed that including slag in asphalt mixes instead of basalt aggregates showed environmental and cost advantages while also providing better mechanical performance of the asphalt pavement structure while under load [17]. You et al. [18] studied the integration of recycled plastic waste into asphalt pavements. This study showed that using recycled plastic waste in asphalt pavements instead of the original plastic and applying the right process to reduce the generation of toxic exhaust gasses that further pollute the climate can prove to be beneficial to the environment. It should be noted, however, that despite some improvement in rutting, aging resistance, and tensile strength of asphalt pavements, fatigue life and resistance to cracking at low temperatures are significantly lower when plastic waste is used. White et al. [19] considered climate change and proposed a method for estimating CO_2 emissions and the cost of using alternative materials for road construction. LCA research of bituminous mixtures containing recycled materials such as crumb rubber found significant environmental impact and energy savings benefits when wet technology was used, but showed almost no benefits when the dry technology was used, which the authors attributed to the lack of data on the maintenance and life cycle of rubber-reinforced asphalt [20]. Practical applications of the use of plastic waste have shown that it is possible to build durable roads from plastic and that the construction of such roads reduces CO_2 emissions compared to conventional road pavements [21–23]. The research suggests that waste material is an important building stone in pavement design and should be considered when constructing a new pavement. This paper further evaluates the impact of the quality of the waste material and virgin material mixture on the optimal pavement design and its CO_2 equivalent.

Methods for designing pavements vary, with some being purely empirical and others mechanistic-empirical. Base course thickness can be reduced with geosynthetics. The performance of geosynthetic-reinforced flexible pavements can be evaluated using field tests, laboratory tests, and numerical simulations. To compare conventional flexible pavements with geosynthetic-reinforced flexible pavements, both unreinforced and reinforced pavement types should be optimized at a minimum cost. For this purpose, optimization models were developed and advanced algorithms were applied [24].

The main objective of this study is to investigate the feasibility of integrating waste materials into bound and unbound layers of the road pavement while evaluating the impact of these waste materials on the overall pavement design. This investigation relies on the utilization of existing empirical methods for pavement design in conjunction with an in-depth analysis of research efforts aimed at describing the material properties of these waste constituents. Such an approach considers the complex interplay between the inclusion of waste materials and the impact on the economic viability and environmental sustainability of road pavements. Exploration of this multifaceted area seeks to unravel the intricate link between waste utilization, pavement design optimization, and the broader considerations of cost-effectiveness and ecological responsibility. A genetic algorithm was employed in this study to optimize pavement structure cost and CO_2 emissions in accordance with pavement design guidelines. Genetic algorithms have proven to be very effective because they do not require differentiable functions, which is advantageous for complex nonlinear functions such as those often used in pavement optimization. The main novelty of this work is the optimization model, which allows us to obtain an optimal design of the pavement structure and is able to take into account different material properties affected by the inclusion of waste materials, both in terms of cost and CO_2 emissions. Such an approach demonstrates the potential of using waste materials in flexible pavements in terms of cost and CO_2 emissions. To represent the effects of the material properties of the asphalt, base layer, sub-base layer, and subgrade, all of which are affected by the inclusion of waste materials, a parametric study was conducted to determine the optimal design of flexible pavement structures using the developed optimization model. Due to the large number of combinations of design parameters (material properties, traffic loads, and geosynthetic reinforcement) involved in the optimization process, manual execution of the algorithm was not possible. Therefore, an optimization model was developed that includes a loop that performs optimizations for each combination. In addition, a sensitivity analysis was performed on the importance of each design parameter based on all optimal solutions.

2. The Performance of Asphalt and Unbound Layers Containing Waste Materials

The performance of asphalt mixtures containing waste materials has been investigated in Marshall tests, and numerous studies have concluded that waste materials, particularly plastic waste, increase the stiffness modulus and thus the strength of the pavement. However, researchers have also found that the stiffness modulus decreases above a certain ratio of waste material to virgin material [25–28]. The Slovenian technical specifications for the design of new asphalt pavements, based on the empirical AASHTO method, contain a series of diagrams showing a spectrum of values for Marshall test results of asphalt mixtures and the CBR values of base and sub-base mixtures [29]. These diagrams were considered in the analyses in this paper (see Figure 2). The Marshall test diagram and a range of CBR values for the base and sub-base are paired with equivalence factors that affect pavement design.

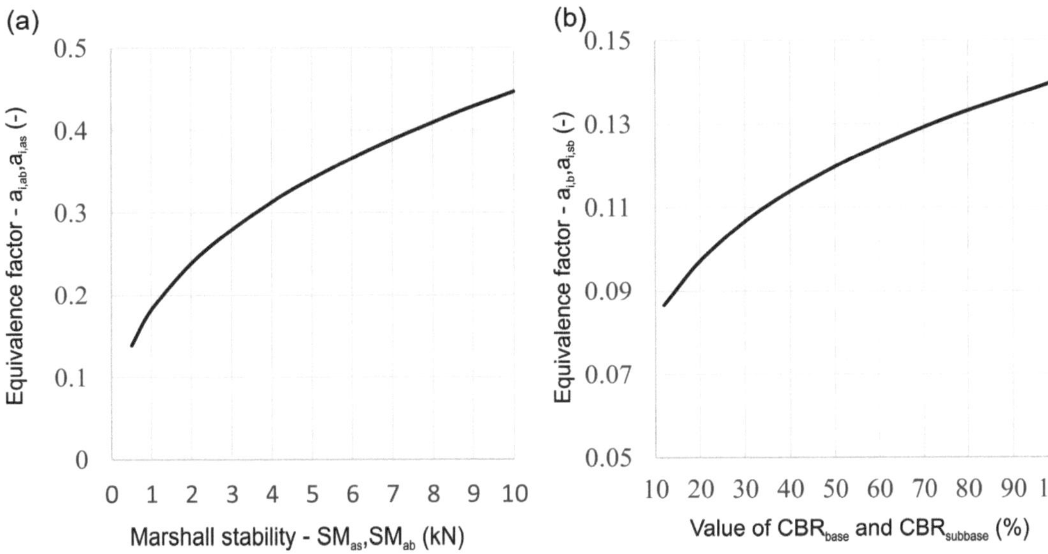

Figure 2. Equivalence factors for given values of (**a**) Marshall stability of asphalt and (**b**) CBR of base and sub-base material.

To show the Marshall stability for asphalt pavements with different waste materials, a diagram (Figure 3) was prepared based on several authors. The diagram includes the ratio of waste materials to virgin materials and the effects on Marshall stability. Fly ash was analyzed and showed an increase of 4 kN in the Marshall stability up to a certain point of the waste ratio when mixed with bitumen, but the addition of additional fly ash, above 7% of the mixture, was found to be suboptimal [30]. Plastic waste in the form of PE carry bags also showed a similar effect on the Marshall stability, but the inflection point was 12% of the bitumen weight, indicating a larger amount of waste that can be added to the asphalt pavement and make it more environmentally friendly [31]. To further study plastic waste, PET bottles were used in combination with bitumen. The test showed a steady increase in Marshall stability with the turning point being at around 12% of PET bottles added [32]. The results are consistent with the previous study on PE carry bags, thus proving that plastic waste is a reliable way to replace virgin material in asphalt pavements. Reinforcing asphalt mixtures with Electric Arc Furnace Dust (EAFD), a hazardous waste generated in the metallurgical industry, proved to be efficient for reinforcing asphalt pavements and could be a cost-effective and sustainable way to reduce the potential environmental impact of this waste material [33].

The inclusion of waste material in base and sub-base layers is rarely discussed, and how mixing waste and gravel in various proportions affects CBR values is rarely stated. The paper considers the experience of the authors of previous studies. An analysis of permanent deformations of the base course of pavements and their dependence on the proportion of accessors in the stone mixture had been carried out [34]. Žlender and Trauner [35] presented the use of the electro-filter granular (EFG) for the base course of pavements. Different types of EFG specimens were studied, first without reinforcement and then with one, three, and five reinforcements. The failure and deformation envelope curves are given separately for all types of EFG specimens and compared with the results performed for standard ballast base courses. The use of geosynthetic geocells as base course reinforcement was studied. The results show that the proper use and positioning of geocells filled with supplementary material can significantly reduce the thickness of asphalt layers [36,37]. The Cinderella project [38] aimed to develop a new Circular Economy Business Model for the use of secondary raw materials in urban areas. It included the use of secondary raw materials created by recycling construction, industrial, mining, and some municipal wastes. One of the project objectives was to create a circular model for resource use that would reduce negative impacts on the environment. This would be achieved by introducing circular supply chains in urban construction. The URGE project [39] accelerates the transition to a circular economy. It aims to develop integrated urban circular economy strategies in the construction sector as the main consumer of raw materials. Research on lightweight materials as a possible solution to improve the low-bearing-capacity subgrade of pavements has concluded that, due to their low density, lightweight materials can be used to reduce weight in areas with the low-bearing-capacity subgrade. Due to their high porosity and the resulting good thermal insulation properties, lightweight materials can be used as frost protection layers in cold regions or as thermal insulation layers under road pavements. From a mechanical point of view, this material can provide good compressive strength and stiffness due to the relatively thick pore walls compared to the diameter of the voids [40].

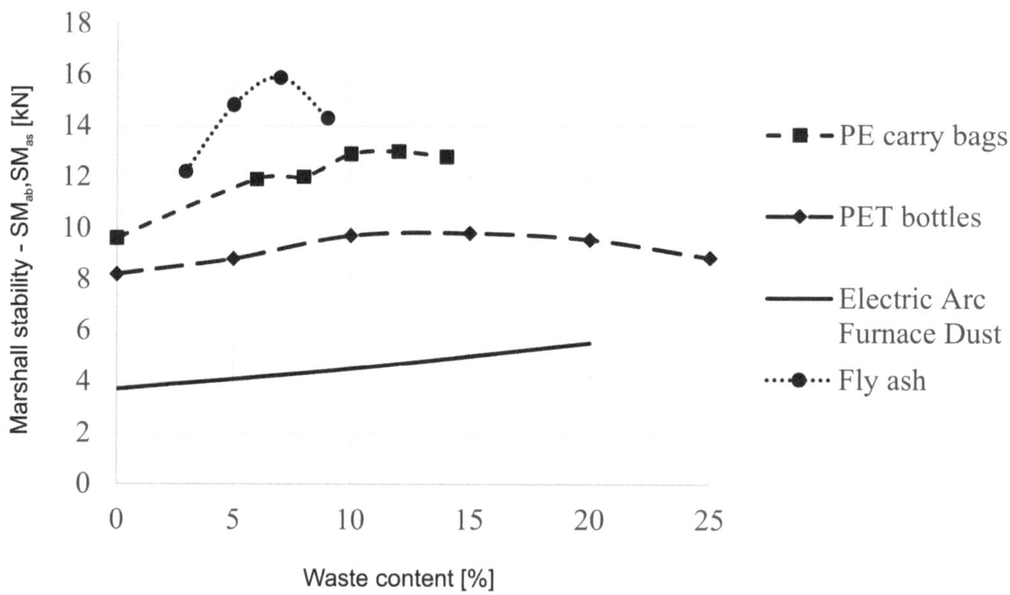

Figure 3. Various test results of SM for different waste materials and their content in the bituminous mix.

3. Optimization Model of Pavement Structure

The design and construction of pavement structures are of great importance due to several interrelated factors. First, it is important to create a pavement structure that can withstand the expected traffic loads during its lifespan. To achieve this, the criteria set forth in the design specifications must be followed. By carefully considering these criteria, engineers can ensure that the pavement structure will withstand the expected traffic loads and maintain its structural integrity. Furthermore, another significant objective is to construct the pavement structure while minimizing construction costs. To achieve this, a cost objective function is defined for the pavement structure, which can be minimized using optimization algorithms. By using such algorithms, engineers can determine the most cost-effective pavement design and materials, thereby reducing the financial burden associated with pavement construction.

Moreover, it is also crucial to address environmental concerns during the pavement construction process. Alongside minimizing construction costs, efforts should be made to decrease the carbon footprint of the pavement structure. This can be achieved by implementing sustainable practices, utilizing environmentally friendly materials, and optimizing construction processes to reduce energy consumption and emissions. In addition, it is critical that environmental considerations be taken into account when constructing pavements. In addition to minimizing construction costs, efforts should be made to reduce the carbon footprint of pavement construction. This can be achieved by adopting sustainable practices, minimizing pavement thickness, using environmentally friendly materials, and optimizing construction processes to reduce energy consumption and emissions. To effectively achieve the aforementioned objectives, the development of an optimization model is required. This model should include a cost objective function that considers construction costs and a carbon footprint function that considers environmental impacts. In addition, the optimization model should include various constraints that limit the objective function to ensure that the design of a pavement structure meets the required standards and specifications. By incorporating these elements into the optimization model, engineers can aim for a pavement structure that is both structurally sound and environmentally sustainable while minimizing construction costs.

Such an optimization model is shown in Table 1, where the cost objective function of the pavement design is developed, and the CO_2 emissions generated during construction are evaluated. The optimized pavement structure meets all required standards and specifications based on an empirical pavement design method.

Table 1. Optimization model.

Objective function	$COST_{pav} = C_{exc} + C_{gc} + C_{fill,b} + C_{as,subs} + C_{fill,sb} + C_{as} + C_{ab} + C_{geo}$ $C_{exc} = c_{exc} \cdot h_{total} \cdot (B_{ve} + B_{as}) \cdot L$ $C_{gc} = c_{gc} \cdot (B_{ve} + B_{as}) \cdot L$ $C_{fill,b} = c_{fill,b} \cdot (B_{ve} + B_{as}) \cdot d_b \cdot L$ $C_{as,subs} = c_{as,subs} \cdot B_{as} \cdot L$ $C_{fill,sb} = c_{fill,sb} \cdot (B_{ve} + B_{as}) \cdot d_{sb} \cdot L$ $C_{as} = c_{as} \cdot B_{as} \cdot d_{as} \cdot L$ $C_{ab} = c_{ab} \cdot B_{as} \cdot d_{ab} \cdot L$ $C_{geo} = c_{geo} \cdot (B_{ve} + B_{as}) \cdot L$
CO_2 emissions	$CO_{2,\,total} = CO_{2,exc} + CO_{2,fill,b} + CO_{2,as,subs} + CO_{2,fill,sb} + CO_{2,as} + CO_{2,ab} + CO_{2,geo}$ $CO_{2,exc} = ci_{exc} \cdot h_{total} \cdot (B_{ve} + B_{as}) \cdot L$ $CO_{2,fill,b} = ci_{fill,b} \cdot (B_{ve} + B_{as}) \cdot d_b \cdot L \cdot \rho_{base}$ $CO_{2,\,as,subs} = ci_{as,subs} \cdot B_{as} \cdot L$ $CO_{2,fill,sb} = ci_{fill,sb} \cdot (B_{ve} + B_{as}) \cdot d_{sb} \cdot L \cdot \rho_{sub-base}$ $CO_{2,as} = ci_{as} \cdot B_{as} \cdot d_{as} \cdot L \cdot \rho_{as}$ $CO_{2,ab} = ci_{ab} \cdot B_{as} \cdot d_{ab} \cdot L \cdot \rho_{ab}$ $CO_{2,geo} = ci_{geo} \cdot (B_{ve} + B_{as}) \cdot L$
Condition 1	$D_{total} \geq D_{req}$ $D_{total} = d_{as} \cdot a_{i,as} + d_{ab} \cdot a_{i,ab} + d_b \cdot a_{i,b}$ $D_{req} = d_{asb,0} \cdot 0.38 + d_{b,CBR_{mod}} \cdot 0.14$ $d_{asb,0} = a_1 \cdot T_n{}^{a_2}$ $d_{b,CBR_{mod}} = \left((c_1 - c_2 \cdot CBR_{mod}) \cdot \ln(T_n) - c_3 + e^{(c_4 \cdot CBR_{mod}) \cdot c_5} \right) / \gamma_{geo,b}$
Condition 2	$D_{total,AC} \geq D_{req,AC}$ $D_{total,AC} = d_{as} \cdot a_{i,as} + d_{ab} \cdot a_{i,ab}$ $D_{req,AC} = d_{asb,0} \cdot 0.38$

Table 1. *Cont.*

Condition 3	$D_{total,base} \geq D_{req,base}$ $D_{total,base} = d_b \cdot a_{i,b}$ $D_{req,base} = d_{b,CBR_{mod}} \cdot 0.14$
Condition 4	$d_{prov} \geq d_{asb,0}$ $d_{prov} = d_{as} + d_{ab}$
Condition 5	$d_b \geq d_{b,req}$ $d_{b,req} = d_{b,CBR_{mod}}$
Condition 6	$h_{total} \geq h_{req}$ $h_{total} = d_{as} + d_{ab} + d_b + d_{sb}$ $h_{req} = h_m \cdot f_{fr}$
Condition 7	$d_{sb} \geq d_{sb,CBR_{mod}}$ $d_{sb,CBR_{mod}} = \left(b_1 \cdot \left(\frac{b_2 \cdot (CBR_{mod} - CBR)}{b_3 - CBR} + b_4\right) \cdot \left(\frac{0.14}{a_{i,sb}}\right)\right) / \gamma_{geo,sb}$
Condition 8	$d_{as} \geq d_{as,min}$
Condition 9	$d_{ab} \geq d_{ab,min}$
Condition 10	$d_b \geq d_{b,min}$
Condition 11	$d_{sb} \geq d_{sb,min}$

Figure 4 shows the flexible pavement structure, which contains two bound layers (asphalt surface layer and asphalt binder layer) and two unbound layers (base course and sub-base course), as well as geosynthetic reinforcement.

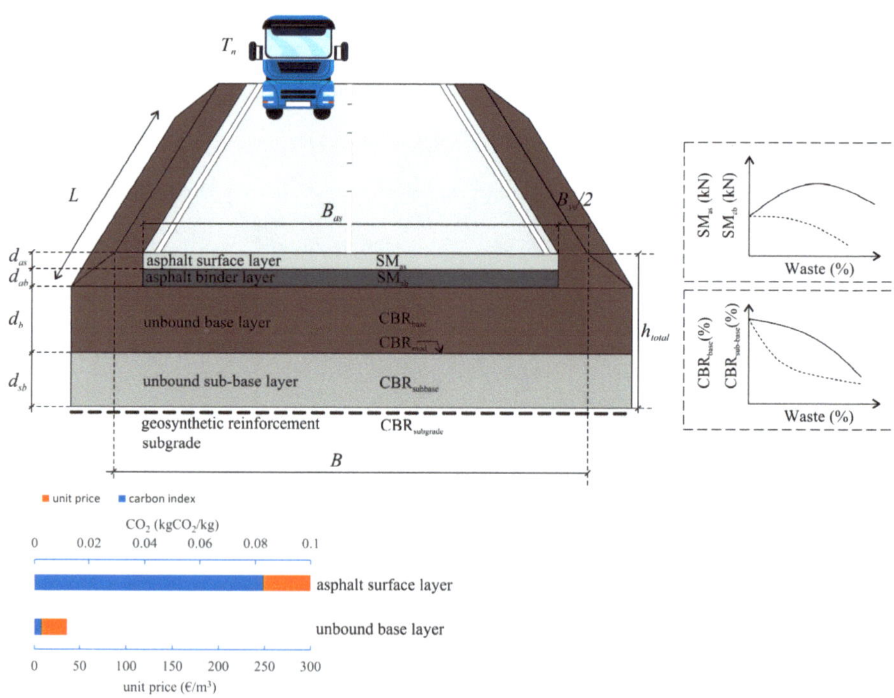

Figure 4. Cross-section of the road pavement structure.

The optimization model comprises four variables that correspond to distinct layers within a pavement structure. The variable d_{as} (m) represents the asphalt surface layer, d_{ab} (m) represents the binder layer of the asphalt, d_b (m) represents the unbound base layer, and d_{sb} (m) represents the unbound sub-base layer. Additionally, the construction cost of the pavement structure is defined as $COST_{pav}$ (EUR). The input data provided are used to represent various parameters and characteristics relevant to the design properties, which are used in the optimization process of the pavement structure. The definitions of the individual input data and the associated values are listed in Table 2. The unit prices (c_{exe}, c_{gc}, $c_{fill,sb}$, $c_{fill,b}$, $c_{as,subs}$, c_{as}, c_{ab}, and c_{geo}) of each material used in the pavement structure were assigned, as well as carbon indices (ci_{exe}, $ci_{fill,sb}$, $ci_{fill,b}$, $ci_{as,subs}$, ci_{as}, ci_{ab}, and ci_{geo}) that allow the calculation of the total CO_2 emissions of the materials used in the pavement structure. While the empirical pavement design method is included in the optimization model, where the pavement thickness is limited by the pavement thickness index, it is of utmost importance to determine the equivalence factors $a_{i,as}$, $a_{i,ab}$, $a_{i,b}$, and $a_{i,sb}$ based on the material properties of the individual pavement layers. In this way, it was possible to take into account the inclusion of waste material affecting the stability according to the Marshall stability test (SM_{as} and SM_{ab}) in the asphalt concrete layer and the CBR value (CBR_{base} and $CBR_{subbase}$) for the unbound base and the sub-base layer.

$$a_{i,as} = t_1 \cdot SM_{as}{}^{t_2} \tag{1}$$

$$a_{i,ab} = t_3 \cdot SM_{ab}{}^{t_4} \tag{2}$$

$$a_{i,b} = t_5 \cdot CBR_{base}{}^{t_6} \tag{3}$$

$$a_{i,sb} = t_7 \cdot CBR_{subbase}{}^{t_8} \tag{4}$$

To satisfy the first five conditions, the required thickness of the asphalt, $\mathrm{d_{asb,0}}$ (cm), and the required thickness of the unbound base layer, $\mathrm{d_{b,CBR,mod}}$ (cm), which depend on the CBR of the subgrade, must be determined by a parameterized function. Therefore, the parameters a_1, a_2, c_1, c_2, c_3, c_4, and c_5 were determined based on an approximation of the charts included in the technical specifications for roads, which relate to the required thickness and the number of ESALs T_n. Since the main objective of placing the sub-base layer is to improve the CBR value of the subgrade, the thickness of the required unbound sub-base layer $\mathrm{d_{sb,CBR,mod}}$ (cm) is calculated to provide the modified CBR value at the top of the sub-base layer $\mathrm{CBR_{mod}}$. The correlation between the original CBR value of the subgrade, the thickness of the sub-base, and the modified CBR value at the top of the sub-base is determined by the parameters b_1, b_2, b_3, and b_4. Furthermore, the thickness of the sub-base layer can be reduced by a factor of $\gamma_{\mathrm{geo,sb}}$ if the geosynthetic reinforcement is installed in the contact between the subgrade and the sub-base layer. Condition 6 ensures that the overall thickness of the pavement is sufficient to be frost resistant. While frost depth depends on geographic location and hydrologic conditions, factors $\mathrm{f_{fr}}$ and $\mathrm{h_m}$ are determined using Table 3. Hydrological conditions are favorable if the total thickness of the pavement structure is at least 1.5 m, the water table is constantly below freezing, and drainage is ensured without water inflow within the pavement. Otherwise, the factors for unfavorable conditions must be considered. The last four conditions (conditions 8–11) ensure that the thickness of each pavement layer is of a sufficient minimum thickness according to conventional pavement construction techniques.

Table 2. Explanation of input data in optimization model along with determined values.

Symbol	Value	Description
c_{exe} (€/m^3)	9	unit price of the ground excavation
c_{gc} (€/m^2)	2.5	unit price of the ground compaction
$c_{fill,sb}$ (€/m^3)	24	unit price of the unbound sub-base fill
$c_{fill,b}$ (€/m^3)	36	unit price for unbound base fill
$c_{as,subs}$ (€/m^2)	1.5	unit price of the asphalt substrate
c_{as} (€/m^3)	300	unit price of the asphalt surface layer
c_{ab} (€/m^3)	200	unit price of the asphalt binder layer
c_{geo} (€/m^2)	3.2	unit price of the geosynthetics
B_{ve} (m)	1	width of the verge
B_{as} (m)	8	width of the asphalt surface
L (m)	1000	length of pavement sections
ci_{exe} (kgCO$_2$/m^3)	1.38	carbon index for the ground excavation
$ci_{fill,b}$ (kgCO$_2$/kg)	0.00248	carbon index for unbound sub-base fill
$ci_{as,subs}$ (kgCO$_2$/m^2)	0.35	carbon index for unbound base fill
$ci_{fill,sb}$ (kgCO$_2$/kg)	0.00248	carbon index for asphalt substrate
ci_{as} (kgCO$_2$/kg)	0.08278	carbon index for asphalt surface layer
ci_{ab} (kgCO$_2$/kg)	0.08278	carbon index for asphalt binder layer
ci_{geo}(kgCO$_2$/m^2)	0.396	carbon index for geosynthetics
ρ_{base} (kg/m^3)	1800	density of the unbound base fill
$\rho_{\text{sub-base}}$ (kg/m^3)	1800	density of the unbound sub-base fill
ρ_{as} (kg/m^3)	2400	density of the asphalt surface layer
ρ_{ab} (kg/m^3)	2400	density of the asphalt binder layer
$t_1 = t_3$ (-)	0.182104767	parameter for $a_{i,as}$ and $a_{i,ab}$ determination
$t_2 = t_4$ (-)	0.389702035	parameter for $a_{i,as}$ and $a_{i,ab}$ determination
$t_5 = t_7$ (-)	0.049219606	parameter for $a_{i,b}$ and $a_{i,sb}$ determination
$t_6 = t_8$ (-)	0.227144669	parameter for $a_{i,b}$ and $a_{i,sb}$ determination
a_1 (-)	0.6567	parameter for required thickness of the asphalt
a_2 (-)	0.2175	parameter for required thickness of the asphalt
b_1 (-)	8.382	parameter for required thickness of sub-base
b_2 (-)	−0.791	parameter for required thickness of sub-base
b_3 (-)	1.975	parameter for required thickness of sub-base
b_4 (-)	1.912	parameter for required thickness of sub-base
c_1 (-)	6.239	parameter for required thickness of base layer
c_2 (-)	0.376	parameter for required thickness of base layer
c_3 (-)	26.64	parameter for required thickness of base layer
c_4 (-)	0.141	parameter for required thickness of base layer
c_5 (-)	4.882	parameter for required thickness of base layer
CBR_{mod} (%)	15.0	modified CBR value at the top of the sub-base layer
$\gamma_{geo,sb}$(-)	2.0	reduction factor for the consideration of the geosynthetic
h_m (cm)	80	depth of frost penetration
f_{fr} (-)	0.8	Factor for the conditions of the material at the site
$\text{d}_{\text{as,min}}$ (m)	0.04	minimum thickness of the asphalt surface layer
$\text{d}_{\text{ab,min}}$ (m)	0.06	minimum thickness of the asphalt binder layer
$\text{d}_{\text{b,min}}$ (m)	0.25	minimum thickness of the unbound base layer
$\text{d}_{\text{sb,min}}$ (m)	0.20	minimum thickness of the unbound sub-base layer

Table 3. Minimum required thicknesses of the road pavement structures h_{req} [29].

Resistance of the Material under the Pavement Structure against the Effects of Freezing and Thawing	Hydrological Conditions	Minimum Thickness of Pavement Structure $h_{req} = (f_{fr})\cdot h_m$ h_m Is Depth of Frost Penetration	
		to an Altitude of 600 m	from an Altitude of 600 m
resistant	favorable	$(0.6)\cdot h_m$	$(0.7)\cdot h_m$
	unfavorable	$(0.7)\cdot h_m$	$(0.8)\cdot h_m$
not resistant	favorable	$(0.7)\cdot h_m$	$(0.8)\cdot h_m$
	unfavorable	$(0.8)\cdot h_m$	$(0.9)\cdot h_m$

4. Multi Parametric Optimization

The main objective of this work is to analyze how the material properties of each layer of the pavement affect the design thickness of the pavement and consequently the construction costs and CO_2 emissions. An optimization model was used to determine the minimum thickness of each layer that still meets all conditions and consequently ensures sufficient performance over the intended 20-year period. The ESAL is the most important design parameter in pavement design, so a parametric analysis was also performed for this parameter. Therefore, the optimal designs of the pavement structure were determined for 450 combinations of the following parameters:

- Total number of ESALs: T_n (1×10^4; 1×10^5; 1×10^6; 1×10^7; 1×10^7).
- California Bearing Ratio of subgrade: *CBR* (3%; 4%; 5%; 6%; 7%).
- Marshall stability of asphalt layers: $SM_{as} = SM_{ab}$ (2 kN; 4 kN; 6 kN; 8 kN; 10 kN).
- California Bearing Ratio of unbound layers: $CBR_{base} = CBR_{subbase}$ (100%; 60%; 30%).

By performing such a multiparametric analysis, it is possible to show how the thickness of each pavement layer increases as the material properties decrease due to the incorporation of waste material into each layer of the pavement structure under different traffic loads. The optimization model was developed so that different values of Marshall stability could be applied to the asphalt surface layer and the asphalt binder layer. However, in the parametric analysis, the Marshall stability takes the same value for the asphalt surface and asphalt binder layer ($SM_{as} = SM_{ab}$). The same applies to the unbound base and the sub-base layer ($CBR_{base} = CBR_{subbase}$).

The results of the optimal solution are presented in several steps. First, the parallel coordinate plot is used to present multidimensional data on Marshall stability, CBR of unbound layers, pavement cost, and CO_2 emissions. In a parallel coordinate plot, the data points are represented as contiguous lines, and the parallel axes represent the different variables (Marshall stability, CBR of unbound layers, values of optimal pavement cost, and CO_2 emissions).

The data points are grouped based on Marshall stability and plotted in different colors. Figure 5a shows fifteen different optimal costs and CO_2 emissions for different combinations of Marshall stability and CBR of the unbound layers for an ESAL of $T_n = 1 \times 10^6$ and for subgrade $CBR_{subgrade} = 3\%$. While Figure 5b shows the results for pavements with geosynthetics, Figure 5a presents the results without geosynthetic reinforcement. The parallel plot can be read as in Figure 5a (see the blue lines) as follows: For a Marshall stability of SM = 2 kN and a $CBR_{base} = 30\%$, the optimal construction cost is 95.5 €/m^2 and CO_2 emissions are 41 kgCO_2/m^2, while for the same SM = 2 kN and a better $CBR_{base} = 100\%$, the optimal construction cost is 85 €/m^2 and CO_2 emissions are 39.3 kgCO_2/m^2. In this case, reducing the CBR of the unbound layer from 100% to 30% means an increase of 12% in costs and 4% in CO_2 emissions. Similarly, for a pavement structure with geosynthetic reinforcement, and the reduction in the quality of the unbound layers increases the cost by 9% and the CO_2 emissions by 3%. It was also found that Marshall stability has the largest impact on both cost and CO_2 emissions.

From the results shown in Figure 6, it can be seen that costs increase by 25% and CO_2 emissions by 48% when the Marshal stability of the asphalt layer is reduced from 10 kN to 2 kN. This reduction was calculated for a pavement with geosynthetics and a CBR value of 30% for unbound layers. The use of geosynthetics in most of the cases discussed in the parametric analysis reduces the cost of pavement structure and the amount of CO_2 emissions. The largest reduction in COST and CO_2 is given in the case where T_n is in the range of 1×10^4, CBR = 3%, SM = 10 kN, and the CBR of the base and sub-base layers is 30%. In this case, the use of geosynthetics results in a 15% reduction in COST and a 9% reduction in CO_2 due to the reduced thickness of the unbound pavement structure. Figure 6 shows that the use of geosynthetics is economically justified when the CBR of the subgrade is less than 5%, and that the use of geosynthetics is environmentally justified when the $CBR_{subgrade}$ is less than 6% if the properties of the base and sub-base layer are assumed to be $CBR_{base,subbase} = 100\%$. If the CBR value of the base and sub-base layer is low ($CBR_{base,subbase}$ = 30%), the use of geosynthetics is justified from an economic and environmental point of view if the CBR value of subgrade is less than 7%. It was found that the use of geosynthetics is particularly important in the case where the base and subgrade layers partially contain waste materials.

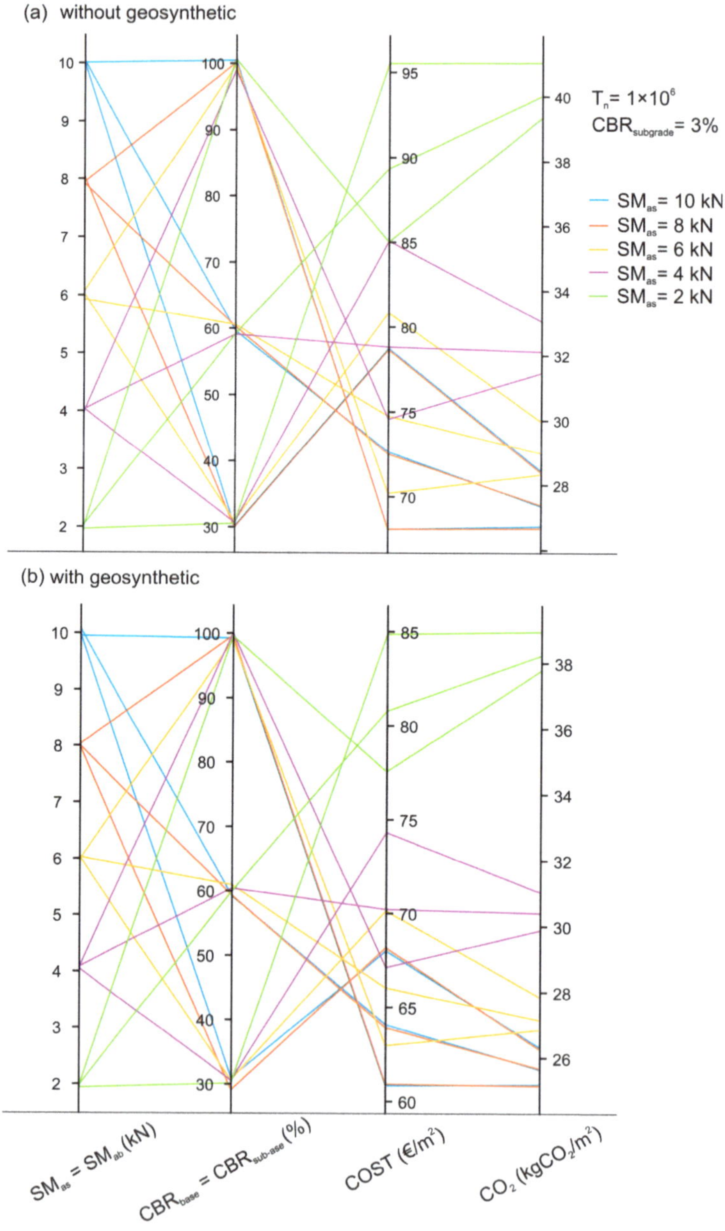

Figure 5. Parallel coordinate plot of optimal pavement cost and CO_2 emissions for different Marshall stability of asphalt layer and CBR of unbound layers: (**a**) without geosynthetic reinforcement and (**b**) with geosynthetic reinforcement.

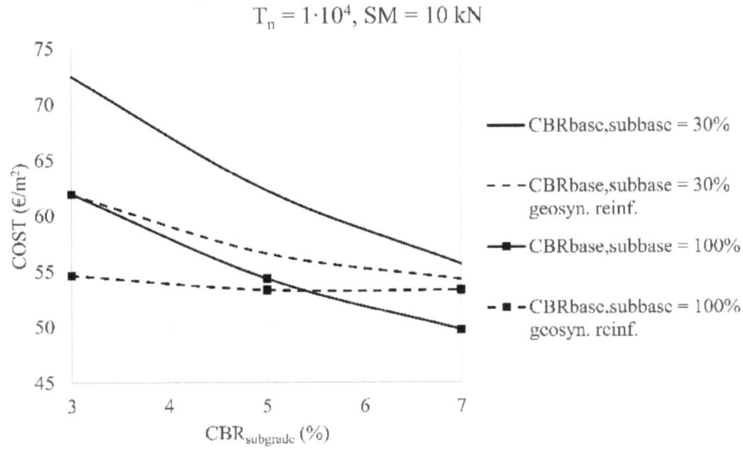

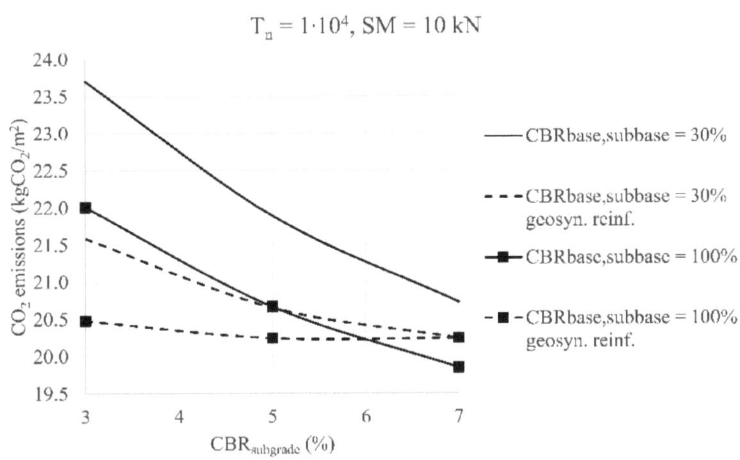

Figure 6. Costs and CO_2 emissions of road pavements depend on the quality of the subgrade and the use of geosynthetics.

Table 4 (without geosynthetics) and Table 5 (with geosynthetics) show the optimal pavement design including the thickness of asphalt layers, the base layer, and the sub-base, as well as pavement costs and CO_2 emissions. It should be noted that for all parameter combinations, the optimal thickness of the asphalt surface layer was calculated as d_{as} = 4 cm, which corresponds to the minimum value specified. Based on these two tables, it was possible to evaluate the effects of the Marshall stability and the CBR value of the unbound layer on the design of flexible pavements. The thickness of the unbound layers increased from 91 cm to 120 cm when the CBR value of the base and sub-base layers decreased from 100% to 30%. For real-world pavement projects, Tables 4 and 5 can help engineers and designers select the most appropriate materials, layer thicknesses, and construction methods for pavements where the design is based on the minimum cost and CO_2 emissions. The model was developed in a general form that allows an optimal design to be obtained for any input data based on real site conditions, material properties, and traffic loads.

Table 4. Optimal pavement design for subgrade $CBR_{subgrade}$ = 3%, without geosynthetic.

		$CBR_{base} = CBR_{subbase}$ = 100%					$CBR_{base} = CBR_{subbase}$ = 30%				
T_n (ESAL)	SM	$d_{as} + d_{ab}$	d_b	d_{sb}	COST	CO_2	$d_{as} + d_{ab}$	d_b	d_{sb}	COST	CO_2
(-)	(kN)	(cm)	(cm)	(cm)	(€/m^2)	(kgCO_2/m^2)	(cm)	(cm)	(cm)	(€/m^2)	(kgCO_2/m^2)
1.0×10^4	10	10	25	66	61.9	22.0	10	33	87	72.5	23.7
1.0×10^5	10	10	25	66	61.9	22.0	10	33	87	72.5	23.7
1.0×10^6	10	13	25	66	68.2	26.7	13	33	87	78.7	28.4
1.0×10^7	10	22	25	66	87.0	40.9	22	33	87	97.5	42.6
1.0×10^8	10	36	25	66	116.3	62.9	36	33	87	126.8	64.6
1.0×10^4	8	10	25	66	61.9	22.0	10	33	87	72.5	23.7
1.0×10^5	8	10	25	66	61.9	22.0	10	33	87	72.5	23.7
1.0×10^6	8	13	25	66	68.2	26.7	13	33	87	78.7	28.4
1.0×10^7	8	22	25	66	87.0	40.9	22	33	87	97.5	42.6
1.0×10^8	8	36	25	66	116.3	62.9	36	33	87	126.8	64.6
1.0×10^4	6	10	25	66	61.9	22.0	10	33	87	72.5	23.7
1.0×10^5	6	10	25	66	61.9	22.0	10	33	87	72.5	23.7
1.0×10^6	6	14	25	66	70.3	28.3	14	33	87	80.8	30.0
1.0×10^7	6	23	25	66	89.1	42.4	23	33	87	99.6	44.1
1.0×10^8	6	38	25	66	120.5	66.0	38	33	87	131.0	67.7
1.0×10^4	4	10	25	66	61.9	22.0	10	33	87	72.5	23.7
1.0×10^5	4	10	25	66	61.9	22.0	10	33	87	72.5	23.7
1.0×10^6	4	16	25	66	74.5	31.4	16	33	87	85.0	33.1
1.0×10^7	4	27	25	66	97.5	48.7	27	33	87	108.0	50.4
1.0×10^8	4	44	25	66	133.0	75.4	44	33	87	143.5	77.1
1.0×10^4	2	13	25	66	68.2	26.7	13	33	87	78.7	28.4
1.0×10^5	2	13	25	66	68.2	26.7	13	33	87	78.7	28.4
1.0×10^6	2	21	25	66	84.9	39.3	21	33	87	95.5	41.0
1.0×10^7	2	36	25	66	116.3	62.9	36	33	87	126.8	64.6
1.0×10^8	2	58	25	66	162.3	97.4	58	33	87	172.8	99.1

Table 5. Optimal geosynthetic reinforced pavement design for subgrade CBR = 3%.

		$CBR_{base} = CBR_{subbase}$ = 100%					$CBR_{base} = CBR_{subbase}$ = 30%				
T_n (ESAL)	SM	$d_{as} + d_{ab}$	d_b	d_{sb}	COST	CO_2	$d_{as} + d_{ab}$	d_b	d_{sb}	COST	CO_2
(-)	(kN)	(cm)	(cm)	(cm)	(€/m^2)	(kgCO_2/m^2)	(cm)	(cm)	(cm)	(€/m^2)	(kgCO_2/m^2)
1.0×10^4	10	10	25	33	54.6	20.5	10	33	44	61.9	21.6
1.0×10^5	10	10	25	33	54.6	20.5	10	33	44	61.9	21.6
1.0×10^6	10	13	25	33	60.9	25.2	13	33	44	68.1	26.3
1.0×10^7	10	22	25	33	79.7	39.3	22	33	44	87.0	40.4
1.0×10^8	10	36	25	33	109.0	61.3	36	33	44	116.2	62.4
1.0×10^4	8	10	25	33	54.6	20.5	10	33	44	61.9	21.6
1.0×10^5	8	10	25	33	54.6	20.5	10	33	44	61.9	21.6
1.0×10^6	8	13	25	33	60.9	25.2	13	33	44	68.1	26.3
1.0×10^7	8	22	25	33	79.7	39.3	22	33	44	87.0	40.4
1.0×10^8	8	36	25	33	109.0	61.3	36	33	44	116.2	62.4
1.0×10^4	6	10	25	33	54.6	20.5	10	33	44	61.9	21.6
1.0×10^5	6	10	25	33	54.6	20.5	10	33	44	61.9	21.6
1.0×10^6	6	14	25	33	63.0	26.8	14	33	44	70.2	27.9
1.0×10^7	6	23	25	33	81.8	40.9	23	33	44	89.0	42.0
1.0×10^8	6	38	25	33	113.2	64.5	38	33	44	120.4	65.6
1.0×10^4	4	10	25	33	54.6	20.5	10	33	44	61.9	21.6
1.0×10^5	4	10	25	33	54.6	20.5	10	33	44	61.9	21.6
1.0×10^6	4	16	25	33	67.2	29.9	16	33	44	74.4	31.0
1.0×10^7	4	27	25	33	90.2	47.2	27	33	44	97.4	48.3
1.0×10^8	4	44	25	33	125.7	73.9	44	33	44	132.9	75.0
1.0×10^4	2	13	25	33	60.9	25.2	13	33	44	68.1	26.3
1.0×10^5	2	13	25	33	60.9	25.2	13	33	44	68.1	26.3
1.0×10^6	2	21	25	33	77.6	37.8	21	33	44	84.9	38.9
1.0×10^7	2	36	25	33	109.0	61.3	36	33	44	116.2	62.4
1.0×10^8	2	58	25	33	155.0	95.9	58	33	44	162.2	97.0

Different input data were utilized to determine the optimal configuration of the pavement structure. The four primary inputs comprise the overall count of ESAL (T_n), the Marshall stability of asphalt layers ($SM_{as} = SM_{ab}$), the California Bearing Ratio of the base and sub-base layer ($CBR_{base} = CBR_{subbase}$), and the subgrade conditions ($CBR_{subgrade}$). Through this multiparametric analysis, the primary goal was to employ these key attributes to anticipate other continuous characteristics, including the minimum cost of pavement structure (COST) and the minimum CO_2 emissions (CO_2). Prior to employing the predictive model, the dataset was split into a training dataset (odd-indexed samples) and a checking dataset (even-indexed samples). The "exhsrch" function in MATLAB (R2021a) was utilized to exhaustively search among the available inputs and determine the set of inputs that have the greatest impact on the optimal cost of the pavement structure and layer thickness. The "exhsrch" function involved building predictive models for each parameter combination, training them for an epoch, and subsequently reporting their achieved performance. In Figure 7, the leftmost input variable is the most pertinent in terms of the output, as it exhibits the lowest root-mean-square error (RMSE). The RMSE is defined as follows:

$$\text{RMSE} = \sqrt{\frac{\sum_{i=1}^{n}(\hat{x}_i - x_i)^2}{n}} \tag{5}$$

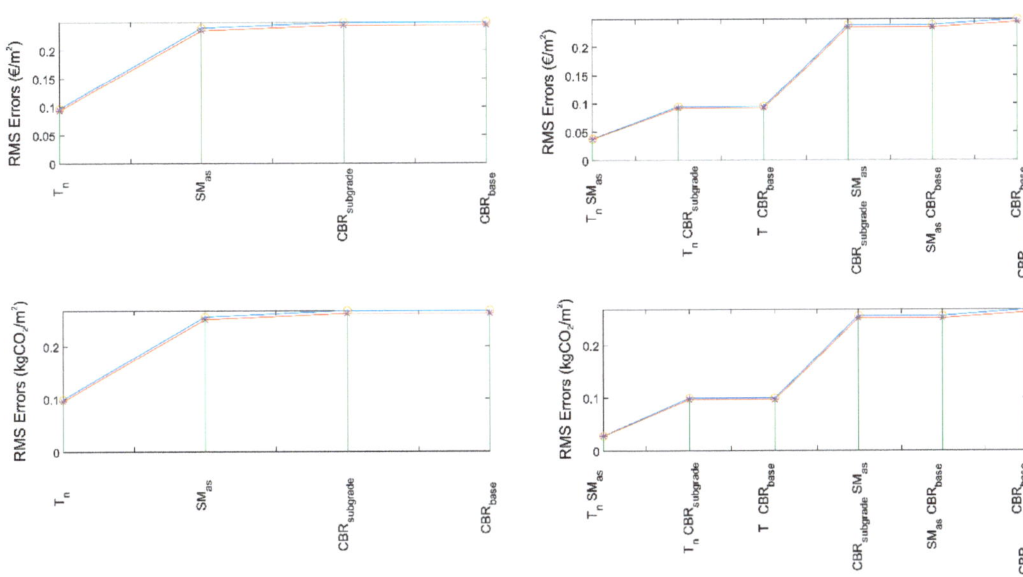

Figure 7. Influence of each input variable on the optimal cost of a pavement structure and the CO_2 emissions (training data in blue line, test data in red line).

Here, $\hat{x}_i$ represents the predicted values, while x_i represents the values obtained through the optimization procedure (COST, CO_2). Prediction models often face the challenge of overfitting. However, in this simple prediction model, the training and checking errors are comparable, indicating the absence of overfitting. It is important to note that the primary objective of this prediction model is to identify the inputs that exert the greatest influence on the output, rather than constructing a prediction model with minimal training error. To enhance the accuracy of the prediction model, it is advisable to incorporate more neurons in the neural networks. However, an increase in neurons may potentially lead to overfitting issues. The analysis also examines the combination of two inputs that hold the greatest influence over the output. The results of the parametric analysis unmistakably indicate that the total number of ESALs (T_n) is the most crucial parameter for achieving the optimal cost of a pavement structure. Subsequently, the Marshall stability ($SM_{as} = SM_{ab}$), CBR of the subgrade ($CBR_{subgrade}$), and CBR of unbound layers ($CBR_{base} = CBR_{subbase}$) follow suit in terms of their significance.

Based on Figure 8, regarding the CO_2 emissions for the pavement structure, the bound layers (asphalt layers) are responsible for 96% of the CO_2 emissions, while the unbound layers account for the remaining 4%. This distribution of CO_2 emissions is valid for $T_n = 1 \times 10^8$, $CBR_{subgrade}$ = 7%, CBR_{base} = 100%, and SM_{as} = 2 kN. The analysis shows that the fraction of CO_2 emissions caused by asphalt layers is much more sensitive to design parameters, while the fraction of pavement costs caused by asphalt layers is less sensitive to design parameters.

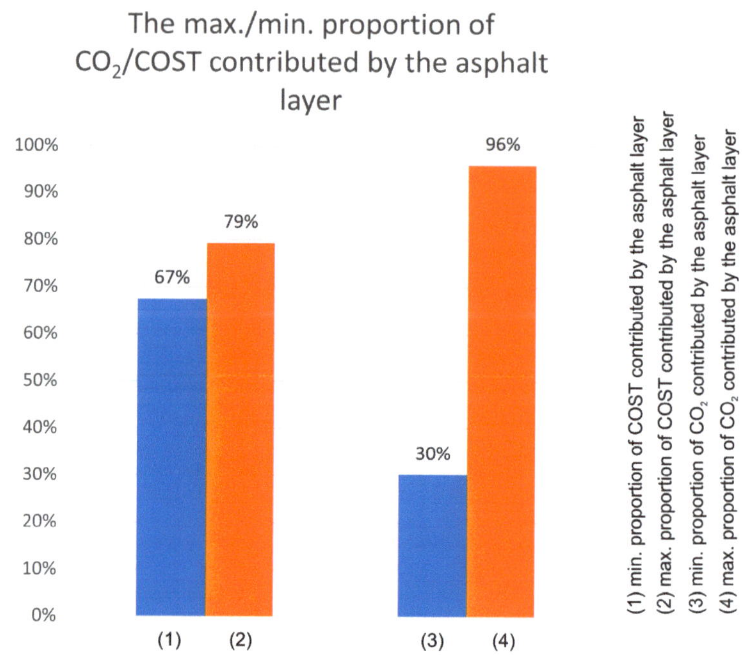

i	(1)	(2)	(3)	(4)
T_n (-)	1×10^4	1×10^8	1×10^4	1×10^8
$CBR_{subgrade}$ (%)	3	7	3	7
SM_{as} (kN)	10	2	10	2
CBR_{base} (%)	30	100	30	100
Geosyn.	reinforced	reinforced	reinforced	un-reinforced

Figure 8. Proportion of COST and CO_2 emissions contributed by the asphalt layer.

5. Conclusions

The present work deals with the aspect of the material quality of the pavement structure, provided that other legal, organizational, and logistical conditions are also met. It examines how the incorporation of waste materials in bound and unbound pavement layers affects layer thicknesses and consequently on costs and CO_2 emissions. The inclusion of waste materials was accounted for via equivalence factors used in the empirical pavement design method. The geosynthetic reinforced and unreinforced pavement design was optimized for different traffic loads and material properties. The proportion of costs and CO_2 emissions of the asphalt layers were also calculated. The main conclusions are the following:

- For the most unfavorable design parameters examined in the parametric analysis, the thickness of the unbound layers increased from 91 cm to 120 cm (32% increase in thickness) when the CBR value of the base and sub-base layers decreased from 100% to 30%.
- For the most unfavorable design parameters examined in the parametric analysis, the thickness of the asphalt layer increased from 36 cm to 58 cm (61% increase in thickness) when the Marshall stability value of the asphalt layer decreased from 10 kN to 2 kN.
- The analysis shows that the proportion of CO_2 emissions caused by asphalt layers can vary from 30% to 96% depending on the design parameters, while the proportion of costs caused by asphalt layers only ranges from 67% to 79% for the same design parameters. This is due to the fact that the ratio of CO_2 emissions between the asphalt layer and the unbound layer is higher than the ratio of prices.
- The results of the parametric analysis show that the total number of ESALs (T_n) is the most important parameter for achieving the optimal cost of a pavement structure. This is followed by the Marshall stability (SM), the CBR value of the subgrade ($CBR_{subgrade}$), and the CBR value of the unbound layers ($CBR_{base} = CBR_{subbase}$) in terms of their importance.
- The use of geosynthetics could result in a 15% reduction in pavement structure cost and a 9% reduction in CO_2 emissions due to the reduced thickness of unbound layers. However, the use of geosynthetics could also result in an increase in road pavement structure cost and CO_2 emissions under favorable site conditions (e.g., with a CBR subgrade of 7%).
- The empirical design method for pavements limits the Marshall stability to approximately 10 kN, although the stability of asphalt concrete could be higher. Therefore, the mechanical-empirical design method could further improve the optimization model by considering even larger Marshall stability values.

The optimization model was developed in a general form that can provide an optimal solution for various design parameters including different traffic loads, site conditions, and material properties that depend on specific real project data. Further research is needed to evaluate how the properties of asphalt and unbound layers are altered by the addition of waste in various percentages and by the type of waste included. Once these relationships are known, waste reduction could also be determined in terms of cost and CO_2 emissions while achieving a reduction in waste deposition. Since this optimization model and, consequently, the results presented are based on an empirical pavement design method, further investigations could be investigated by semi-empirical pavement design methods or methods based on finite element modeling.

Author Contributions: Conceptualization, P.J., B.Ž., B.M., S.G. and C.G.; methodology, P.J., B.Ž., B.M., S.G. and B.M.; software, P.J. and R.V.; validation, P.J., B.Ž., B.M., S.G., C.G., R.V., T.B., Ş.Y., M.V.T. and B.E.K.; writing—original draft preparation, P.J., B.Ž., B.M., S.G. and C.G.; writing—review and editing, P.J., B.Ž., B.M., S.G., C.G., R.V., T.B., Ş.Y., M.V.T. and B.E.K. All authors have read and agreed to the published version of the manuscript.

Funding: This research was funded by the Slovenian Research Agency (ARIS) and Scientific and Technological research Council of Türkiye (TÜBİTAK) by supporting a bilateral project (grant numbers BI-TR/22-24-06 and 122N273) and by the EU project GEOLAB (grant number 101006512).

Institutional Review Board Statement: Not applicable.

Informed Consent Statement: Not applicable.

Data Availability Statement: The data presented in this study are available on request from the corresponding author.

Conflicts of Interest: The authors declare no conflict of interest.

References

1. Hale, S.E.; Roque, A.J.; Okkenhaug, G.; Sørmo, E.; Lenoir, T.; Carlsson, C.; Kupryianchyk, D.; Flyhammar, P.; Žlender, B. The Reuse of Excavated Soils from Construction and Demolition Projects: Limitations and Possibilities. *Sustainability* **2021**, *13*, 6083. [CrossRef]
2. Huang, Y.; Bird, R.N.; Heidrich, O. A review of the use of recycled solid waste materials in asphalt pavements. *Resour. Conserv. Recycl.* **2007**, *52*, 58–73. [CrossRef]
3. Bamigboye, G.O.; Bassey, D.E.; Olukanni, D.O.; Ngene, B.U.; Adegoke, D.; Odetoyan, A.O.; Kareem, M.A.; Enabulele, D.O.; Nworgu, A.T. Waste materials in highway applications: An overview on generation and utilization implications on sustainability. *J. Clean. Prod.* **2021**, *283*, 124581. [CrossRef]
4. Li, J.; Xiao, F.; Zhang, L.; Amirkhanian, S.N. Life cycle assessment and life cycle cost analysis of recycled solid waste materials in highway pavement: A review. *J. Clean. Prod.* **2019**, *233*, 1182–1206. [CrossRef]
5. Patel, S.; Pai, R.R.; Bakare, M.D.; Shahu, J.T. Field evaluation of road pavement constructed with waste materials through nondestructive testing. *Mater. Today Proc.* **2020**, *28*, 1254–1260. [CrossRef]
6. Abukhettala, M.; Fall, M. Geotechnical characterization of plastic waste materials in pavement subgrade applications. *Transp. Geotech.* **2021**, *27*, 100472. [CrossRef]
7. Khan, S.; Ashish, P.K.; Kannelli, V.; Hossain, K.; Nagabhushana, M.N.; Tiwari, D. Potential application of over-burnt brick and fly ash for sustainable inverted pavement structure. *Constr. Build. Mater.* **2022**, *345*, 128298. [CrossRef]
8. Zimar, Z.; Robert, D.; Zhou, A.; Giustozzi, F.; Setunge, S.; Kodikara, J. Application of coal fly ash in pavement subgrade stabilisation: A review. *J. Environ. Manag.* **2022**, *312*, 114926. [CrossRef]
9. Arifuzzaman, M.; Qureshi, H.J.; Al Fuhaid, A.F.; Alanazi, F.; Javed, M.F.; Eldin, S.M. Novel ensemble modelling for prediction of fundamental properties of bitumen incorporating plastic waste. *J. Mater. Res. Technol.* **2023**, *24*, 3334–3351. [CrossRef]
10. Kumar, B.; Kumar, N. Assessment of Marshall Stability of asphalt concrete with plastic waste using soft computing techniques. *Multiscale Multidiscip. Model. Exp. Des.* **2023**. [CrossRef]
11. Lee, S.; Le, T.H.M. Feasibility of Sustainable Asphalt Concrete Materials Utilizing Waste Plastic Aggregate, Epoxy Resin, and Magnesium-Based Additive. *Polymers* **2023**, *15*, 3293. [CrossRef]
12. White, G.; Reid, G. Recycled waste plastic for extending and modifying asphalt binders. In Proceedings of the 8th Symposium on Pavement Surface Characteristics (SURF), Brisbane, Australia, 2–4 May 2018; p. 12.
13. Vasudevan, R.; Sekar, A.R.C.; Sundarakannan, B.; Velkennedy, R. A technique to dispose waste plastics in an ecofriendly way –Application in construction of flexible pavements. *Constr. Build. Mater.* **2012**, *28*, 311–320. [CrossRef]
14. Picado-Santos, L.G.; Capitão, S.D.; Neves, J.M.C. Crumb rubber asphalt mixtures: A literature review. *Constr. Build. Mater.* **2020**, *247*, 118577. [CrossRef]
15. Wen, Y.; Wang, Y.; Zhao, K.; Chong, D.; Huang, W.; Hao, G.; Mo, S. The engineering, economic, and environmental performance of terminal blend rubberized asphalt binders with wax-based warm mix additives. *J. Clean. Prod.* **2018**, *184*, 985–1001. [CrossRef]
16. Wang, T.; Li, K.; Liu, D.; Yang, Y.; Wu, D. Estimating the Carbon Emission of Construction Waste Recycling Using Grey Model and Life Cycle Assessment: A Case Study of Shanghai. *Int. J. Environ. Res. Public Health* **2022**, *19*, 8507. [CrossRef]
17. Bosurgi, G.; Celauro, C.; Pellegrino, O.; Ruggeri, A.; Sollazzo, G. Mechanical and environmental performance comparisons of improved asphalt pavement wearing courses with high quality aggregates, steel slags, and polymeric compound. *Constr. Build. Mater.* **2023**, *382*, 131252. [CrossRef]
18. You, L.; Long, Z.; You, Z.; Ge, D.; Yang, X.; Xu, F.; Hashemi, M.; Diab, A. Review of recycling waste plastics in asphalt paving materials. *J. Traffic Transp. Eng.* **2022**, *9*, 742–764. [CrossRef]
19. White, P.; Golden, J.S.; Biligiri, K.P.; Kaloush, K. Modeling climate change impacts of pavement production and construction. *Resour. Conserv. Recycl.* **2010**, *54*, 776–782. [CrossRef]
20. Farina, A.; Zanetti, M.C.; Santagata, E.; Blengini, G.A. Life cycle assessment applied to bituminous mixtures containing recycled materials: Crumb rubber and reclaimed asphalt pavement. *Resour. Conserv. Recycl.* **2017**, *117*, 204–212. [CrossRef]
21. Macrebur. Available online: https://www.macrebur.com/#projects (accessed on 8 August 2023).
22. VolkerWessels. Available online: https://www.volkerwessels.com/en/projects/plasticroad (accessed on 8 August 2023).
23. TechniSoil Industrial. Available online: https://technisoilind.com/technisoil-roads.html (accessed on 8 August 2023).
24. Jelušič, P.; Varga, R.; Žlender, B. Parametric analysis of the minimum cost design of flexible pavements. *Ain Shams Eng. J.* **2023**, *14*, 101840. [CrossRef]
25. Naveed, M.; Raza, M.A.; Mehmood, R. Performance analyses of conventional hot mix asphalt with waste additives. *Case Stud. Constr. Mater.* **2022**, *16*, e00850. [CrossRef]
26. Fatemi, S.; Imaninasab, R. Performance evaluation of recycled asphalt mixtures by construction and demolition waste materials. *Constr. Build. Mater.* **2016**, *120*, 450–456. [CrossRef]
27. Hake, S.L.; Damgir, R.M.; Awsarmal, P.R. Utilization of Plastic waste in Bitumen Mixes for Flexible Pavement. *Transp. Res. Procedia* **2020**, *48*, 3779–3785. [CrossRef]
28. Heydari, S.; Hajimohammadi, A.; Javadi, N.H.S.; Khalili, N. The use of plastic waste in asphalt: A critical review on asphalt mix design and Marshall properties. *Constr. Build. Mater.* **2021**, *309*, 125185. [CrossRef]
29. *Projektiranje Dimenzioniranje Novih Asfaltnih Vozišcnih Konstrukcij*; TSC 06.520:2009; Ministrstvo za promet—Direkcija RS za ceste: Ljubljana, Slovenia, 2009.

30. Bohara, N. Study of the Influence of Fly Ash and Its Content in Marshall Properties of Asphalt Concrete. *J. Sustain. Constr. Mater. Technol.* **2018**, *3*, 261–270. [CrossRef]
31. Rajput, P.S.; Yadav, R.K. Use of Plastic Waste in Bituminuous Road Construction. *Int. J. Sci. Technol. Eng.* **2016**, *2*, 509–513.
32. Shah, A.M.; Lodhi, R.H.; Javed, M.F.; Jasiński, M.; Jasińska, E.; Gono, M. Structural Performance of Waste Plastic Bottles Modified Asphalt: A Review. *Resources* **2023**, *12*, 10. [CrossRef]
33. Loaiza, A.; Colorado, H.A. Marshall stability and flow tests for asphalt concrete containing electric arc furnace dust waste with high ZnO contents from the steel making process. *Constr. Build. Mater.* **2018**, *166*, 769–778. [CrossRef]
34. Ficko, G.; Žlender, B. The analysis of permanent deformations of repeatedly loaded gravels from the Mura region. *Acta Geotech. Slov.* **2005**, *2*, 24–37.
35. Jing, P.; Chazallon, C. Hydro-Mechanical Behaviour of an Unbound Granular Base Course Material Used in Low Traffic Pavements. *Materials* **2020**, *13*, 852. [CrossRef] [PubMed]
36. Orduz-Duarte, E.J.; Pineda-Jaimes, J.A. Performance of a Reinforced Pavement Structure with Geocells via Numerical Simulation, Considering a Discontinuous Subgrade. In Proceedings of the 2017 Congreso Internacional de Innovacion y Tendencias en Ingenieria (CONIITI), Bogota, Colombia, 3–5 October 2018; pp. 1–6.
37. Ghanizadeh, A.R.; Ghaderi, F.; Tavassoti, P. Numerical Investigation of the Performance of Geocell-Reinforced Granular Base in Inverted Pavement Systems Using Nonlinear Finite Element Modeling. *Can. J. Civ. Eng.* **2022**, *50*, 395–407. [CrossRef]
38. EU Project CINDERELA—New Circular Economy Business Model for More Sustainable Urban Construction. Available online: https://www.cinderela.eu/The-project (accessed on 8 August 2023).
39. EU Project URGE—Circular Building Cities. Available online: https://urbact.eu/networks/urge (accessed on 8 August 2023).
40. Øiseth, E.; Aabøe, R.; Hoff, I. Field Test Comparing Frost Insulation Materials in Road Construction. In *Current Practices in Cold Regions Engineering*; American Society of Civil Engineers: Reston, VA, USA; pp. 1–11.

sustainability

Article

Performance Measurement Framework for Prediction and Management of Construction Investments

Kristina Galjanić [1,2], Ivan Marović [2,*] and Tomaš Hanak [3]

[1] GP Krk d.d., Stjepana Radića 31, 51500 Krk, Croatia; kgaljanic@student.uniri.hr
[2] Faculty of Civil Engineering, University of Rijeka, Radmile Matejčić 3, 51000 Rijeka, Croatia
[3] Faculty of Civil Engineering, Brno University of Technology, Veveri 95, 602 00 Brno, Czech Republic; hanak.t@vutbr.cz
* Correspondence: ivan.marovic@uniri.hr

Abstract: Despite good ideas, great efforts, and high investments, many projects do not end with success. Projects often fail due to a lack of understanding of the project requirements and constraints necessary for overall success. Five selected projects were analyzed in detail through the multiple case study method followed by semi-structured interviews with 56 experts to develop a pattern for the future prediction of project success. This paper aims to identify key factors for project performance in a multi-stakeholder environment, define a performance measurement framework for construction investments, and establish a link between performance measurement and prediction of project performance. The findings could help researchers in modeling performance measurement tools for project managers to achieve their designated project goals, reach better decisions, and achieve full potential in their future projects.

Keywords: performance measurement; project management; multiple case study; performance indicators; prediction; stakeholder management

Citation: Galjanić, K.; Marović, I.; Hanak, T. Performance Measurement Framework for Prediction and Management of Construction Investments. *Sustainability* **2023**, *15*, 13617. https://doi.org/10.3390/su151813617

Academic Editors: Uroš Klanšek and Tomaž Žula

Received: 11 August 2023
Revised: 9 September 2023
Accepted: 10 September 2023
Published: 12 September 2023

1. Introduction

Despite good ideas, great efforts, and high investments, many projects do not end with success. Although there could be several reasons for such, a pivotal task in the study of project management remains the same, as Chen [1] stated, "to identify the critical determinants of project management performance". Therefore, over the years, many researchers and practitioners have examined and identified a wide variety of approaches, tools, and techniques to describe and measure project management performance focusing on input characteristics that affect project outcomes [1–4]. Those studies often focus on the overall project life cycle [2,3,5], with relatively few focused on the perspective level of the project phases [1], especially how various stakeholders will perceive project success [6–8]. The paradigm of focusing solely on technical and economic aspects and areas over the years has shifted towards the integration of social and behavioral areas as well, thus focusing on the interaction between project stakeholders and the project team [8–11]. Both practitioners and academics have difficulties coping with such problems, clearly showing that there is still room to investigate and contribute. Therefore, this article builds on the previous research and stakeholder management theories [12–14] and multi-criteria decision tools [15–18], thus proposing such a framework that considers complex project environment, especially in construction projects, and a multi-stakeholder view to enable reaching full success in future projects.

The complex surroundings and the goal for overall betterment, often viewed as sustainability, have a specific imperative that project performance is recognized and measured on long-term strategic objectives instead of short-term tactical performance [18–21]. While the authors, in general, often focus on achieving short-term project targets as long-term benefits management, especially in public projects [20], there are "significant variations in the levels of success", as Flyvbjerg reported [22]. The PMI's report [23] claimed "only 70% of projects successfully met their original goals and business intent", so, there is still much room for improving performance.

Another aspect is the multi-dimensionality of success, as the interests of different stakeholders imply that they will sometimes have quite different perceptions of the project's success [24–28]. Therefore, project failure is often seen as a lack of understanding of the project requirements and the constraints necessary for overall success, emphasizing the early stages of the project. Such is most evident in the architecture, engineering, and construction (AEC) industry as construction costs are one of the main criteria for decision making in the early stages and of interest to all project participants, i.e., stakeholders [29,30]. Very often, there are discrepancies between the estimated (e.g., planned or contracted) costs concerning the realized (e.g., actual) costs of the construction project [31]. Usually, discrepancies occur due to a lack of data and information in the conceptual phase [32–37]. Therefore, monitoring performance and reward depending on outcomes is increasingly common.

While the studies mentioned above indicate how to approach project performance measurement, they do not address how performance determinants (as both performance areas (PAs) and performance indicators (PIs)) influence project success from a multi-stakeholder viewpoint. Therefore, we aim to define a performance measurement framework that could transparently involve different stakeholders in defining a set of PIs. It could help manage projects based on their current performance to predict future success.

The main goal of this research is to develop a performance measurement framework as a conceptual framework that takes into account different project stakeholders' points of view (POV) as well as projects' performance criteria, i.e., key performance indicators (KPIs), to help project managers to make their decisions in the best possible way to reach project goals and outcomes. In a construction project environment, the stakeholders' POV is represented by clients, contractors, consultants, and their project representatives, i.e., project managers. Therefore, the proposed framework to create a pattern for the future prediction of project success will be based on five case studies, i.e., real construction investment projects, and analyzed with the multiple case study method followed by semi-structured interviews with identified stakeholders. At the same time, the performance management areas are derived from previously performed, detailed bibliographical research analysis [38] and multiple case study that takes into account pre-defined performance measures and their outcomes in selected construction investment projects. The proposed framework takes project data as inputs for successfully managing performance during the whole life-cycle of construction projects. The contribution of this research would be to the better understanding and improvement of project performance in construction projects by providing such a framework that offers project managers the opportunity to evaluate the current state of the project, regardless of its stage, and provides a reasonable possibility of prediction to meet project constraints. Therefore, to achieve the main goal, this research intends to deal with the following three objectives, namely, (1) to create a procedure that can be used in a multi-stakeholder environment to identify project performance indicators for measuring performance, (2) to define a performance measurement framework for construction investments, and (3) to establish a link between performance measurement and prediction of project performance.

2. Theoretical Background

2.1. Project Management, Success, and Performance of the Project

Project management theory initially defines project success based on three core criteria: delivery on time, within budget, and to an agreed quality [39]. Such an approach gained popularity thanks to the good measurability of the criteria. However, later studies have greatly criticized this concept as these three criteria are insufficient to capture the project's success from a broader point of view [40]. Accordingly, the required level of performance can only be achieved if other aspects are observed [41]. The project's success is affected, among others, also by its complexity, which may increase the level of cost and time risks [42]. In this context, a breakdown of project success criteria into 29 categories was proposed [43] supplementing traditional criteria (time, cost, and quality) with other macro-level categories covering stakeholders-, deliverables-, and project organization- and management-related criteria.

The theory recognizes project complexity as the number and heterogeneity of different inter-related elements [44]. Vidal and Marle [45] highlighted that complexity renders the project difficult to understand, foresee, and keep under control. The multi-dimensionality of project complexity is seen in the literature from technological and organizational views, while, in our paper, we mostly focus on the organizational complexity relating to both in terms of the complexity of project objectives and related tasks as well as to interactions between a high number of people and stakeholders involved [46]. It is also believed that a higher number of inter-related elements that have to be co-ordinated causes greater exposure to delays and cost overruns [42]. In addition, it is argued that when the scope and complexity of the project increases, the need for a more comprehensive portfolio of criteria increases as well [47].

The later studies have further conceptualized performance management on a project level in a wide range of areas, such as supply chain management [48–50], risk considerations [51–54], safety [55,56], and sustainability aspects [57–59]. In such a way, it is possible to capture a broader range of data necessary for effectively managing the project and evaluating its performance. Such an approach becomes pivotal, especially in an unstable business environment characterized by changes in competition, technologies, and customer preferences and requirements [60]. As ascertained by Ward and Chapman [14], stakeholders represent the main source of uncertainty in the project due to the multiplicity of their objectives, which can be conflicting. From this perspective and in line with performance management efforts, analyzing various stakeholders' POVs on the project's success becomes pivotal. Accordingly, in our study, we investigate the POV on performance management issues of these three central stakeholders of any construction project.

Many root causes of cost and time overruns have already been identified in the literature, including project complexity, price increases, slow decision making, rework, or shortage of equipment [61]. Many scholars have incorporated risk aspect into their performance management approaches in terms of particular KPIs, such as overtime work rate and rework rate [62] or time–cost predictability [33,63–66]. Accordingly, risk performance indexes and measurement systems have been developed [52], mainly covering cost and schedule over-run-related risk.

Available literature suggests numerous models, systems, and frameworks developing performance management issues. As Lin and Shen [67] discussed, the need for so many models arises from the fact that they look at the various facets of performance from different points of view. Furthermore, they argue that multi-perspective indicators are essential for performance measurement and applying the balance scorecard approach [68–71] should help improve overall performance. However, these models are often criticized for grouping causes and effects together as an overall performance indicator [72].

Hence, researchers have generally focused on providing advances (1) for the overall performance measurement and (2) by developing fragmentized forecasting models and models addressing specific aspects of the performance. In relation to (1), several approaches have been built, e.g., to predict project failure at completion by considering seven variables (communication, team, creativity, technology, risk, quality, and materials; as suggested by [1]), in terms of the total performance score that has been developed in order to quantify project performance indicator system based on 18 KPIs covering eight PAs [62], or by a system dynamics approach to predict construction project performance [73]. Regarding (2), the following models can be noted: the operational research model has been developed to predict contractor performance [74], the decision support model for construction supply chain performance management was introduced by Yildiz and Ahi [48], while Kim [52] presented a risk performance management model based on cost and schedule risk considerations.

Stakeholder perspectives and their POVs have been widely studied. Prior analyses have shown that the perception of specific KPIs differs across stakeholders [41], similar to the perception of particular attributes that influence cost performance [28]. That is why engaging stakeholders already at the early stages of the project is of high importance [75–77] in as much as many projects disagree on the priority of particular criteria across individual stakeholders [47]. Previous research also revealed performance objectives and indicators of stakeholder management [75,78,79] and pinpointed collaborative management, which could produce positive effects such as increased cost performance of the project [80]. Considering the various concepts raised, it is desirable to reflect on how much uncertainty exists in managers' predictive models [81], which can adversely affect achieving project success. Therefore, the choice of PAs to be monitored and measured is crucial.

2.2. Stakeholders Management

As ascertained by Ward and Chapman [14], stakeholders represent the main source of uncertainty in the project due to the multiplicity of their objectives, which can be conflicting. From this perspective and in line with performance management efforts, it becomes pivotal to analyze various stakeholders' POVs on the project's success. Stakeholders are defined usually as "groups or individuals who have a stake in, or expectation of, the project's performance". The origins of the stakeholder concept have been described by Freeman [12], highlighting its dynamic aspect as every stakeholder role is temporary and issue-specific. The further development of stakeholder theory has included, among others, the approach of Mitchell et al. [13] regarding the identification (normative theory), salience (descriptive theory), and establishing the typology of stakeholders. It should be mentioned that stakeholder identification belongs to the main challenges of project managers [75,78]. Once stakeholders are identified, Mitchell's theory [13] further facilitates the determination of stakeholders' salience based on three main elements of typology: power, legitimacy, and urgency, and their assignment to one of the nine classes. Accordingly, managers can decide on the priority they give to competing stakeholders' claims. One of the prime project management goals is to support a balance between the needs and expectations of individual stakeholders [79].

A high number of stakeholders raises the need for careful strategic considerations in buyer–supplier relationships. Previous theoretical findings pointed out that there is no single and ideal way to manage these relationships ([82] Kim and Choi, 2015). Deep and long-term relations might benefit from the mutual trust of the parties involved, which is important as trust can influence the success of the project ([83] Cerić et al., 2021). Since the buyer has to control the relationship with its suppliers and is in line with the effort to avoid poor performance, an incentive/disincentive mechanism might be considered as a suitable managerial approach [49,84]. From the construction industry's point of view, private and public projects have to be differentiated. As for public projects, relationships are often limited to a single contract [49]. In this context, supply chain management in construction becomes more complicated. Additionally, available literature recognizes, e.g., in the supply chain operations reference model, its metrics were used to manage the performance of the construction supply chain [48].

While the spectrum of construction project stakeholders is broad, e.g., clients, project managers, designers, subcontractors, supplies, funding bodies, users, community, local authorities, environmentalists [85], project management as well as construction management, the literature recognizes three key stakeholders, namely, clients, contractors, and consultants [75,86]. Especially, as the stakeholders being seen [84] as "one of the underestimated factors of project success". Accordingly, in our study, we investigate the POV on performance management issues of these three central stakeholders of any construction project.

2.3. Project Performance Areas

Previous research has shown that investigations into individual aspects of performance management on the project level have been widely conducted. More specifically, available literature suggests a wider spectrum of areas (apart from time, cost, and quality) that can be suggested as subjects of performance measurement. By conducting an extensive literature review [38], we have identified eight common PAs (namely, profitability, productivity, quality, time/schedule, cost, safety, team satisfaction, and client satisfaction) used to evaluate the project's success (see Table 1). The typical PA list has been developed based on 56 relevant publications and their distribution into individual PAs. Our study uses them to analyze them from a multi-stakeholder's point of view.

Table 1. Overview of common performance areas to evaluate project success.

Decade	Reference	Profitability	Productivity	Quality	Time/Schedule	Cost	Safety	Team Satisfaction	Client Satisfaction
1980s	[81]	+	+						
	[40]				+			+	+
1990s	[73]			+	+	+	+		
	[85]	+	+	+					
	[39]			+	+	+		+	+
2000s	[62]	+	+	+	+	+	+		
	[87]			+	+	+	+	+	+
	[88]		+	+	+	+	+		
	[67]		+	+	+	+	+		+
	[89]	+	+	+	+	+	+		+
	[90]	+	+	+	+	+			+
	[91]		+		+	+	+		+
	[92]		+	+	+	+	+		
	[93]	+	+		+	+	+		+
	[66]					+	+	+	+
	[94]	+	+			+			
	[95]			+	+	+			
	[96]				+	+			
	[97]		+	+		+	+	+	+
	[2]	+	+	+	+	+	+	+	+
	[45]			+	+	+	+	+	+
	[14]				+	+		+	
	[68]	+	+	+	+	+		+	+
	[98]		+	+	+	+	+	+	+
	[99]	+	+	+	+	+	+		+
	[100]	+	+	+	+	+	+		
	[4]		+						+
2010s	[101]		+	+	+	+			+
	[41]	+	+	+	+	+	+	+	+
	[102]		+	+	+	+			
	[61]		+	+	+	+	+	+	+
	[103]		+	+		+			
	[104]		+	+					+
	[6]			+	+	+		+	+
	[105]			+	+	+	+		+
	[31]				+	+			
	[69]	+	+	+	+	+	+	+	+
	[106]			+			+		
	[107]		+	+	+	+	+		+
	[108]			+	+	+	+		
	[3]				+	+			
	[109]	+	+	+	+	+	+		+
	[1]			+	+	+	+		
	[19]	+	+	+	+				
	[110]		+	+	+	+	+		
	[111]	+	+	+	+	+	+	+	+
	[27]			+	+	+	+	+	+
	[60]				+	+			
	[112]		+	+	+	+	+		
	[57]							+	+
	[72]			+	+	+		+	+
	[113]			+		+	+	+	
	[113]			+		+	+	+	

Table 1. *Cont.*

Decade	Reference	Profitability	Productivity	Quality	Time/Schedule	Cost	Safety	Team Satisfaction	Client Satisfaction
2010s	[78]			+	+	+	+	+	+
	[58]	+	+			+			
	[114]	+				+			
	[115]			+	+	+	+	+	
	[116]			+	+	+	+	+	+
	[117]	+	+	+	+	+	+	+	+
	[79]				+	+			
	[118]			+		+			+
2020s	[7]			+	+	+			+
	[8]		+	+	+	+	+	+	+
	[119]		+	+	+	+	+		+
	[43]			+	+	+	+	+	+
	[9]		+	+	+	+		+	+
	[49]			+	+	+	+		
	[38]			+	+	+	+	+	
	[76]		+	+	+	+	+		
	[10]		+	+	+	+	+	+	+

As previously shown in the above theoretical studies, there is a consensus that the improvement of project performance represents a difficult task in a complex construction environment. Diversity of projects, variability of stakeholders, differences in external influences or exposure to various risks, among others, complicates the easy implementation of performance management systems. Recognizing this challenge and given its importance in the broader project literature, we would contribute to a better understanding of performance management in construction investment projects by proposing a framework built on the combination of stakeholder theory and PAs.

Therefore, the performance measurement framework is developed and described in the following section to offer not only an insight into how PAs mutually interact and have an impact on the project's success in a multi-stakeholder environment but also be a framework for the prediction and management of construction investments based on accumulated past project performance and knowledge.

3. Research Methodology

3.1. Developing Performance Measurement Framework for Predicting and Managing Construction Investments

A performance measurement framework for predicting and managing construction investments is proposed to address the previously defined main goal. It consists of three parts (see Figure 1), each defined with a particular added value to the decision-maker. Firstly, defining common PAs to evaluate project success is essential. Such can be achieved through an extensive literature review. In this particular case, an extensive literature review has previously been done by Marović et al. [38], as the research query was focused on performance management in civil engineering. This resulted in 1240 documents published in Scopus and Web of Science databases from 2000 to 2021. The results were extended with the theoretical background previously stated, resulting in 8 common PAs shown in Table 1. Such provided a level playing field for the following multi-stakeholder analysis of performance criteria. Once stakeholders are defined and start with project-related communications, PAs serve as well-needed constraints to define performance criteria. Therefore, performance criteria are defined transparently, in the stakeholders' hands, and their exclusive responsibility. As it can be conducted in different ways and using different techniques, the hierarchical goal structure procedure [75,120] showed promising results in dealing with multi-stakeholders in the construction project environment. Therefore, we are keen to use it in this particular framework as well to not just define performance criteria for each PA but also to develop a hierarchical criteria structure to evaluate projects' performance.

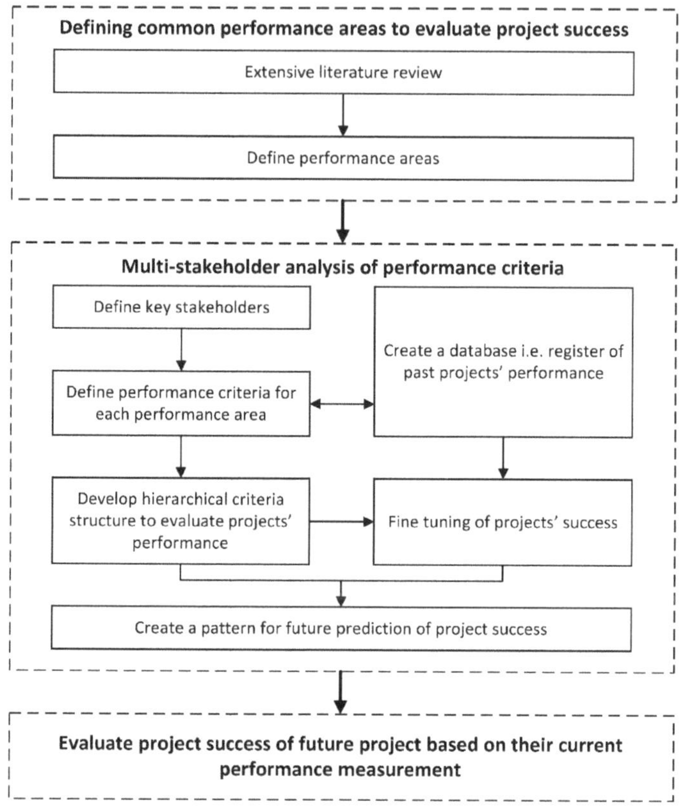

Figure 1. A performance measurement framework for predicting and managing construction investments.

A register of past projects' performance can also be of use to help stakeholders to develop it. As such development is an iterative process, an additional fine-tuning of projects' success can be done either quantitatively or qualitatively. It is an open loop, so applied in different constraints and project environments will bring additional needed aspects to the framework and, therefore, new added value. The aforementioned enables us to perceive how PAs mutually interact and the potential impact of the project's success from the different stakeholders' points of view. As herein, the proposed framework is qualitative and will, undoubtedly, provide insight to create a pattern for a future prediction of project success. Therefore, addressing more and more project cases to create a pattern using multiple case study is expected to bring more precision to future predictions. Such a pattern could undoubtedly serve as a valuable tool to evaluate project success of future projects based on their current performance measurements.

A multiple case study research approach [121] is adopted to understand and facilitate the identification of projects' performance criteria and, based on them, to develop a pattern as a decision-maker's tool for future prediction of project success. The central issue is developing a theoretical performance model based on a detailed literature review and stakeholder inputs on one side and construction project data on the other to help project managers decide the best way to reach project goals and outcomes. To differ between case study and multiple case study approaches, it is necessary to highlight adequate terminology that is used in this manuscript. Therefore, as stated by Yin [121], we adopted the notions that the case study research stands for the mode of inquiry, case studies for the method of inquiry, and the case is related to the unit of inquiry in a particular case study. Additionally, illustrative case studies used herein serve primarily "to make the unfamiliar familiar and to give readers a common language about the topic in question". Therefore, our multiple case study protocol consists of three phases (1) Define and Design Phase, (2) Prepare, Collect, and Analyze Phase, and (3) Conclude Phase, as [121] suggested. During the first phase, some additional methods were used to develop the theory, such as review methods for analyzing the existing literature [122], selecting a representative case sample, and designing a data collection protocol. Throughout case selection, it is essential to ensure a valid variation on the dimensions of theoretical interest [123].

In order to identify projects' PAs (defined as objective no. 1), several review methods for analyzing the existing literature, such as critical review, literature review, meta-analysis, and systematic search and review, were used. This was predominantly used to develop a theory as the first step of the protocol mentioned above (research background is given in Section 2). For this purpose, bibliometrics [38,124,125] was used, as a systematic quantitative literature review, followed by a transparent and systematic method for reviewing collected bibliography and systematization of information. Therefore, by combining the quantitative and qualitative approaches, the goal is to identify performance areas for project performance and achieve its success. This approach can be used particularly for transdisciplinary and interdisciplinary research to identify the literature's geographic, scalar, theoretical, and methodological gaps [126].

3.2. Brief Description of Analyzed Cases

The boundaries/restrictions for selecting cases were the following: (1) public investments; (2) in the area of Primorje Gorski Kotar and Istria County (Republic Croatia); (3) contract value in the range of 15–25 mil. EUR; (4) contracted and being active from 2016 onwards; (5) executed by the same contractor. In order to collect data from selected cases, the data collection protocol is developed. The interview guide has been prepared and used as a tool during semi-structured interviews with the experts involved. Each case study was separately discussed with clients, contractors, and consultants to gather different points of view regarding the success of a particular project. As previously published works mainly adopt collecting data from various companies/institutions [61,88,108,112,115,116,127], we focused just on construction projects executed by the same contractor in a geographically limited locality with a more significant volume of work. In order to ensure the diversity of analyzed projects, this study purposefully examines public projects of different natures as well as under various supplier arrangements (single contractor vs. consortium). Therefore, it is possible to document how the approach to performance management may differ across the projects despite having an identical entity responsible for carrying out the construction works. The selected company represents the biggest contractor with 65 years of tradition in the surveyed area.

The second phase was conducted in detailed case studies and their analysis based on those above. Data derived from these case studies are robust in underpinning the case analysis, which consists of information for (1) Case A—multipurpose logistics center; (2) Case B—water utility infrastructure; (3) Case C—clinical hospital facility; (4) Case D—road infrastructure (highway); and (5) Case E—road infrastructure (state road). Each case was analyzed in detail (Section 4), resulting in writing an individual case report according to a defined protocol. In addition, once the case studies were analyzed, a dash–dotted-line feedback loop is given the possibility to update or redesign the approach if it is a situation where an important discovery occurs during the study of one of the individual cases. One of the essential aspects of Yin's case study research [121] is having a strict procedure or protocol that enables later investigators to arrive at the same findings and conclusions.

3.3. Sample and Data Collection Procedure

Once the theory had been developed, it was necessary to select cases and design a data collection protocol to conclude the "Define and Design Phase" as the first phase of the multiple case study protocol. As previously mentioned, five cases were selected based upon five restrictions and studied in detail. Knowing that the construction cost is one of the main criteria for decision making in the early stages of the construction process, and, therefore, their prediction is of interest to all project participants [29], the focus was placed on all time–cost related documentation related to selected projects throughout the projects' life-cycle. Therefore, to identify discrepancies between the estimated time–cost and the project's realized time–cost and avoid or minimize time–cost overruns, importance was placed on collecting planning data, i.e., contracted and realized values. Such was performed throughout the project documentation from the initiation and planning stage (main contracts), execution stage (monthly reports of planned and realized works, annexes, internal communication within the contractor team, and official communication between project stakeholders), and closure stage (records of handover of the facility).

In addition to the project documentation, semi-structured interviews were performed with all identified stakeholders to gather their POVs. Stakeholders of all selected projects were identified according to their connection to the projects and grouped as clients, contractors, and consultants. There were several experts in each group reflecting on the project performance. The overall list of interviewed stakeholders consisted of 65 people, i.e., experts involved in all phases of a particular project. In the end, 56 experts were involved (8 clients, 28 contractors, and 20 consultants; see Table 2) in the interviews, which represents a relatively high response rate (86%).

Table 2. Overview of the experts involved in this study.

Stakeholders	Case A	Case B	Case C	Case D	Case E
Clients	2	2	1	2	1
Contractors	5	6	5	7	5
Consultants	3	4	3	4	6

During the second part of the interviews, an additional 13-question questionnaire was given to each stakeholder group so they could reflect upon their project with a Likert scale of 1–7. In addition to scoring the statements, there was a conversation with the participants about the problems and challenges of the project. The gathered attitudes of clients, contractors, and consultants served for fine-tuning of projects' success but also served as insights into the dynamic of a particular project and problems that occur. All gathered information has been normalized to each stakeholder group to have comparable insight into stakeholders' POV throughout the particular projects and across other projects. Based on those mentioned above, the second phase of the multiple case study protocol was executed to achieve this study's second and third objectives.

4. Results and Discussion

The following results are presented according to the defined protocol. As the projects' data are bulky, herein are only presented the necessary ones to validate the proposed performance measurement framework.

4.1. Conducting Multiple Case Study Analyses

To perform the "Prepare, Collects, and Analyze Phase" of multiple case study protocol, all collected data were systematically analyzed for a particular case, focusing on its execution stage, and presented below. During interviews, stakeholders were asked to reflect upon defined PAs and give their POV regarding the case project performance by assigning "+" (i.e., green) to those that have been taken into account to manage project performance successfully, with "±" (i.e., yellow) to those that have been partially taken, and with "-" (i.e., red) to those that have not been taken into account.

4.1.1. Case A—Multipurpose Logistics Center

The Case A project is a public investment of 15.2 million EUR for the construction of a multipurpose logistics center. The project commenced in August 2016, with a planned completion date of November 2017. While the contractor claimed that the project was completed within the contractual deadline, which was regulated by annexes on several occasions, the project was finished in February 2019 (time over-run approx. 94%). The construction costs were also 88% higher compared to the original contract.

The major challenge encountered in the project was the bankruptcy of one of the bidders in the contractor consortium during the first half of the planned project duration. This led to the remaining construction works being divided between the two contractor companies, resulting in additional tension between them. Another challenge faced by the project was financing. The client faced difficulties in continuing to finance the project, which necessitated significant alterations to the project scope to achieve most project goals and outcomes. Several unforeseen works and significant shortcomings were encountered during the project. Some works were not designed at all, the correct fitting into the existing condition was not foreseen, and there were shortcomings in the design and planning stages. During the construction works, geological problems were discovered that were not documented during the design stage, leading to several months of delays.

Despite these difficulties, all stakeholders were satisfied with the completion of the project. However, it is important to note that better planning, communication, and coordination could have prevented or minimized many of the issues encountered during the project.

4.1.2. Case B—Water Utility Infrastructure

Case B is a public investment of 18.7 million EUR for the construction of new water supply and sewerage systems and rehabilitation of the existing ones. The project started in November 2017 and was scheduled to be completed in April 2020. However, the project was completed in March 2021 (time over-run approx. 37%), with a 5% increase in construction costs compared to the original contract. Although the increase in cost was only 5%, significant changes occurred during the execution phase of the project. The contracted cost was initially reduced by 30%, but with the addition of new infrastructure network, the contractor and the client agreed on a new contract cost, which was similar to the original.

During the construction works, historically valuable remains were discovered multiple times, which required conservation surveys as unforeseen works. This resulted in the interruption of the works and extension of the deadline for 11 months which was initially contracted. All stakeholders involved in the project expressed dissatisfaction with the contract documentation, and additional and unforeseen works arose frequently, requiring constant changes and refinements of project documentation. Poor communication and a bad atmosphere among clients, contractors, and consultants significantly affected the resolution of project problems. Moreover, the availability of materials was impaired by the COVID-19 pandemic, and there were significant changes in the prices of construction products and services during the project. The stakeholders concluded that the cohesion of project participants could have been higher, and some project participants were considered insufficiently expert for the positions they held. Overall, the project faced significant challenges, but despite these difficulties, it was completed within the contractual deadline, and all stakeholders were satisfied with the final outcome.

4.1.3. Case C—Clinical Hospital Facility

Case C is a public investment project worth 22.7 million EUR, aimed at constructing a clinical hospital facility. The project commenced in September 2019, and the anticipated completion date was December 2021. The construction was completed on time, and the costs were regulated with annexes on several occasions, resulting in an increase of 13% in construction costs compared to the original contract.

One of the primary issues encountered during the project was the fluctuation in construction product and service prices due to the COVID-19 pandemic. However, the contract does not provide for a sliding scale to accommodate such changes. Therefore, the contractor was not able to charge the difference or seek compensation for this unforeseen challenge.

During the interviews, the stakeholders revealed that the project documentation was frequently problematic, resulting in poor cost estimates due to the discrepancies between the contracted and actual quantities of work. The design and contracting of unnecessary items that were not consumed resulted in additional and unforeseen works, thereby increasing the project's cost. The challenge of the project was certainly the location and space constraints. Strong gusts of wind occasionally limited or stopped working on site. Since other facilities bounded the construction site, the spatial organization of works, ware–houses, and construction site communications was challenging. Despite the space constraints, a large number of workers, and the necessary performance methods, monitoring occupational safety regulations was also highlighted as very demanding.

4.1.4. Case D—Road Infrastructure (Highway)

The project, Case D—road infrastructure (highway), is a public investment of 19.5 million EUR. Construction started in April 2019, with a planned completion date in March 2021. From the contractor's POV, the project was completed within the contractual deadline that was regulated by annexes on several occasions. From a time perspective, the project was completed in August 2021 (time over-run approx. 22%), with construction costs 19% higher compared to the original contract.

The client has signed a contract with a consortium of two companies. The interviewed representatives of the contractors were only from one company of the consortium (7 examinees). They mentioned that communication within the consortium was a big problem throughout the whole construction phase. As the other company did not follow the dynamic plans satisfactorily, the client requested a redistribution of works. The problem was in their productivity, or better to say, stress in productivity, as they bear equal responsibilities and high penalties for non-compliance towards the client. Since the time frame remained the same, the contractor found himself in a situation wherein such a short period, they had to do additional major construction work (e.g., contracted work from consortium partner) to meet the agreed percentage of work to be done, and face the potential loss of income. The examinees from the contractor group highlighted these challenges and assigned them to a "turnkey" type of contract. Several survey participants from various stakeholder groups complained about the incompetence of the other participants and the inadequate atmosphere among the teams of clients, contractors, and consultants. Also, all stakeholders highlighted that the commitment and demands of the client in monitoring safety at work and the quality of work were significantly higher than usual.

4.1.5. Case E—Road Infrastructure (State Road)

The project, Case E—road infrastructure (state road), is a public investment of 14.95 million EUR. Construction started in October 2020, with a planned completion date in March 2023. From the contractor's POV, the project will be completed with several annexes as they build their assumption based on already made changes and signed annexes.

At the very beginning of the execution phase, historically valuable remains were discovered on the construction site, which slowed down the works according to the baseplan for at least two months. Also, some additional and unforeseen works occur that comprise geomorphological characteristics of the terrain and the need for updating the initial designs. Additionally, the project documentation did not correspond to the actual situation on-site in several places, so among others, in the position of the future road viaduct, the existing buildings still existed (the private house that needed to be demolished) thus becoming part of the works needed to be done but was not part of the project. This and similar problems resulted in additional time overruns. Interviewed stakeholders also commented on the state of the project in which they are currently engaged. They agreed that communication in the project has been solid, so far, but that cohesion between teams of clients, contractors, and consultants is only partially satisfactory. They highlighted that the problems are solved extremely slowly and that something needs to be changed in this regard as soon as possible to meet project deadlines. The project's performance and potential overruns will need to be calculated once the project is finished, but, at the moment, it is evident that they will be present in both time and cost.

4.2. Cross-Case Analysis and Discussion

The "Conclude phase" of the protocol starts with a cross-case conclusion and modifying the theory, followed by implications. By analyzing the data collected from each case and performing analysis throughout the protocol, it is possible to conclude that the time over-run was on average 38%, and the cost over-run on average 32% for all finished projects. As the set theory is found adequate, the changes that occur during the analysis can be seen as driving fine-tuning factors. Although most of the interviewed experts answered that the projects were completed successfully in the end, the problems that occurred during the projects can be seen in more detail through the conducted case study analysis. Table 3 gives an overview of the time and cost overruns of five cases based on their detailed project documentation, contracts, drafts, S-curves, etc. On the other hand, the conducted analysis gives insight into each project throughout the time of execution, and the drivers for the overruns are mostly paced in unforeseen works and, sometimes, in additional work as a result of clients' changes.

Table 3. Overview of cases regarding their time and cost overruns.

Project	Status of Completion	Time Over-Run	Cost Over-Run
Case A	Completed	94%	88%
Case B	Completed	37%	5%
Case C	Completed	0%	13%
Case D	Completed	22%	19%
Case E	At the very end	N/A	N/A

Although the aforementioned time–cost overruns, calculated by MacDonald's equation [128], give insight into the past performance of the projects it is interesting to overlap information from each previously analyzed case (Section 4.1). Overlaying data from stakeholders' POV of defined PAs regarding projects' successful performance (Table 4) with time–cost overruns (Table 3) provides an additional level of information, not only how stakeholders see PAs to measure for project performance, but also how the performance measurement and project performance is interdependent.

From the data presented in Table 4, it is possible to draw up several interesting insights. For instance, for Case A, even though stakeholders mostly took into consideration "cost" as a vital PA, similarly to "time" and "quality" (i.e., iron triangle), actual results were achieved in terms of cost and time overruns are not satisfactory at all. Therefore, the cause of the problems must be hidden elsewhere, that is, in another PA and their interferences. In Case A, we attribute the occurrence cost and time overruns predominantly to the "team satisfaction" PA, as the realization has been planned in a consortium of contractors. Thus, the structure of stakeholders was even more complex than usual, and, therefore, neglecting team satisfaction has a high potential to negatively impact the project's success. On the other hand, for Case C, resulting time and cost overruns provide relatively satisfactory outputs despite omitting the "cost" PA. Notwithstanding, stakeholders involved emphasized both "team" and "client satisfaction" PAs which contributed positively to the completion of the project close to its cost constraints and on time.

Table 4. Created pattern for the future prediction of project success.

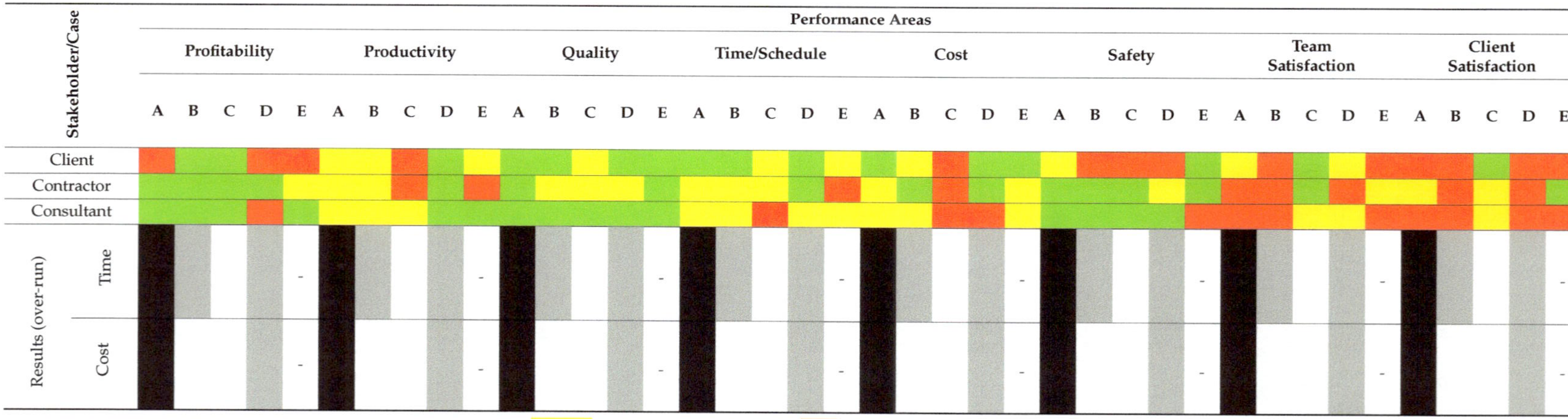

Legend: green taken into account, yellow partially taken into account, red not taken into account; black color—time–cost over-run is over 75%, dark grey—time–cost over-run is between 50–75%, grey color—time–cost over-run is between 25–50%, light grey color —time–cost over-run is between 1–25%, white color—there is no time–cost over-run,—N/A.

These findings highlight that it is not desirable to limit performance measurement to selected (e.g., well-measurable) PAs and that a broader scope of performance management is necessary due to their interconnectedness. Suppose these findings are confronted with data presented in Table 1. In that case, one may suggest that preferred PAs of "quality", "time/schedule", "cost", and "safety" are more frequently addressed in the available literature, should be supplemented with other areas taking into account also the surrounding factors affecting the project success. Presented findings, therefore, reflect on the complexity of construction projects and the importance of a multi-stakeholder environment. Based on the analyzed cases, it can be concluded that the participants were moderately satisfied with the communication on the project. In general, they considered that solving problems was slow, inefficient, and unsuccessful. In addition to the professional knowledge that is necessary, and seldom highlighted by various stakeholders, co-operation between clients, contractors, and consultants is extremely important. If co-operation is improved and better business relations are established, it is expected that the problems of the construction site will be solved easier and faster. For sure, one possible direction in order to overcome the aforementioned limitation of the qualitative approach is to perform detailed quantitative assessment focusing on, among other things, correlation between the performance areas.

Participants in the analyzed cases were moderately satisfied with project communication, but considered problem solving to be slow, inefficient, and unsuccessful. Improved co-operation between clients, contractors, and consultants was identified as a key factor in resolving construction site issues more easily and quickly. Participants were also generally dissatisfied with project documentation, leading to extensions of deadlines, discussions, and financial claims from contractors. The quantities of work foreseen by the project were seen as too large and not in compliance with the foreseen deadlines, which was compounded by documentation issues. The impact of the COVID-19 pandemic was noted in terms of availability and delivery of construction products, and changes in their prices. Finally, all 56 stakeholders involved in the interviews ranked the eight PAs, and the results are shown in Table 5.

Table 5. Ranking of performance areas based on stakeholders' POV.

Rank	Clients	Contractors	Consultants
1	quality	profitability	quality
2	productivity	safety	safety
3	time/schedule	quality	profitability
4	team satisfaction	cost	cost
5	cost	client satisfaction	productivity
6	safety	productivity	time/schedule
7	client satisfaction	team satisfaction	team satisfaction
8	profitability	time/schedule	client satisfaction

Contractors and consultants share similar opinions on project success factors, with the top four being the same, due to their expertise and responsibility for the project's performance. However, their opinions can differ based on their specific business goals. Surprisingly, "team satisfaction" is not ranked highly, despite participants stating its importance in interviews. Clients prioritize the final product's quality, longevity, and timely delivery, rather than team satisfaction. The participants' perspectives on project success extend beyond the traditional iron triangle model but not excessively so.

5. Conclusions

5.1. Theoretical and Practical Implications

The goal of this study was to develop a performance management framework that can be used for predicting project outcomes and facilitating advanced management. The proposed framework utilizes pattern creation through multiple case study to forecast future project success. While the framework has been adopted by academics and practitioners in the AEC industry, the results from the multiple case study have provided additional insights into different stakeholders' views on project performance and management. The qualitative approach used in this study has exposed gaps between stakeholders' expectations and realities in managing performance in construction projects.

As a key theoretical contribution to the extant project management literature, the current research demonstrates the importance of multi-stakeholders' POV on performance measurement and perception of the project success, even in a specific set of cases where the contractor is represented by the same company. The multiple case study protocol, combined with other scientific methods, has enabled the identification of performance areas and criteria for project performance, as well as the development of a performance measurement model that offers a conceptual framework for linking performance measurement and prediction of project performance. This study's significant theoretical contribution is demonstrating the importance of multi-stakeholder perspectives on performance measurement and project success, even when a single company represents the contractor. Additionally, this research provides new insights into using patterns for predicting future performance in construction projects, enriching understanding of the performance management challenges associated with a project's complexity and uniqueness.

5.2. Limitation and Future Research

The findings of this study can benefit project management practices by proposing an innovative managerial tool that enables the prediction of future project outcomes in their early stages. Combining well-known performance areas with the proposed performance measurement framework allows a multi-stakeholder environment to be adaptive and open to different viewpoints while having a consistent process in managing project performance. However, there are some limitations to this framework, such as the influence of stakeholders' expertise on the quality of the prediction pattern and the potential for inconsistencies in defining particular performance criteria. One of the most important limitations is that the stakeholders' expertise's greatly influence the quality of the prediction pattern. Therefore, it is important, especially in public investments, to build up the pattern on a large number of past projects in order to have better predictions. Also, as the whole framework is open to stakeholders to define particular performance criteria freely, they want to use on their projects, it could bring some inconsistencies to it. Such is solved by having rigidly defined performance areas that are based on previous knowledge. Therefore, it can be seen as a benefit to the stakeholders because they are not limited with what to particularly measure but at the same time have a clear structure of performance areas.

This conceptual framework also suggests avenues for future research, such as quantifying performance areas and criteria for project performance and creating a quantitative performance measurement model that can be linked to other prediction time–cost models that serve to predict project performance. Such research could be supported by intelligent Industry 4.0 tools such as AI and big data analytics. The emergence of this evolving conceptual framework calls for further interdisciplinary collaborations.

Author Contributions: Conceptualization, K.G., I.M. and T.H.; methodology, K.G., I.M. and T.H.; software, K.G. and I.M.; validation, K.G., I.M. and T.H.; formal analysis, K.G., I.M. and T.H.; investigation, K.G., I.M. and T.H.; resources, K.G. and I.M.; data curation, K.G. and I.M.; writing—original draft preparation, K.G., I.M. and T.H.; writing—review and editing, K.G., I.M. and T.H.; visualization, K.G., I.M. and T.H.; supervision, I.M. and T.H.; project administration, I.M.; funding acquisition, I.M. All authors have read and agreed to the published version of the manuscript.

Funding: This research received no external funding.

Institutional Review Board Statement: Not applicable.

Informed Consent Statement: Not applicable.

Data Availability Statement: All research data are available on request to authors.

Acknowledgments: The authors would like to thank all involved experts from private and public sector for their engagement during conducted interviews and workshops. This research has been fully supported by the University of Rijeka under the project number uniri-pr-tehnic-19-18 and research project Brno University of Technology no. FAST-S-23-8253—Life cycle cost analysis of construction objects.

Conflicts of Interest: The authors declare no conflict of interest.

References

1. Chen, H.L. Performance measurement and the prediction of capital project failure. *Int. J. Proj. Manag.* **2015**, *33*, 1393–1404.
2. Scott-Young, C.; Samson, D. Project success and project team management: Evidence from capital projects in the process industries. *J. Oper. Manag.* **2008**, *26*, 749–766.
3. Chen, H.L. Innovation stimulants, innovation capacity, and the performance of capital projects. *J. Bus. Econ. Manag.* **2014**, *15*, 212–223.
4. Tabassi, A.A.; Bakar, A.H.A. Training, motivation, and performance: The case of human resource management in construction projects in Mashhad, Iran. *Int. J. Proj. Manag.* **2009**, *27*, 471–480.
5. El-Sayegh, S.M. Risk assessment and allocation in the UAE construction industry. *Int. J. Proj. Manag.* **2008**, *26*, 431–438. [CrossRef]
6. Turner, R.; Zolin, R. Forecasting success on large projects: Developing reliable scales o predict multiple perspectives by multiple stakeholders over multiple time frames. *Proj. Manag. J.* **2012**, *43*, 87–99. [CrossRef]
7. Montenegro, A.; Dobrota, M.; Todorović, M.; Slavinski, T.; Obradović, V. Impact of construction project managers' emotional intelligence on project success. *Sustainability* **2021**, *13*, 10804. [CrossRef]
8. Korhonen, T.; Jaaskelainen, A.; Laine, T.; Saukkonen, N. How performance measurement can support achieving success in project-based operations. *Int. J. Proj. Manag.* **2023**, *41*, 102429. [CrossRef]
9. Bukoye, O.T.; Ejohwomu, O.; Roehrich, J.; Too, J. Using nudges to realize project performance management. *Int. J. Proj. Manag.* **2022**, *40*, 886–905.
10. Pavez, I.; Gomez, H.; Liu, C.; Gonzalez, V.A. Measuring project team performance: A review and conceptualization. *Int. J. Proj. Manag.* **2022**, *40*, 951–971.
11. Blais, C.; St-Pierre, J.; Bergeron, H. Performance measurement in new product development projects: Findings from successful small and medium enterprises. *Int. J. Proj. Manag.* **2023**, *41*, 102451. [CrossRef]
12. Freeman, R. *Strategic Management: A Stakeholder Approach*; Pitman: Boston, MA, USA, 1984.
13. Mitchell, R.K.; Agle, B.R.; Wood, D.J. Toward a Theory of Stakeholder Identification and Salience: Defining the Principle of Who and What Really Counts. *Acad. Manag. Rev.* **1997**, *22*, 853. [CrossRef]
14. Ward, S.; Chapman, C. Stakeholders and uncertainty management in projects. *Constr. Manag. Econ.* **2008**, *26*, 563–577. [CrossRef]
15. Behzadian, M.; Kazemzadeh, R.B.; Albadvi, A.; Aghdasi, M. PROMETHEE: A comprehensive literature review on methodologies and applications. *Eur. J. Oper. Res.* **2010**, *200*, 198–215.
16. Darko, A.; Chan, A.P.C.; Ameyaw, E.E.; Owusu, E.K.; Parn, E.; Edwards, D.J. Review of application of analytic hierarchy process (AHP) in construction. *Int. J. Constr. Manag.* **2019**, *19*, 436–452. [CrossRef]
17. Marović, I.; Tijanić, K.; Šopić, M.; Car-Pušić, D. Group decision-making in civil engineering based on AHP and PROMETHEE methods. *Sci. Rev. Eng. Environ. Sci.* **2020**, *29*, 474–484. [CrossRef]
18. Mladenovic, G.; Vajdic, N.; Wündsch, B.; Temeljotov Salaj, A. Use of key performance indicators for PPP transport projects to meet stakeholders' performance objectives. *Built Environ. Proj. Asset Manag.* **2013**, *3*, 228–249. [CrossRef]
19. Serra, C.E.M.; Kunc, M. Benefits realization management and its influence on project success and on the execution of business strategies. *Int. J. Proj. Manag.* **2015**, *33*, 53–66. [CrossRef]
20. Williams, T.; Vo, H.; Bourne, M.; Bourne, P.; Cooke-Davies, T.; Kirkham, R.; Masterton, G.; Quattrone, P.; Valette, J. A cross-national comparison of public project benefits management practices–the effectiveness of benefits management frameworks in application. *Prod. Plan. Control.* **2020**, *31*, 644–659. [CrossRef]
21. Hughes, D.L.; Dwivedi, Y.K.; Rana, N.P. Mapping is failure factors on PRINCE2 stages: An application of interpretive ranking process (IRP). *Prod. Plan. Control.* **2017**, *28*, 776–790. [CrossRef]
22. Flyvbjerg, B. From Nobel prize to project management: Getting risks right. *Proj. Manag. J.* **2006**, *37*, 5–15. [CrossRef]
23. PMI (Project Management Institute). Success Rates Rise–Transforming the High Cost of Low Performance. Newton Square, PA: Pulse of the Profession. 2017. Available online: https://www.pmi.org/-/media/pmi/documents/public/pdf/learning/thought-leadesrhip/pulse/pulse-of-the-profession-2017.pdf (accessed on 23 September 2021).
24. Ika, L.A. Project success as a topic in project management journals. *Proj. Manag. J.* **2009**, *40*, 6–19. [CrossRef]
25. Davis, K. Different stakeholder groups and their perceptions of project success. *Int. J. Proj. Manag.* **2014**, *32*, 189–201. [CrossRef]
26. McLeod, L.; Doolin, B.; MacDonell, S.G. A perspective-based understanding of project success. *Proj. Manag. J.* **2012**, *43*, 68–86. [CrossRef]
27. Williams, T. Identifying success factors in construction projects: A case study. *Proj. Manag. J.* **2016**, *47*, 97–112. [CrossRef]
28. Yang, R.Y.; Jayasuriya, S.; Gunarathna, C.; Arashpour, M.; Xue, X.; Zhang, G. The evolution of stakeholder management practices in Australian mega construction projects. *Eng. Constr. Archit. Manag.* **2018**, *25*, 690–706. [CrossRef]

29. Ambrule, V.R.; Bhirud, A.N. Use of artificial neural network for pre design cost estimation of building projects. *Int. J. Recent Innov. Trends Comput. Commun.* **2017**, *5*, 173–176.
30. Galjanić, K.; Marović, I.; Jajac, N. Decision support systems for managing construction projects: A scientific evolution analysis. *Sustainability* **2022**, *14*, 4977. [CrossRef]
31. Peško, I.; Trivunić, M.; Cirović, G.; Mučenski, V. A preliminary estimate of time and cost in urban road construction using neural networks. *Tech. Gaz.* **2013**, *20*, 563–570.
32. Al-Zwainy, F.M.; Aidan, I.A.A. Forecasting the cost of structure of infrastructure projects utilizing artificial neural network model (highway projects as case study). *Indian J. Sci. Technol.* **2017**, *10*, 1–12. [CrossRef]
33. Car-Pušić, D.; Tijanić, K.; Marović, I.; Mlađen, M. Predicting buildings construction cost overruns on the basis of cost overruns structure. *Sci. Rev. Eng. Environ. Sci.* **2020**, *29*, 366–376. [CrossRef]
34. Mrak, I.; Ambruš, D.; Marović, I. A Holistic Approach to Strategic Sustainable Development of Urban Voids as Historic Urban Landscapes from the Perspective of Urban Resilience. *Buildings* **2022**, *12*, 1852. [CrossRef]
35. Marović, I. Possible Applications of Neural Networks in Managing Urban Road Networks. In *Current Topics and Trends on Durability of Building Materials and Components*; Serrat, C., Casas, J.R., Gibert, V., Eds.; International Center for Numerical Methods in Engineering (CIMNE): Barcelona, Spain, 2020. Available online: https://www.scipedia.com/public/Marovic_2020a (accessed on 17 October 2022).
36. Biolek, V.; Hanak, T.; Marović, I. Data flow in relation to life-cycle costing of construction projects in the Czech Republic. *IOP Conf. Ser. Mater. Sci. Eng.* **2017**, *245*, 072032. [CrossRef]
37. Mandičák, T.; Spišáková, M.; Mésároš, P.; Kozlovská, M. Design of Economic Sustainability Supported by Enterprise Resource Planning Systems in Architecture, Engineering, and Construction. *Buildings* **2022**, *12*, 2241. [CrossRef]
38. Marović, I.; Hanak, T.; Plaum, S. Performance management in civil engineering: A systematic literature review. *Adv. Civ. Archit. Eng.* **2022**, *13*, 47–58. [CrossRef]
39. Atkinson, R. Project management: Cost, time and quality, two best guesses and a phenomenon, its time to accept other success criteria. *Int. J. Proj. Manag.* **1999**, *17*, 337–342. [CrossRef]
40. de Wit, A. Measurement of project success. *Int. J. Proj. Manag.* **1988**, *6*, 164–170. [CrossRef]
41. Toor, S.-R.; Ogunlana, S.O. Beyond the "iron triangle": Stakeholder perception of key performance indicators (KPIs) for large-scale public sector development projects. *Int. J. Proj. Manag.* **2010**, *28*, 228–236. [CrossRef]
42. Bjorvatn, T.; Wald, A. Project complexity and team-level absorptive capacity as drivers of project management performance. *Int. J. Proj. Manag.* **2018**, *36*, 876–888. [CrossRef]
43. Radujković, M.; Sjekavica Klepo, M.; Bosch-Rekveldt, M. Breakdown of Engineering Projects' Success Criteria. *J. Constr. Eng. Manag.* **2021**, *147*, 04021144. [CrossRef]
44. Burke, C.M.; Morley, M.J. On temporary organizations: A review, synthesis and research agenda. *Hum. Relat.* **2016**, *69*, 1235–1258. [CrossRef]
45. Vidal, L.; Marle, F. Understanding project complexity: Implications on project management. *Kybernetes* **2008**, *37*, 1094–1110. [CrossRef]
46. Eriksson, P.E.; Larsson, J.; Pesämaa, O. Managing complex projects in the infrastructure sector—A structural equation model for flexibility-focused project management. *Int. J. Proj. Manag.* **2017**, *35*, 1512–1523. [CrossRef]
47. Bryde, D.J.; Robinson, L. Client versus contractor perspectives on project success criteria. *Int. J. Proj. Manag.* **2005**, *23*, 622–629. [CrossRef]
48. Yildiz, K.; Ahi, M.T. Innovative decision support model for construction supply chain performance management. *Prod. Plan. Control.* **2020**, *33*, 894–906. [CrossRef]
49. Patrucco, A.S.; Moretto, A.; Knight, L. Does relationship control hinder relationship commitment? The role of supplier performance measurement systems in construction infrastructure projects. *Int. J. Prod. Econ.* **2021**, *233*, 108000. [CrossRef]
50. Mandičák, T.; Mésároš, P.; Kanáliková, A.; Špak, M. Supply Chain Management and Big Data Concept Effects on Economic Sustainability of Building Design and Project Planning. *Appl. Sci.* **2021**, *11*, 11512. [CrossRef]
51. Floricel, S.; Miller, R. Strategizing for anticipated risks and turbulence in large-scale engineering projects. *Int. J. Proj. Manag.* **2001**, *19*, 445–455. [CrossRef]
52. Kim, S.G. Risk performance indexes and measurement systems for mega construction projects. *J. Civ. Eng. Manag.* **2010**, *16*, 586–594. [CrossRef]
53. Cerić, A. Minimizing communication risk in construction: A Delphi study of the key role of project managers. *J. Civ. Eng. Manag.* **2014**, *20*, 829–838. [CrossRef]
54. Sanni-Anibire, M.O.; Mahmoud, A.S.; Hassanain, M.A.; Salami, B.A. A risk assessment approach for enhancing construction safety performance. *Saf. Sci.* **2019**, *121*, 15–29. [CrossRef]
55. Cheng, E.W.L.; Ryan, N.; Kelly, S. Exploring the perceived influence of safety management practices on project performance in the construction industry. *Saf. Sci.* **2012**, *50*, 363–369. [CrossRef]
56. Winge, S.; Albrechtsen, E.; Arnesen, J. A comparative analysis of safety management and safety performance in twelve construction projects. *J. Saf. Res.* **2019**, *71*, 139–152. [CrossRef] [PubMed]
57. Chang, R.-D.; Zuo, J.; Soebarto, V.; Zhao, Z.-Y.; Zillante, G.; Gan, X. Discovering the Transition Pathways toward Sustainability for Construction Enterprises: Importance-Performance Analysis. *J. Constr. Eng. Manag.* **2017**, *143*, 04017013. [CrossRef]

58. Sertyesilisik, B. A preliminary study on the regenerative construction project management concept for enhancing sustainability performance of the construction industry. *Int. J. Constr. Manag.* **2017**, *17*, 293–309. [CrossRef]
59. Danneels, E. The dynamics of product innovation and firm competences. *Strateg. Manag. J.* **2002**, *23*, 1095–1121. [CrossRef]
60. Adam, A.; Josephson, P.-E.B.; Lindahl, G. Aggregation of factors causing cost overruns and time delays in large public construction projects. *Eng. Constr. Archit. Manag.* **2017**, *24*, 393–406. [CrossRef]
61. Cha, H.S.; Kim, C.K. Quantitative approach for project performance measurement on building construction in South Korea. *KSCE J. Civ. Eng.* **2011**, *15*, 1319–1328. [CrossRef]
62. Kagioglou, M.; Cooper, R.; Aouad, G. Performance management in construction: A conceptual framework. *Constr. Manag. Econ.* **2001**, *19*, 85–95. [CrossRef]
63. Radujković, M.; Vukomanović, M.; Burcar Dunović, I. Application of Key Performance Indicators in South-Eastern European construction. *J. Civ. Eng. Manag.* **2010**, *16*, 521–530. [CrossRef]
64. Petruševa, S.; Žileska-Pancovska, V.; Car-Pušić, D. Implementation of process-based and data-driven models for early prediction of construction time. *Adv. Civ. Eng.* **2019**, *2019*, 7405863. [CrossRef]
65. Tijanić, K.; Car-Pušić, D.; Čulo, K. Impact of funding on cost-time aspects of public and social buildings. *Gradjevinar* **2019**, *71*, 21–32. [CrossRef]
66. Lin, G.; Shen, Q. Measuring the Performance of Value Management Studies in Construction: Critical Review. *J. Manag. Eng.* **2007**, *23*, 2–9. [CrossRef]
67. Bassioni, H.A.; Price, A.D.F.; Hassan, T.M. Performance Measurement in Construction. *J. Manag. Eng.* **2004**, *20*, 42–50. [CrossRef]
68. Bou-Llusar, J.C.; Escrig-Tena, A.B.; Roca-Puig, V.; Beltran-Martin, I. An empirical assessment of the EFQM Excellence Model: Evaluation as a TQM framework relative to the MBNQA Model. *J. Oper. Manag.* **2009**, *27*, 1–22. [CrossRef]
69. Vukomanović, M.; Radujković, M. The balanced scorecard and EFQM working together in a performance management framework in construction industry. *J. Civ. Eng. Manag.* **2013**, *19*, 683–695. [CrossRef]
70. Dwivedi, R.; Prasad, K.; Mandal, N.; Singh, S.; Vardhan, M.; Pamučar, D. Performance evaluation of an insurance company using an integrated Balanced Scorecard (BSC) and Best-Worst Method (BWM). *Decis. Mak. Appl. Manag. Eng.* **2021**, *4*, 33–50. [CrossRef]
71. Jonas, D.; Kock, A.; Gemünden, H.G. Predicting Project Portfolio Success by Measuring Management Quality—A Longitudinal Study. *IEEE Trans. Eng. Manag.* **2013**, *60*, 215–226. [CrossRef]
72. Leon, H.; Osman, H.; Georgy, M.; Elsaid, M. System dynamics approach for forecasting performance of construction projects. *J. Manag. Eng.* **2017**, *34*, 04017049. [CrossRef]
73. Tam, C.M.; Harris, F. Model for assessing building contractors' project performance. *Eng. Constr. Archit. Manag.* **1996**, *3*, 187–203.
74. Doloi, H. Cost overruns and failure in project management: Understanding the roles of key stakeholders in construction projects. *J. Constr. Eng. Manag.* **2013**, *139*, 267–279. [CrossRef]
75. Marović, I.; Perić, M.; Hanak, T. A multi-criteria decision support concept for selecting the optimal contractor. *Appl. Sci.* **2021**, *11*, 1660. [CrossRef]
76. Pamučar, D.; Macura, D.; Tavana, M.; Božanić, D.; Knežević, N. An integrated rough group multicriteria decision-making model for the ex-ante prioritization of infrastructure projects: The Serbian Railways case. *Socio-Econ. Plan. Sci.* **2022**, *79*, 101098. [CrossRef]
77. Yang, J.; Shen, G.Q.; Drew, D.S.; Ho, M. Critical success factors for stakeholder management: Construction practitioners' perspectives. *J. Constr. Eng. Manag.* **2010**, *136*, 778–786. [CrossRef]
78. Oppong, G.D.; Chan, A.P.C.; Dansoh, A. A review of stakeholder management performance attributes in construction projects. *Int. J. Proj. Manag.* **2017**, *35*, 1037–1051. [CrossRef]
79. Xue, H.; Zhang, S.; Su, Y.; Wu, Z.; Yang, R.Y. Effect of stakeholder collaborative management on off-site construction cost performance. *J. Clean. Prod.* **2018**, *184*, 490–502. [CrossRef]
80. Franco-Santos, M.; Otley, D. Reviewing and theorizing the unintended consequences of performance management systems. *Int. J. Manag. Rev.* **2018**, *20*, 696–730. [CrossRef]
81. Miller, D.M. Profitability = Productivity + Price Recovery. *Harv. Bus. Rev.* **1984**, *62*, 145–153.
82. Kim, Y.; Choi, T.Y. Deep, Sticky, Transient, and Gracious: An Expanded Buyer-Supplier Relationship Typology. *J. Supply Chain. Manag.* **2015**, *51*, 61–86. [CrossRef]
83. Cerić, A.; Vukomanović, M.; Ivić, I.; Kolarić, S. Trust in megaprojects: A comprehensive literature review of research trends. *Int. J. Proj. Manag.* **2021**, *39*, 325–338. [CrossRef]
84. Klaus-Rosińska, A.; Iwko, J. Stakeholder Management—One of the Clues of Sustainable Project Management—As an Underestimated Factor of Project Success in Small Construction Companies. *Sustainability* **2021**, *13*, 9877. [CrossRef]
85. Bernolak, I. Effective measurement and successful elements of company productivity: The basis of competitiveness and world prosperity. *Int. J. Prod. Econ.* **1997**, *52*, 203–213. [CrossRef]
86. Xia, N.; Zou, P.X.W.; Griffin, M.A.; Wang, X.; Zhong, R. Towards integrating construction risk management and stakeholder management: A systematic literature review and future research agendas. *Int. J. Proj. Manag.* **2018**, *36*, 701–715. [CrossRef]
87. Chan, A.P.C.; Scott, D.; Lam, E.W.M. Framework of Success Criteria for Design/Build Projects. *J. Manag. Eng.* **2002**, *18*, 120–128. [CrossRef]
88. Cox, R.F.; Issa, R.R.A.; Ahrens, D. Management's Perception of Key Performance Indicators for Construction. *J. Constr. Eng. Manag.* **2003**, *129*, 142–151. [CrossRef]

89. Chan, A.P.C.; Chan, A.P.L. Key performance indicators for measuring construction success. *Benchmarking Int. J.* **2004**, *11*, 203–221. [CrossRef]
90. Tangen, S. Demystifying productivity and performance. *Int. J. Product. Perform. Manag.* **2005**, *54*, 34–46. [CrossRef]
91. Dawood, N.; Sikka, S.; Marasini, R.; Dean, J. Development of key performance indicators to establish the benefits of 4D planning. In Proceedings of the 22nd Annual ARCOM Conference, Birmingham, UK, 4–6 September 2006.
92. Park, H.-S. Conceptual framework of construction productivity estimation. *KSCE J. Civ. Eng.* **2006**, *10*, 311–317. [CrossRef]
93. El- Mashaleh, M.; Minchin, R.E.; O'Brien, W.J. Management of Construction Firm Performance Using Benchmarking. *J. Manag. Eng.* **2007**, *23*, 10–17. [CrossRef]
94. Bottazzi, G.; Secchi, A.; Tamagni, F. Productivity, profitability and financial performance. *Ind. Corp. Change* **2008**, *17*, 711–751. [CrossRef]
95. Ling, F.Y.Y.; Low, S.P.; Wang, S.; Egbelakin, T. Models for Predicting Project Performance in China Using Project Management Practices Adopted by Foreign AEC Firms. *J. Constr. Eng. Manag.* **2008**, *134*, 983–990. [CrossRef]
96. Nasirzadeh, F.; Afshar, A.; Khanzadi, M. System dynamics approach for construction risk analysis. *Int. J. Civ. Eng.* **2008**, *6*, 120–131.
97. Rankin, J.; Fayek, A.R.; Meade, G.; Haas, C.; Mandeau, A. Initial metrics and pilot program results for measuring the performance of the Canadian construction industry. *Can. J. Civ. Eng.* **2008**, *35*, 894–907. [CrossRef]
98. Ling, F.Y.Y.; Low, S.P.; Wang, S.Q.; Lim, H.H. Key project management practices affecting Singaporean firms' project performance in China. *Int. J. Proj. Manag.* **2009**, *27*, 59–71. [CrossRef]
99. Nassar, N.K. An integrated framework for evaluation of performance of construction projects. In Proceedings of the PMI Global Congress 2009—North America, Orlando, FL, USA, 10–13 October 2009.
100. Skibniewski, M.J.; Ghosh, S. Determination of Key Performance Indicators with Enterprise Resource Planning Systems in Engineering Construction Firms. *J. Constr. Eng. Manag.* **2009**, *135*, 965–978. [CrossRef]
101. Ali, A.S.; Rahmat, I. The performance measurement of construction projects managed by ISO-certified contractors in Malaysia. *J. Retail. Leis. Prop.* **2010**, *9*, 25–35. [CrossRef]
102. Wang, Q.; El-Gafy, M.; Zha, J. Bi-Level Framework for Measuring Performance to Improve Productivity of Construction Enterprises. In Proceedings of the Construction Research Congress 2010: Innovation for Reshaping Construction Practice, Banff, Canada, 8–10 May 2010.
103. Pekuri, A.; Haapasalo, H.; Herrala, M. Productivity and performance management-Managerial practices in the construction industry. *Int. J. Perform. Meas.* **2011**, *1*, 39–58.
104. Rezaei, A.R.; Celik, T.; Baalousha, Y. Performance measurement in a quality management system. *Sci. Iran.* **2011**, *18*, 742–752. [CrossRef]
105. Chovichien, V.; Nguyen, T.A. List of indicators and criteria for evaluating construction project success and their weight assignment. In Proceedings of the 4th International Conference on Engineering, Project, and Production Management, Bangkok, Thailand, 23–25 October 2013; pp. 130–150. [CrossRef]
106. Wanberg, J.; Harper, C.; Hallowell, M.; Rajendran, S. Relationship between Construction Safety and Quality Performance. *J. Constr. Eng. Manag.* **2013**, *139*, 04013003. [CrossRef]
107. Xiao, H.; Proverbs, D. Factors influencing contractor performance: An international investigation. *Eng. Constr. Archit. Manag.* **2013**, *10*, 322–332. [CrossRef]
108. Auma, E. Factors Affecting the Performance of Construction Projects in Kenya: A Survey of Low-Rise Buildings in Nairobi Central Business District. *Int. J. Bus. Manag.* **2014**, *2*, 115–140.
109. Nassar, N.; AbouRizk, S. Practical Application for Integrated Performance Measurement of Construction Projects. *J. Manag. Eng.* **2014**, *30*, 04014027. [CrossRef]
110. Omar, M.N.; Fayek, A.R. Modeling and evaluating construction project competencies and their relationship to project performance. *Autom. Constr.* **2016**, *69*, 115–130. [CrossRef]
111. Silva, G.A.; Warnakulasuriya, B.N.F.; Arachchige, B. Criteria for Construction Project Success: A Literature Review. *SSRN Electron. J.* **2016**. [CrossRef]
112. Cha, H.S.; Kim, K.H. Measuring Project Performance in Consideration of Optimal Best Management Practices for Building Construction in South Korea. *KSCE J. Civ. Eng.* **2017**, *22*, 1614–1625. [CrossRef]
113. Molwus, J.J.; Erdogan, B.; Ogunlana, S. Using structural equation modelling (SEM) to understand the relationships among critical success factors (CSFs) for stakeholder management in construction. *Eng. Constr. Archit. Manag.* **2017**, *24*, 426–450. [CrossRef]
114. Wibowo, M.A.; Astana, I.N.Y.; Rusdi, H.A. Dynamic Modelling of the Relation between Bidding Strategy and Construction Project Performance. *Procedia Eng.* **2017**, *171*, 341–347. [CrossRef]
115. Demirkesen, S.; Ozorhon, B. Impact of integration management on construction project management performance. *Int. J. Proj. Manag.* **2017**, *35*, 1639–1654. [CrossRef]
116. Demirkesen, S.; Ozorhon, B. Measuring Project Management Performance: Case of Construction Industry. *Eng. Manag. J.* **2017**, *29*, 258–277. [CrossRef]
117. Tripathi, K.K.; Jha, K.N. An Empirical Study on Performance Measurement Factors for Construction Organizations. *KSCE J. Civ. Eng.* **2018**, *22*, 1052–1066. [CrossRef]

118. Keenan, M.; Rostami, A. The impact of quality management systems on construction performance in the North West of England. *Int. J. Constr. Manag.* **2019**, *21*, 1–13. [CrossRef]
119. Moradi, S.; Ansari, R.; Taherkhani, R. A Systematic Analysis of Construction Performance Management: Key Performance Indicators from 2000 to 2020. *Iran. J. Sci. Technol. Trans. Civ. Eng.* **2021**, *46*, 15–31. [CrossRef]
120. Marović, I.; Završki, I.; Jajac, N. Ranking zones model—A multicriterial approach to the spatial management of urban areas. *Croat. Oper. Res. Rev.* **2015**, *6*, 91–103. [CrossRef]
121. Yin, R.K. *Case Study Research and Applications: Design and Methods*; SAGE Publications: Thousand Oaks, CA, USA, 2018.
122. Grant, M.J.; Booth, A. A typology of reviews: An analysis of 14 review types and associated methodologies. *Health Inf. Libr. J.* **2009**, *26*, 91–108. [CrossRef] [PubMed]
123. Seawright, J.; Gerring, J. Case selection techniques in case study research: A menu of qualitative and quantitative options. *Political Res. Q.* **2008**, *61*, 294–308. [CrossRef]
124. Pollack, J.; Adler, D. Emergent trends and passing fads in project management research: A scientometric analysis of changes in the field. *Int. J. Proj. Manag.* **2015**, *33*, 236–248. [CrossRef]
125. Norouzi, M.; Chàfer, M.; Cabeza, L.F.; Jiménez, L.; Boer, B. Circular economy in the building and construction sector: A scientific evolution analysis. *J. Build. Eng.* **2021**, *44*, 102704. [CrossRef]
126. Pickering, C.; Byrne, J. The benefits of publishing systematic quantitative literature reviews for PhD candidates and other early-career researchers. *High. Educ. Res. Dev.* **2014**, *33*, 534–548. [CrossRef]
127. Hanak, T.; Marović, I. Performance management in Czech construction: Public investors' perspective. *Teh. Glas.* **2022**, *16*, 113–120. [CrossRef]
128. MacDonald, M. *Review of Large Public Procurement in the UK*; HM Treasury: London, UK, 2002.

sustainability

MDPI

Article

A Roof Refurbishment Strategy to Improve the Sustainability of Building Stock: A Case Study

María J. Ruá *, Ángel M. Pitarch, Inés Arín and Lucía Reig

Mechanical Engineering and Construction Department, Universitat Jaume I, 12071 Castellón de la Plana, Spain; pitarcha@uji.es (Á.M.P.); arin@uji.es (I.A.); lreig@uji.es (L.R.)
* Correspondence: rua@uji.es

Abstract: The aging of the building stock in most cities highlights the relevance of refurbishment to achieve sustainability. Current refurbishment practices are often short-sighted and do not encompass holistic strategies beyond energy saving. This research study aims to analyze the factors involved in roof refurbishment versus current decision-making determinants. The objective is to identify the barriers that hinder their implementation and to find arguments to support roof renovations. A multicriteria analysis, which considered environmental, economic and performance factors, was employed to select optimal roof refurbishment solutions. This study evaluated five solutions. With interviews held with construction professionals and a survey of experts and homeowners, the preferences and criteria for making decisions about roof refurbishments were analyzed. Simulation tools were then used to estimate the energy savings, payback periods and environmental impact for a representative building in the study area. The results were extrapolated to a neighborhood level. The results highlight the importance of considering factors, such as weight, cost and user preferences when selecting suitable refurbishment solutions. The findings not only estimate the potential energy savings and carbon emission reductions in the area but also underscore the relevance of roof refurbishments for prolonging a building's life span to contribute to sustainability.

Keywords: urban sustainability; roof refurbishment; multicriteria analysis; simulation tools; energy performance

check for updates

Citation: Ruá, M.J.; Pitarch, Á.M.; Arín, I.; Reig, L. A Roof Refurbishment Strategy to Improve the Sustainability of Building Stock: A Case Study. *Sustainability* **2024**, *16*, 2028. https://doi.org/10.3390/su16052028

Academic Editors: Uroš Klanšek and Tomaž Žula

Received: 29 January 2024
Revised: 20 February 2024
Accepted: 22 February 2024
Published: 29 February 2024

1. Introduction

Energy efficiency in buildings has been a key objective in recent years because of its contribution to low-carbon economies. In 2018, buildings accounted for more than 40% of Europe's energy use. Buildings have been put forward as an emission reduction target alongside other sectors, such as transport, agriculture, waste and industry. As part of the European Green Deal strategy, European Union members agreed to increase the greenhouse gas emission reduction target for 2030 from 40% to 55% by means of the "Fit for 55" initiative. This commitment is set out in the Climate Law Regulation and is, thus, becoming a requisite. To reach the new target, the European Commission presented a package of legislative proposals to revise and update EU legislation on land use, taxation, transport and energy to ensure that the EU achieves climate neutrality by 2050. Regarding buildings, the Council and the Parliament reached a provisional political agreement on a proposal to revise the Energy Performance of Buildings Directive (EPBD) from 2010 [1]. The revised directive sets more ambitious energy performance requirements for new and renovated buildings in the EU, and promotes the building stock's renovation. The main objectives of this revision are that all new buildings should be zero-emission by 2030, and the existing building stock should be transformed into zero-emission buildings by 2050. Zero-emission buildings are defined in Article 2 as buildings with very high energy performance using a small amount of energy that is still needed and fully covered by renewable energies. These buildings will set a new standard for new constructions. Major renovations will need to reach this level as of 2030, and the entire stock must comply with it by 2050. Article 2 also clarifies that "nearly zero-energy buildings" remain the standard for new buildings until the application of the zero-emission building standard and will be the level to be met by profound renovations until 2030.

When looking at energy classes, according to the data provided by the European Union (EU), 51% of the existing residential building stock is in an energy class below Class D (Classes E, F and G when on the A–G scale), and only 3% of this stock is in energy Class A. Currently, the energy refurbishment rate of the stock in Europe is 1% per year, which is below the rate of 3% recommended by the European Commission [2].

In Spain, the situation is significantly worse because 81% of the building stock is in an energy class below Class D, and only 0.3% is in Class A. The National Energy and Climate Plan aims to increase this rate to 1.2% by 2030, with progressive increases over the years. This is because the refurbishment rate in Spain is 0.1%, which is still much lower than the European average of 1% [3], and is mainly due to the predominant property regime in Spain being home ownership, which makes it very difficult to renovate buildings where several owners with different situations and perspectives co-exist. This often makes it difficult to reach agreements to carry out renovation interventions. Many buildings in Spain are multifamily homes with flat roofs and are typically located in temperate climate zones. Flat roofs are systematically seen in most buildings that were built before 1979, when the first regulation on thermal conditions was approved in Spain.

Refurbishment interventions on roofs usually contribute poorly to overall energy performance because a roof accounts for a low percentage of the thermal envelope. So, it is quite possible that intervention on the roof will not yield the most cost-efficient results. However, as highlighted by Morgado et al. [4], beyond energy improvement, a proper roof is crucial to maintaining the whole building and prolonging its service life because these interventions can prevent diseases that derive from construction deterioration. The proposal for revising the EPBD focuses on reducing operational greenhouse gas emissions [1]. However, initial measures are being taken to address carbon emissions throughout a building's entire life cycle. This highlights the importance of roof refurbishment in enhancing a building's durability and overall performance. Furthermore, given the high cost of the investments needed to adapt buildings' energy performance to currently required standards, roof interventions would be aligned with the philosophy of the Electronic Building Book, whose purpose is to program buildings' partial renovations with realistic, affordable and adequate planning, rather than incurring the excessive cost inherent to comprehensive interventions [5,6]. The proposal to update the EPBD presents "staged renovation" as a solution to the high upfront costs, which may act as an obstacle when renovating "in one go". Moreover, such renovation needs to be thoroughly planned to avoid a situation in which one stage excludes the following stages. The Renovation Passport has been suggested as a voluntary tool for owners and investors, and it provides a roadmap for planned renovation. However, according to current conditions, interventions on roofs, which are not a very noticeable element of the thermal envelope, are not perceived as a priority, and maintenance works are carried out only when a disease or problem is detected. Additionally, the most economical intervention to solve the problem in the short term is almost always selected.

This research study focuses on the potential contribution of roof refurbishment as the most exposed part of the thermal envelope. This study was conducted on a multifamily building with a flat roof that was built before 1979 in a medium-sized Spanish Mediterranean city (Castellón de la Plana). This pattern is probably similar in many cities and municipalities in Southern Europe and North Africa [7]. The first stage of this work analyzed the most commonly used roof restoration solutions based on a multicriteria analysis (environmental, economic and performance) and selected the most appropriate ones. The second phase of this work aimed to quantify the area of different roof typologies and the potential improvement that renovation would entail in a neighborhood located on the coastline of Castellón de la Plana (East Spain), as presented in Figure 1. For this purpose, the energy savings of some roof rehabilitation solutions were estimated, and the theoretical savings for a statistically representative building of the area, with no insulation on the roof, were determined depending on the refurbishment roof solutions.

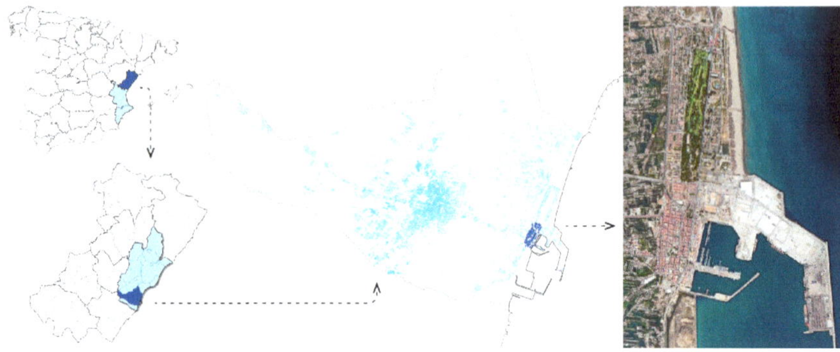

Figure 1. Location of the case study neighborhood: Grao de Castellón.

2. Materials and Methods

This research was conducted in two stages. Figure 2 summarizes the objectives, methods and main results for each stage in accordance with the way the paper is organized. This study focused on flat roofs without thermal insulation, typically used in cities with a temperate climate in Spain. Then, suitable construction solutions for flat roof refurbishment were analyzed by considering current regulatory standards. Five different refurbishment solutions were selected. Three were walkable roof solutions (inverted with raised paving, inverted with adhered paving and inverted with permeable paving (WINVR, WINVAD and WPER, respectively)); two were non-walkable (green roofs and roofs with gravel protection (NWGRE and NWGRA, respectively)).

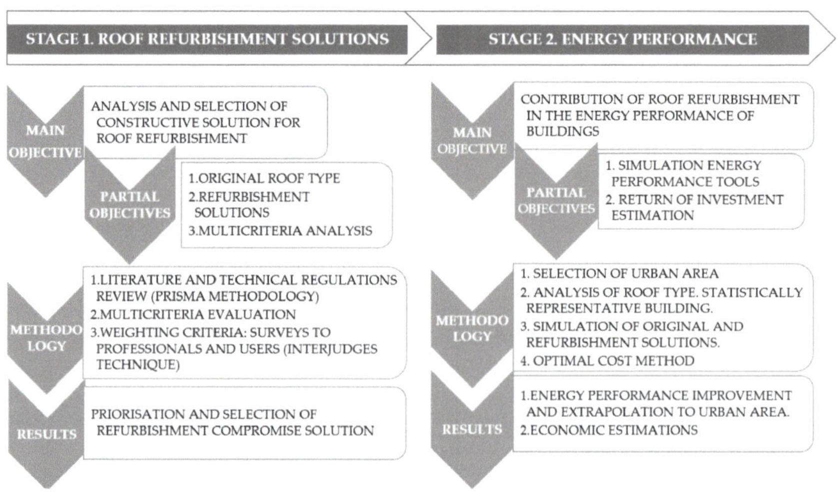

Figure 2. Study stages, followed methodologies and main findings.

The multicriteria analysis used to evaluate these five solutions was based on three criteria categories, namely, environmental (A), economic (E) and performance (P), with nine total evaluation indicators. To properly define the indicators to be considered, two population focus groups—one made up of experts in construction and refurbishment and the other of users/owners of buildings—were asked to answer a survey to analyze their preferences in the decision-making process for building refurbishment and to determine the weighted importance assigned to the different factors. The information provided by planners and builders on the renovation solutions commonly used in actual professional practice also allowed the practical and theoretical solutions to be compared.

The multicriteria analysis established an order of preference for flat roof refurbishment solutions, from the most suitable to the least suitable. Additionally, some suggestions were collected from the obtained results.

In the second study stage, three of the initially proposed solutions were further analyzed to estimate the energy performance of roofs. To do so, simulation values were obtained employing an energy certification tool that requires modeling a complete building. For this reason, a statistically representative building in the urban area that presented the typical non-insulated flat roof solution was selected. This case study selection was based on an exhaustive and detailed study about building typologies, roof types, roof surfaces, etc. This building was simulated with a tool officially approved by the Ministry of Energy certification, and the improvement made with the chosen refurbishment solutions was subsequently tested. The obtained results were used to estimate the potential refurbishment at the neighborhood level by extrapolating to the total area of this roof type. Finally, based on the acquired results, the payback periods for the considered rehabilitation interventions were estimated.

3. Stage 1 Results: Analysis of Refurbishment Solutions

3.1. Thermal Insulation of Roofs in the Current Building Stock

Roof types in Mediterranean cities can be classified according to their thermal insulation. In line with this, the first mandatory regulation in Spain with requirements for thermal conditions for buildings dates back to 1979: the Basic Housing Regulation on Thermal Conditions [8]. Therefore, in mild climate zones in Spain, buildings before that year present no insulated roofs at all. After the 1979 regulation, insulation was required, but it was not until 2006 when the Building Technical Code (CTE, in Spanish), updating previous and obsolete regulations in the building sector, established more ambitious energy restrictions. The part of the CTE that regulates the thermal conditions of buildings is called CTE-HE, and has been updated in the last few years and aligned with the European EPBD [9].

Regarding the building typology classification, the research work previously undertaken by the Valencian Government (East Spain) and the Valencian Building Institute IVE, which has been integrated into the European project Tabula, presented a catalog for the building typologies in different climates in Spain [10]. It characterized the thermal envelope depending on climate, building typology and year of construction in Spain. As described in this study, buildings' estimated thermal transmittance has progressively reduced over the years, which has allowed buildings to be organized into periods according to building regulations: "before 1939", "1937–1959", "1960–1979", "1980–2006" and "after 2006". These periods correspond to a reduction in their roof thermal transmittances from 3.08, 1.67, 1.61, 0.56 to 0.45 W/m^2K, respectively. The research undertaken by Braulio [11] characterized the building thermal envelope of building stock in the Mediterranean climate over a similar time interval. The results show that thermal envelopes can be characterized depending on year of construction due to regulations on thermal conditions at the time and the use of standard constructive systems.

3.2. Identification of Refurbishment Solutions

Multifamily buildings with flat roofs built before 1979, when the first thermal regulation on buildings came into force, were selected as the reference roof to be refurbished. Table 1 summarizes the main roof system layers used as a reference and the five proposed refurbishment solutions, as described in Section 2. These refurbishment systems have been analyzed in previous studies and the corresponding references are included in Table 1. As the first row shows, the reference system presents a roof with no type of thermal insulation.

Table 1. Flat roof type to be refurbished and selected roof refurbishment solutions.

Roof Type		Layers on the Roof	Refs.
Existent	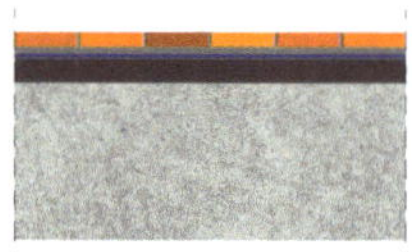	1. Ceramic tiles 2. Layer of mortar 3. Waterproofing membrane 4. Lightweight concrete (Slope formations)	[10,11]
NWGRE Green roof		1. Vegetation 2. Substrate 3. Filtering layer 4. Drainage layer 5. Protective layer 6. Anti-rooting layer 7. Waterproofing membrane (EPDM)	[12–16]
NWGRA Gravel protection	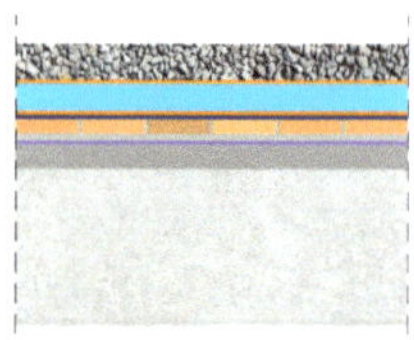	1. Gravel 2. Separation and drainage layer (geotextile) 3. Thermal insulation (XPS) 4. Separation and drainage layer (geotextile) 5. Waterproofing membrane (EPDM)	[17,18]
WINVR Raised paving	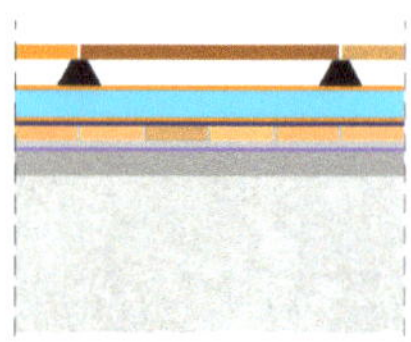	1. Ceramic tiles 2.Polypropylene plots 3. Separation and drainage layer (geotextile) 4. Thermal insulation (XPS) 5. Separation and drainage layer (geotextile) 6. Waterproofing membrane (EPDM)	[18,19]
WINVAD Adhered paving	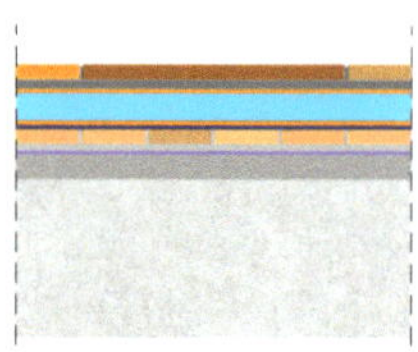	1. Ceramic tiles + Cementitious adhesive 2. Regularization mortar 3. Separation and drainage layer (geotextile) 4. Thermal insulation (XPS) 5. Separation and draining layer (geotextile)	[18]
WPER Permeable paving	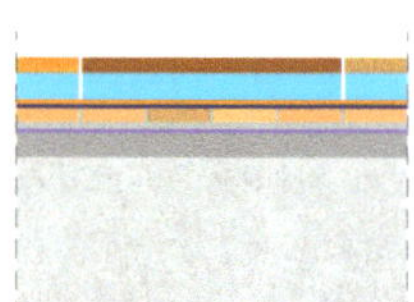	1. Ceramic tiles + thermal insulator 2. Separation and drainage layer (geotextile) 3. Waterproofing membrane (EPDM)	[20,21]

3.3. *Flat Roof Refurbishment Solutions*

3.3.1. Definition of Quantitative and Qualitative Criteria for Multicriteria Analysis

The suitability of different flat roof refurbishment solutions depends on several evaluation criteria. In some cases, they are decisive for selection purposes, such as the weight that the pre-existing building structure can support, but are of less relevance in other cases, such as the aesthetic aspect of the finish. Table 2 shows the indicators proposed in this multicriteria study. Each indicator was rated on a scale from 1 to 5, with 1 being the most unfavorable and 5 the most favorable. Some indicators were qualitatively evaluated based on their technical characteristics, advantages and disadvantages observed in practice, while others were quantitatively assessed by means of measurable variables, whose values were subsequently standardized on a scale from 1 to 5. As the three variables were inverse to the normalized scale (i.e., the heavier the weight, the worse the value), their normalized values were obtained using Equation (1).

$$VNi = 5 \times (Vmin/Vi), \tag{1}$$

where:

VNi is the normalized value of the indicator for constructive solution i;
Vi is the unnormalized value of solution i;
Vmin is the minimum value to be reached of the i solutions.

Table 2. Categories and indicators for the multicriteria analysis.

Category	Indicator	Type of Indicator: Criteria
A	A.1. Thermal insulation—energy savings	Quantitative: the normalized value of thermal transmittance (W/m^2K).
	A.2. Recovery—Recycling	Qualitative: the recovering, reusing and recycling potentials of the materials used in rehabilitation.
E	E.1. Initial investment cost	Quantitative: the normalized unit cost of executing refurbishment solutions (EUR/m^2).
	E.2. Maintenance (durability–cost–periodicity)	Qualitative: the durability and the need for maintenance operations in frequency and cost terms.
P	P.1. Ease of execution	Quantitative: the system's normalized weight (kN/m^2).
	P.2. Acoustic insulation	Qualitative: capacity to prevent roof leaks.
	P.3. Weight of the system	Qualitative: the roof's aesthetic value.
	P.4. Waterproofing—sealing	Quantitative: the system's normalized weight (kN/m^2).
	P.5. Aesthetic	Qualitative: capacity to prevent roof leaks.

Regarding the quantitative indicators, Figure 3 summarizes the thermal transmittance, weight and economic cost of the different analyzed refurbishment solutions. For thermal transmittance, the building located in Castellón de la Plana (climate zone B3 according to CTE, meaning a mild winter and a hot summer) was considered. For each refurbishment solution, the minimum thermal insulation thickness to comply with the limit transmittance value set by the CTE for roofs that come into contact with outside air (U_{LIM} = 0.44 W/m^2K) [10] was determined, and a commercially available thermal insulation thickness was assigned. As presented in Figure 3, all the refurbished systems' thermal transmittances fell within the 0.417 and 0.425 kW/m^2K range.

Besides determining the weight of each refurbishment solution, the expected overloads according to the regulations in force at the time of construction were also calculated [22,23]. The weight criterion indicated that the load-bearing capacity of the pre-existing roof permits an increase in overload of 0.5 kN/m^2. Consequently, the best refurbishment solution would be a roof with a gravel finish (NWGRA, 0.14 kN/m^2), and the worst would be a green roof (NWGREE, 2.12 kN/m^2), with the other solutions somewhere between these values. Despite green roofs' good energy performances [24,25], this solution was ruled out for rehabilitation due to its excessive weight. Although the solutions WINVR and WINVAD slightly exceeded the limitation, they were considered, but with some modifications to fulfill the weight requirement.

The cost of each refurbishment solution was quantified by taking the IVE price base [26] and some commercial solutions as references, and by adopting average values. The most economical solution was permeable paving at 58.31 EUR/m^2, and the most expensive one was the green roof at 92.89 EUR/m^2.

Figure 4 summarizes the adopted qualitative values (from 1 to 5, with the worst values in red and the best ones in green) according to the following criteria:

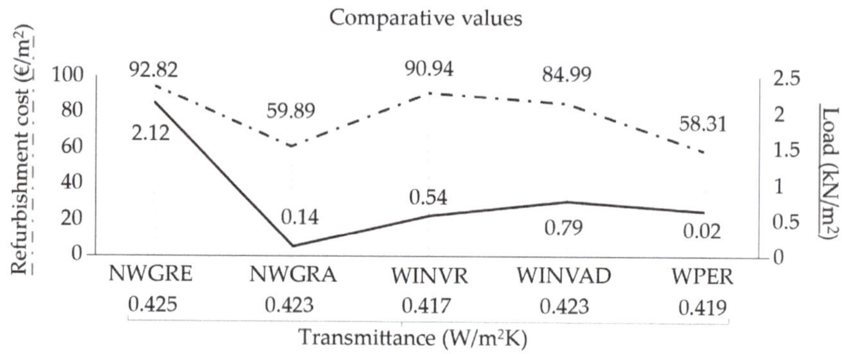

Figure 3. Thermal transmittance, load and refurbishment cost of the analyzed refurbishment solutions.

	WINVR	WINVAD	WPER	NWGRE	NWGRA
A.1. Thermal insulation	5.00	4.93	4.98	4.91	4.93
A.2. Recovery—recycling	5.00	1.00	5.00	4.00	5.00
E.1. Initial investment cost	3.21	3.43	5.00	3.14	4.87
E.2. Maintenance	5.00	3.00	5.00	3.00	5.00
P.1. Ease of execution	4.00	3.00	4.00	1.00	5.00
P.2. Acoustic insulation	4.00	4.00	4.00	5.00	4.00
P.3. Weight of the system	1.30	0.90	1.13	0.33	5.00
P.4. Waterproofing—sealing	5.00	5.00	5.00	5.00	5.00
P.5. Aesthetic	4.00	4.00	4.00	5.00	3.00
Mean values	**4.06**	**3.25**	**4.23**	**3.49**	**4.64**

Tone range: Worst 1 2 3 4 5 Best

Figure 4. Assessment of the qualitative criteria for every refurbishment solution and score obtained before weighting criteria.

A.2 Recovery–recycling: The construction solution that meets the worst refurbishment is adhered paving. This is because this material requires applications with a cementitious adhesive, which makes it impossible to recover the parts to be reused. So, it was assigned the most unfavorable value of 1. This decision is supported by Fayos [27], who analyzed the environmental impact of flat roof construction solutions using the Life Cycle Assessment (LCA) methodology to conclude that the solutions with fixed flooring had the strongest environmental impact, followed by raised flooring and extensive green roofs. Finally, although the green roof was also recoverable, it was assigned a value of 4, which was somewhat lower than the previous ones due to the higher risk of degradation caused by vegetation, such as the waterproof membrane breaking due to roots.

E.2. Maintenance (durability–cost–periodicity): A score of 5 was assigned to the gravel, raised paving and permeable paving solutions because maintenance work on these rehabilitation solutions is minimal. The roofs with adhered paving may present damaged paving pieces due to the tensions generated on pieces given their installation with mortar with green roofs. The main issue was the possible deterioration of any layer in the solution caused by vegetation.

P.1. Ease of execution: The most unfavorable was the green roof, with a score of 1, and the easiest to execute was the gravel roof, with a score of 5. The other two construction solutions were considered similar as regards this parameter, with a score of 4 (they are dry-laid solutions but require cutting and distributing pieces to adapt them to the roof).

P.2. Acoustic insulation: The best suitability was assigned to the green roof due to substrate and vegetation layers, with a score of 5. The other solutions proved similar in terms of composition and the system's mass, with differences in finishing. So, they were considered similar, with a score of 4.

P.4. Waterproofing–sealing: According to the assumption of correct execution, a score of 5 was assigned in all cases because all the solutions were perfectly watertight.

P.5. Aesthetics: The most favorable score was assigned to the green roof, with a score of 5, and the worst to the gravel roof, with a score of 3. The remaining solutions, with a wide range of aesthetic possibilities depending on the tiles selected for the roof's aesthetic finish, were assigned an intermediate score.

In prioritization terms, it was visually observed that the biggest number of favorable criteria corresponded to the gravel finish solution, followed by permeable paving, raised paving and adhered paving. The green roof had the most unfavorable criteria, whose weight was particularly important in rehabilitation. The adhered paving inverted roof obtained the lowest unweighted average value because most scores were medium or low.

3.3.2. Refurbishment Solutions Currently Used: Interviews with Contractors

A semistructured interview was held with active professionals in the construction sector to collect information on the specific roofing systems used in flat roof refurbishments. Interviews were conducted with six construction companies and professionals in the building renovation sector to seek information about the reality of the renovations that had been recently carried out. Interviewees answered the questions that appear below:

1. Identification of the work/project;
2. Location of the construction site (population);
3. Approximate age of the building/roof;
4. Approximate area of the roof (m^2);
5. Initial roof type;
6. Solution applied in renovation;
7. Approximate price of refurbishment (EUR/m^2) or overall budget (EUR).

In general, the professionals highlighted the fact that flat roof refurbishments exclusively aim to repair moisture problems, which originate in the waterproofing membrane and/or at singular points. In most cases, these actions are not used to incorporate thermal insulation or to improve buildings' energy efficiencies, seeing that users do not generally demand all this because this makes interventions more expensive. Moreover, they perceive that such an improvement only benefits the neighbors on the top floor, while the intervention is paid by the whole community of owners. These conclusions fall in line with the observation made by Ramos in his doctoral thesis [28], who observed that adding thermal insulation to roofs impacts the thermal comfort of top-floor dwellings in multifamily multistorey buildings. This fact may compromise the promotion of such measures when a community of owners makes consensual decisions when the economic criterion in refurbishment solution selections is usually the most important one.

So, even though current regulations have made an enormous effort to progressively improve buildings' energy efficiencies and to increase the proportion of buildings with almost zero energy use, their application is almost exclusively relegated to new construction and is optional in refurbishments. This leads to the question of what criteria are used for decision making when carrying out roof refurbishment. This aspect is what the following subsection seeks to clarify.

3.3.3. Survey on Weight Criteria in the Multicriteria Analysis

To estimate the weight of the criteria that can lead to a decision being made about the renovation of a building envelope, a questionnaire was carried out with two population groups: professionals related to the construction sector and homeowners. In both cases, the three above-assessed categories and indicators were used. Each indicator was assessed on a Likert scale, where 1 was considered the most unfavorable and 5 the most favorable. The normalized values were then obtained to evaluate the solutions.

A total of 27 responses were obtained from the professional sector and 55 from homeowners. The collected surveys included an equal number of responses from both men and women. The age range of the homeowners who participated in the survey was consistent with the proportion of homeowners in each age group, which explained the few respondents in their 20s. All the respondents in the expert surveys were professionals in the building sector, with 20 out of 27 respondents having more than 20 years of experience. The detailed breakdown of the survey respondents is as follows: 55 homeowners participated in the study, of whom 24 were women, 30 were men, and 1 person did not answer. The participants' age range was between 24 and 60 years old, with 4 in their 20s, 11 in their 30s, 22 in their 40s, 16 in their 50s and 2 in their 60s. Similarly, 27 experts participated in the survey, of whom 12 were women, 14 were men and 1 person did not answer. All the experts were from the architecture and building engineering field, and their age ranged from 27 to 84 years.

The calculation of the weighting applicable to each criterion was performed by determining the average score assigned by the survey participants to each criterion, and dividing this value by the sum of the average score of all the criteria, as presented in Equation (2). The results are collected in Table 3.

$$\%Pi = Vi/\sum(Vi) \times 100, \quad (2)$$

where:

Pi is the weighting factor for criterion i;
Vi is the value assigned to criterion i.

Table 3. Weighting coefficients.

Indicators	Users	Experts
A.1. Thermal insulation	13.5%	12.7%
A.2. Recovery–recycling	10.48%	12.50%
E.1. Initial investment cost	10.72%	10.47%
E.2. Maintenance	12.00%	10.54%
P.1. Ease of execution	8.90%	10.15%
P.2. Acoustic insulation	11.70%	9.83%
P.3. Weight of the system	9.46%	10.68%
P.4. Waterproofing–sealing	13.58%	13.25%
P.5. Aesthetic	9.68%	9.87%
Total	100%	100%

The values obtained from surveys were used as weighting factors to obtain a composite index, which allows the overall assessment of the suitability index (SI) of each proposed construction solution by considering all the criteria according to Equation (3):

$$SI = (\sum Pi \times Vi)/100, \quad (3)$$

where:

SI is the sustainability index;
Pi is the weighting factor for criterion i;
Vi is the value assigned to criterion i.

As Figure 5 illustrates, all the criteria were generally rated above 3. Two criteria stood out: thermal insulation and watertightness. In contrast, aesthetics was perceived as the least relevant by both users and experts. One particularly noteworthy finding was the little importance that users attached to the system's weight compared to the value given by experts. This was a determining factor for the viability of the system's implementation from a structural stability point of view. Although values were similar, users rather than experts gave slightly higher figures to economic aspects, and it was the other way around for the aspects related to the recovery and recycling of material and to the ease of execution, which were better valued by experts. Figure 5 shows the multicriteria results when considering the weighting coefficients obtained from the corresponding questionnaires for users and experts, with SIu and SIe, respectively. As can be observed, the highest resulting values were for the non-walkable roofing system with gravel, and the lowest values were for the walkable roof finished with adhered paving.

WINVR		WINVAD		WPER		NWGRE		NWGRA	
SIu	SIe	SIu	SIe	SIu	SIe	SIu	SIe	SIu	SIe
4.15	4.12	3.38	3.28	4.33	4.29	3.65	3.55	4.67	4.68

Tone range: Worst 1 2 3 4 5 Best

Figure 5. Suitability index values obtained from users and experts (weighted average of criteria).

4. Stage 2 Results: Energy Performance for Refurbishment Solutions

For this approach, a specific urban area in the Grao neighborhood of Castellón, located on the coastline, was selected. In this area, a previous study by Pitarch et al. [29] was conducted, where the authors identified roof types and measured areas by obtaining surfaces by roof type. The main values are presented in Table 4. They are grouped in the area occupied by census sections 9001, 9002, 9003, 9004, 9005, 9006, 9007 and 9010, at the eastern end of the neighborhood, as presented in Figure 6a. The classification per construction period is summarized in Figure 6b, while Figure 6c depicts the identified roof types, which were sloped roofs, non-walkable flat roofs, walkable flat roofs and inner courtyards. Urban building energy modeling is assisted by Geographical Information System (GIS) maps to easily represent data in the territory [30]. Table 4 shows the distribution of areas per roof type, building typology (Sf—single-family; Mf—multifamily) and construction period. The total amount per type of roof and building typology is also provided. The areas of roofs to be considered in this study are highlighted in bold. They are roofs on high-rise buildings, being mostly flat roofs, which were built during the 1960–1979 period. These buildings were selected because approximately half the building stock in the Mf typologies was built during this construction period (see Figure 6b), when poor-performance energy solutions were employed.

Table 4. Area per building typology, roof type and year of construction, m^2.

Time Period	1840–1936		1937–1959		1960–1979		1980–2006		2007–2012		TOTAL
Typology	Mf	Sf	Mf	Sf	Mf	Sf	Mf	Sf	Mf	Sf	
Sloped	2003	2622	452	771	612	466	18,717	10,293	4355	110	40,400
Flat—non-walkable	758	555	1281	786	**5286**	492	3245	2459	71	35	14,969
Flat—walkable	1237	1657	3379	1742	**30,837**	2271	16,392	1091	2498	127	61,231
Inner Courtyard	390	672	510	712	7609	650	14,119	8657	1993		35,314
TOTAL	4388	5507	5622	4011	44,344	3878	52,474	22,501	8917	272	151,913

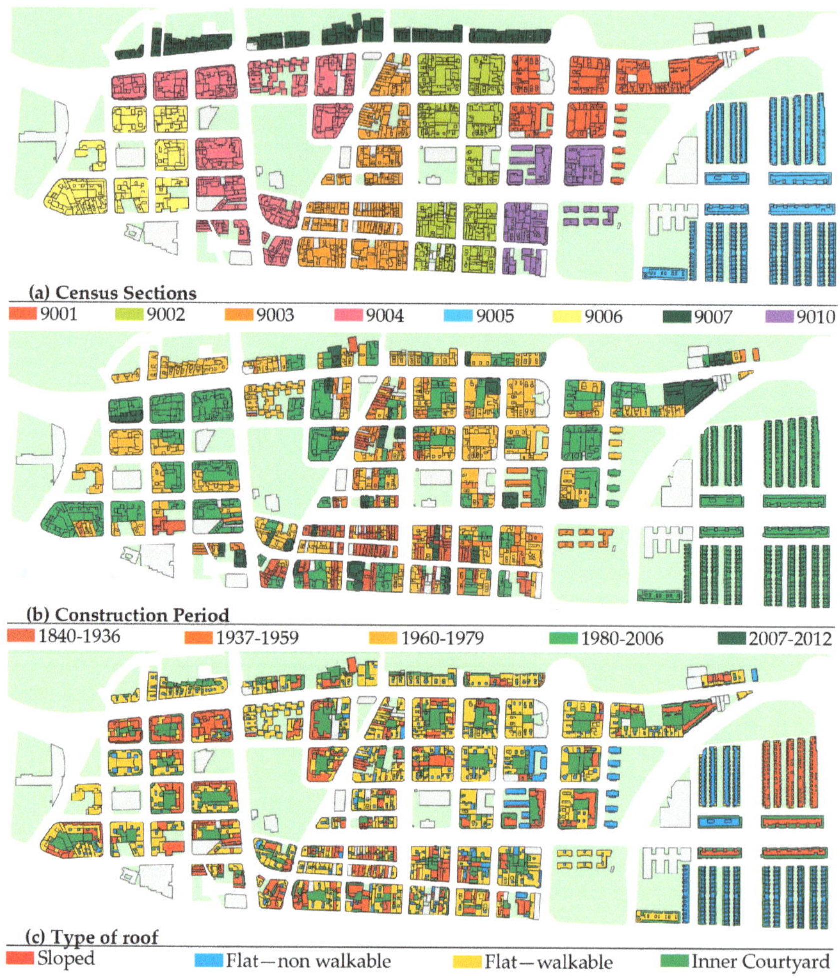

Figure 6. Geographical Information System maps representing the building stock in the area by (**a**) census sections, (**b**) construction period and (**c**) roof type.

To select a statistically representative building of the building stock in the urban area under study, the Cadastral data of the buildings included in the research area were collected following the process previously used by authors like Martín-Consuegra et al. [31]. For the buildings built during the selected construction period, the following variables were analyzed: the number of floors, the number of dwellings and the roof area. Some visits to the area were made to better collect data. A basic statistical analysis allowed the average number of floors, the number of dwellings and roof areas to be estimated for the buildings built in this area from 1960 to 1979, with the following results: 6.1 floors, 31 dwellings, 298.6 m^2 of roof area, 253.4 m^2 of walkable roof area and 36.7 m^2 of non-walkable roof area.

Most Mf buildings during this period present walkable roofs, while the small area of non-walkable roofs usually corresponds to the stairwell cover. To determine the energy rating with a simulation tool, an entire building has to be considered. Figure 7 shows the representative building of the selected studied neighborhood according to statistical criteria for this study stage. It provides a visual description of the most characteristic building typology in the neighborhood. It is located at 45 Alcocebre Street, was built in 1967 and is located on a rectangular plot covering 262 m^2. It consists of a ground floor, five upper floors and three commercial premises on the ground floors. Its total surface area is 229 m^2, with 4 flats per floor, which totals 20 flats ranging from 64 to 79 m^2, with a total residential surface area of 1399 m^2.

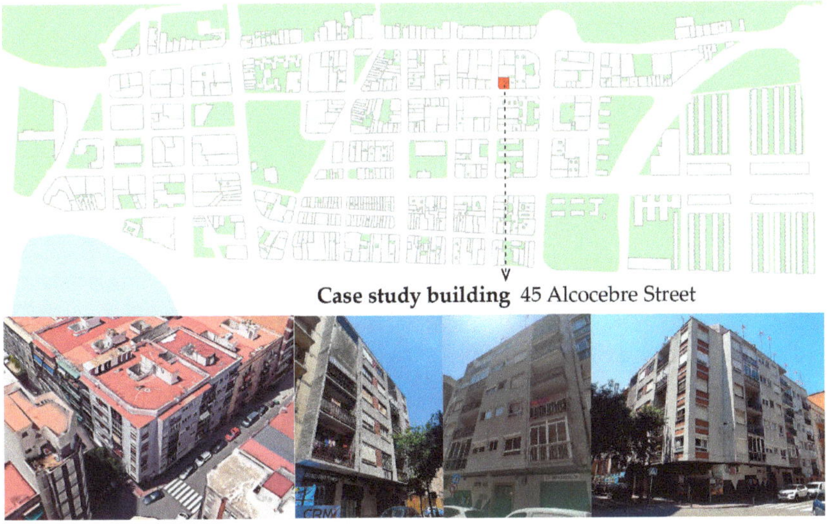

Figure 7. Case study building that is statistically representative of the neighborhood.

After considering the best refurbishment solution for walking roofs obtained in the previous stage by the multicriteria analysis with weighting criteria (WPER—walkable with permeable paving), the next step was to estimate an order of magnitude of energy savings by applying this solution in the urban area wherever the starting solution of the selected period was identified. To do so, the building geometry was checked by using cadastral cartography, the measurements taken at the site and by collecting data about existing constructive solutions. Next, the building was modeled using the CERMA v5.11 tool (Valencian Building Institute, IVE, and Asociación Técnica Española de Climatización y Refrigeración, ATECYR). This tool is one of those officially approved by the Spanish Ministry for Ecological Transition to certify buildings' energy performances (https://energia.gob.es/desarrollo/EficienciaEnergetica/CertificacionEnergetica/DocumentosReconocidos/Paginas/procedimientos-certificacion-proyecto-terminados.aspx, last accessed on 1 December 2023).

The inputs in software, as in other energy performance certification tools, are the climate data of the site, building typology, year of construction, orientation, composition of the thermal envelope and its main measures, together with the main information on building service facilities.

Simulation was conducted for three scenarios: first, the building in its original state (OR); second, the same building with the roof refurbishment by the WPER solution and appropriate thermal insulation to fulfill the requirements set out by the Technical Code for Building in its Energy Savings section (CTE-HE-1); and, finally, improving the thermal insulation of the WPER refurbishment solution by considering commercial formats for the thermal insulation of this permeable paving type available on the market (insulation thicknesses of 40, 50, 60 and 80 mm).

The original building's energy performance class was Class E on a scale from A to G, with emissions of 33.06 $kgCO_2/m^2$ per year. Despite poor insulation and the obsolescence of constructive systems and facilities, the rate is consistent with the climate zone and with mild winters. The study results are summarized in Table 5, where the values for emissions, energy use and heating/cooling demands appear for each scenario, together with the reached rate and the percentage of savings when comparing the original solution to the refurbished one (in brackets).

Table 5. Simulation values and energy class in the representative building.

	CO_2 Emissions ($kgCO_2/m^2$ Year)	Primary Energy Use (kWh/m^2 Year)	Heating Demand of Final Energy (kWh/m^2 Year)	Cooling Demand of Final Energy (kWh/m^2 Year)
Original	33.06 E	168.81 E	84.10 G	14.27 D
40 mm	31.22 E (−5.57%)	159.93 E (−5.26%)	77.81 G (−7.48%)	12.97 C (−9.11%)
50 mm	31.14 E (−5.81%)	159.55 E (−5.49%)	77.54 G (−7.80%)	12.91 C (−9.53%)
60 mm	31.11 E (−5.90%)	159.41 E (−5.70%)	77.44 G (−7.92%)	12.89 C (−9.67%)
80 mm	31.02 E (−6.17%)	158.96 E (−5.83%)	77.13 G (−8.29%)	12.82 C (−10.16%)

As expected, fulfilling the CTE indicated scarce improvement in terms of carbon emissions or energy use when only the roof of a building is refurbished, which implied a decrease in these indicators of around 5–6%. This building's percentage is slightly higher than that published by Abdeen et al. [32], who obtained a 2.3% reduction by improving roof insulation. However, differences grew for heating/cooling demands, at 9% and 10%, respectively. When the tool simulated the commercial formats, the values lowered because insulation thickness increased, which is logical. As the price of paving in those cases could be the decisive factor for selection, the cost efficiency of the solution was analyzed to select the compromise solution [33,34]. To do so, the price of investing in refurbishment was estimated, together with the savings made in energy use. The optimal cost method was applied according to the Commission Delegated Regulation (DR; EU) No. 244/2012 of 16 January 2012, which supplements Directive 2010/31/EU of the European Parliament and the Council on buildings' energy performance. This method allows a comparative methodology framework to be established for calculating cost-optimal levels of minimum energy performance requirements for buildings and building elements. For the representative building, Equation (4) was applied:

$$C_{g(\tau)} = C_I + \sum_j \left[\sum_{i=1}^{\tau} (C_{a,i}(j) R_d(i) + C_{c,i}(j)) - V_{f,\tau}(j)\right], \quad (4)$$

where Cg(τ) is the global cost (referring to the starting year τ 0) over the calculation period to be estimated. Table 6 presents the meaning of each term in Equation (4), as well as the values and starting hypothesis adopted in the analyzed case:

Table 6. Definition of the terms in Equation (4), values and the starting hypothesis.

Term	Definition	Hypothesis
τ	Calculation period.	30 years (as proposed by the DR for residential use).
$C_I(j)$	Initial investment costs for measure or set of measures j.	Obtained using the unit price from the Valencian Building Institute, IVE, construction database (https://bdc.f-ive.es/BDC22/1, accessed on 1 December 2023): EUR 41,373.43, EUR 42,381.84, EUR 45,095.45 and EUR 45,509.18 for 40, 40, 60 and 80 mm, respectively.
$C_{a,i}(j)$	Annual cost during year i for measure or set of measures j.	Energy saving due to the solution by adopting an average energy price of 0.3 EUR/kWh: EUR 3726.94, EUR 3886.42, EUR 3945.18 and EUR 4134.05 for 40, 40, 60 and 80 mm, respectively.
Cci(j)	Carbon cost during year i for measure or set of measures j.	A total of 0 EUR. This term is implemented from a macroeconomic perspective. Spain selected the financial perspective.
$R_d(p)$	$R_d(p) = \left(\frac{1}{1+\frac{r}{100}}\right)^p$ is the discount factor for year i based on discount rate r, where p means the number of years from the starting period.	Three rate factors *r* to undertake a sensitive analysis: 1%, 4% and 6% for the optimistic, medium and pessimistic scenarios, respectively.
$V_{f,\tau}(j)$	Residual value of measure or set of measures j at the end of the calculation period (discounted to starting year τ 0).	A total of 0 EUR. No residual value is considered for the selected solutions.

Table 7 shows the global cost, which considers a net present value of the investment and the payback period (Pp) of the different considered solutions. The last row indicates an additional scenario for the 80 mm solution by assuming an annual increase of 2% in the energy price, which is likely to progressively rise.

Table 7. The global cost and payback period (Pp) of the refurbishment solutions considering several scenarios for the sensitive analysis.

Solution	r1%		r4%		r6%	
	Cg(τ) (EUR)	Pp (Years)	Cg(τ) (EUR)	Pp (Years)	Cg(τ) (EUR)	Pp (Years)
40 mm	53,181.74	12	23,920.24	14	11,644.68	17
50 mm	56,330.30	11	25,779.87	14	12,963.20	17
60 mm	55,147.32	12	24,121.99	15	11,105.94	18
80 mm	59,657.63	11	27,105.84	14	13,448.96	17
80 mm + 2%	95,488.16	10	47,674.82	12	28,132.29	14

Like previous studies, this one used basic statistics to simulate the building stock [33–35]. The magnitude of the potential savings in the neighborhood was estimated by considering the total area of the flat roofs of this building typology and the construction period (see Figure 6 and Table 4). Accordingly, the refurbishment of 36,123 m^2 of the considered roof with the WPER solution of 80 mm thickness would save 339,682.36 $kgCO_2$ and 1,640,132.98 kWh per year. This estimation was obtained by the theoretical simulation of a representative building. Consequently, it must be taken as an order of magnitude rather than as an accurate amount. However, in economic terms, with an investment of EUR 5,416,567.08, and considering the used energy cost, it would lead to savings of EUR 492,039.89 in energy terms. This would imply an estimated return of investment of 11–17 years, depending on the tax rate scenario. This Pp could be reduced to 10–14 years when assuming a scenario with an annual increase in the energy price of 2%. As observed by Jaber [36], these relatively long Pps make the sole rehabilitation of building roofs unattractive from an economic point of view.

5. Discussion

After analyzing the different proposed rehabilitation solutions, and in view of their weaknesses, distinct strategies can be defined to achieve more efficient solutions. Among them, several possibilities can be studied:

- Study the possibility of developing lighter rehabilitation systems by using thin ceramic tiles or sheets. This strategy can be employed with raised flooring by applying a reinforcement layer in direct cladding and in permeable flooring to allow water permeability in tile joints;
- Raise public awareness about the importance of energy efficiency in buildings so that renovations are put to the best possible use to incorporate thermal insulation. This could be performed with existing financial aid so that homeowners' associations could consider it. The cost of the intervention that incorporates thermal insulation is similar or even lower than simply repairing existing damp problems;
- Study the possibility of using recycled aggregate gravel as protection for inverted roofs, which would increase the overall sustainability of the solution.

After analyzing some potential scenarios, possible future considerations can be the following:

- Subsidies and public support for interventions should be maintained to reach an efficient building stock, bearing in mind that low-performance building owners usually face more difficulties that involve making high investments;

- It is necessary to raise more long-term awareness by considering a more holistic view and realizing the benefits of prolonging buildings' life spans through proper maintenance and upgrading;
- Building roofs are normally underrated, perhaps because they cannot be seen from the street level. However, they are important elements in the envelope and are crucial for prolonging the life span of buildings and for promoting proper building functionality. Their proper maintenance will surely avoid future costly repairs.

This work has presented a limited number of refurbishment solutions by basing their environmental performance on improving thermal insulation, which allows the thermal transmittance of the envelope to be reduced. However, this study has not explored other rehabilitation options like those mentioned by Madushika et al. [37], i.e., using paints or highly reflective clay tiles, which could complement the analyzed solutions.

6. Conclusions

This research work assesses the viability of various roof rehabilitation systems by determining the best option based on a multicriteria analysis. To determine the potential energy improvement, the best solutions are simulated in a representative building of a neighborhood located in Castellón de la Plana (East Spain). The energy improvement possibilities at the neighborhood level are determined by extrapolating these refurbishment proposals to the study area. Based on the obtained results, the following conclusions can be drawn:

- The multicriteria assessment indicates the gravel system as the most favorable one when considering the A, E and P aspects together. Raised paving and permeable paving systems obtain an intermediate overall rating, which can be improved by reducing the system's weight or the cost of raised floors;
- Despite the importance of thermally insulating the building envelope and improving the building stock's energy efficiency, the current regulation (DB HE1) is only applicable in certain refurbishment cases to obtain buildings with almost zero energy use. So, owners tend to look for the cheapest solution;
- Users attach more importance to the cost of the investment, and this factor sometimes determines the feasibility of refurbishment. As highlighted by renovation professionals, thermal insulation is generally not incorporated into refurbishments of flat roofs of Mf housing buildings because it makes refurbishment more expensive and is considered to benefit only the top-floor dwellings;
- The fact that roof renovation solutions are overweight is extremely important and conditions their application. However, their importance is not perceived as such by users and experts;
- Although green roofs are highly desirable from a sustainability point of view, they should be ruled out as a refurbishment solution for existing buildings for being overweight;
- Roof refurbishment's cost efficiency does not seem very optimistic. Investments of about 60–90 EUR/m^2 are needed, which means slightly improved energy performance, with savings in overall non-renewable primary energy use of around 5% or 6% for a benign climate zone, such as that herein analyzed. However, considering that this refurbishment means acting on 20% of the building's total thermal envelope, the result is not so low;
- The Pp for the studied cases is 10 years in the best-case scenario and 18 years in the worst-case scenario;
- It is important to emphasize the fact that not only roof insulation is improved with refurbishment but also waterproofing and, therefore, habitability conditions, especially those of the top-floor dwellings. This argument reinforces the convenience of roof refurbishment because it not only improves the building's energy performance but also increases its life span by improving and repairing a very exposed area of the

building envelope, which avoids potential diseases. All these reasons reinforce the alignment of roof refurbishment with buildings' sustainability;

- If the results obtained for the statistically representative building are extrapolated to the neighborhood scale, the total values are quantitatively significant, with an annual saving in atmospheric emissions of almost 340 tons of CO_2, and a reduction in energy use of over 1.6 million kWh. Thus, the potential improvement in the area is quite high.

In the last two decades, there has been a significant increase in the number of scientific studies that explore ways to improve buildings' energy efficiencies. This has been conducted in response to evolving regulations and a global push toward decarbonization of the economy. The new EPBD proposal includes important aspects that provide insights into how sustainability can be promoted effectively in the building industry. Some of these include more ambitious energy performance requirements for both new and renovated buildings, the consideration of carbon emissions throughout a building's entire life cycle, the adoption of "staged renovation" as a solution to the high upfront costs and financial mechanisms to boost the old building stock's renovation. These aspects show the future trend in the building sector, and highlight the importance of roof refurbishment in enhancing a building's durability and overall performance.

Author Contributions: M.J.R., methodology; all authors, formal analysis; M.J.R. and L.R., investigation; all authors, data curation; Á.M.P. and I.A., supervision; L.R., writing—original draft preparation; M.J.R. and L.R., writing—review and editing. All authors have read and agreed to the published version of the manuscript.

Funding: This work was funded by Projects "Rooftiles and Rooftiles II" with a collaboration contract with the research projects with references IMDEEA/2021/34 and IMDEEA/2022/7, respectively, of the Ceramic Industries Research Association (AICE—Asociación de Investigación de las Industrias Cerámicas), in 2022 and 2023. These projects were supported by the Instituto Valenciano de Competitividad Empresarial (IVACE) through FEDER (Fondos Europeos de Desarrollo Regional) funding.

Institutional Review Board Statement: This study did not require ethical approval.

Informed Consent Statement: Informed consent was obtained from all subjects involved in this study.

Data Availability Statement: All data from buildings are public and were retrieved from the Cadastral Office. All respondents to the surveys were informed about the purpose of this study and signed an informed consent form; the responses were anonymized and stored for research purposes, with access restricted to the study's authors.

Conflicts of Interest: The authors declare no conflicts of interest.

References

1. EUR-Lex. Access to European Union Law. Available online: https://eur-lex.europa.eu/legal-content/EN/TXT/?uri=CELEX:52021PC0802&qid=1641802763889 (accessed on 18 February 2024).
2. European Union. Directive (EU) 2018/844 of the European Parliament and of the Council of 30 May 2018 amending Directive 2010/31/EU on the Energy Performance of Buildings and Directive 2012/27/EU on Energy Efficiency. 2018. Available online: https://eur-lex.europa.eu/eli/dir/2018/844/oj (accessed on 29 January 2024).
3. María-Tomé Gil, B. *Propuestas de Escenarios para la Rehabilitación Energética de Viviendas 2030–2050*; Instituto Sindical De Trabajo, Ambiente y Salud (ISTAS): Madrid, Spain, 2021. Available online: https://istas.net/sites/default/files/2021-07/BBPP_REHAB_VIVIENDA.pdf (accessed on 11 January 2024).
4. Morgado, J.; Flores-Colen, I.; de Brito, J.; Silva, A. Maintenance programmes for flat roofs in existing buildings. *Prop. Manag.* **2017**, *35*, 339–362. [CrossRef]
5. Espinoza-Zambrano, P.; Marmolejo-Duarte, C. Hacia un Pasaporte de Renovación de Edificios en España: Necesidades y oportunidades a la luz de experiencias europeas/Towards a Building Renovation Passport in Spain: Needs and opportunities in light of European experiences. *Inf. Construcción* **2022**, *74*, 1988–3234. [CrossRef]
6. Gómez-Gil, M.; Espinosa-Fernández, A.; López-Mesa, B. Review and Analysis of Models for a European Digital Building Logbook. *Energies* **2022**, *15*, 1994. [CrossRef]
7. Maaouane, M.; Chennaif, M.; Zouggar, S.; Krajačić, G.; Amrani, S.; Zahboune, H. Cost-effective design of energy efficiency measures in the building sector in North Africa using Building Information Modeling. *Energy Build.* **2023**, *294*, 113283. [CrossRef]

8. Royal Decree 2429/79, 6 July, Approving Basic Housing Regulation on Thermal Conditions, NBE-CT-79. Available online: https://www.boe.es/eli/es/rd/1979/07/06/2429 (accessed on 29 January 2024).
9. Royal Decree 450/2022 of 14 June Amending the Technical Building Code, Approved by Royal Decree 314/2006 of 17 March. Available online: https://www.boe.es/eli/es/rd/2022/06/14/450 (accessed on 29 January 2024).
10. Valencian Government and Valencian Building Institute. Catálogo de Tipología Edificatoria Residencial. Ámbito España. 2016. Available online: https://episcope.eu/fileadmin/tabula/public/docs/brochure/ES_TABULA_TypologyBrochure_IVE.pdf (accessed on 2 January 2024).
11. Braulio, M. Propuesta Metodológica para la Caracterización del Comportamiento Energético Pasivo del Parque Edificatorio Residencial Existente Considerando su Contexto Urbano. Ph.D. Thesis, Universitat Jaume I, Castellón, Spain, 2016. [CrossRef]
12. Bevilacqua, P.; Bruno, R.; Arcuri, N. Green roofs in a Mediterranean climate: Energy performances based on in-situ experimental data. *Renew. Energy* **2020**, *152*, 1414–1430. [CrossRef]
13. Cascone, S. Green Roof Design: State of the Art on Technology and Materials. *Sustainability* **2019**, *11*, 3020. [CrossRef]
14. Maiolo, M.; Pirouz, B.; Bruno, R.; Palermo, S.A.; Arcuri, N.; Piro, P. The Role of the Extensive Green Roofs on Decreasing Building Energy Consumption in the Mediterranean Climate. *Sustainability* **2020**, *12*, 359. [CrossRef]
15. Saadatian, O.; Sopian, K.; Salleh, E.; Lim, C.H.; Riffat, S.; Saadatian, E.; Arash, T.; Sulaiman, M.Y. A review of energy aspects of green roofs. *Renew. Sustain. Energy Rev.* **2013**, *23*, 155–168. [CrossRef]
16. Zheng, Y.; Chen, L. Modeling the Effect of Green Roofs for Building Energy Savings and Air Pollution Reduction in Shanghai. *Sustainability* **2024**, *16*, 286. [CrossRef]
17. Al-Turki, A.M.; Gari, H.N.; Zaki, G.M. Comparative study on reduction of cooling loads by roof gravel cover. *Energy Build.* **1997**, *25*, 1–5. [CrossRef]
18. Sedbauer, K.; Schunk, E.; Berthel, R. *Flat Roof Construction Manual. Materials Design Applications*; Edition Detail: Munich, Germany, 2012.
19. D'Orazio, M.; Perna, C.D.; Stazi, F. Thermal behaviour of vented roofs. *Struct. Surv.* **2009**, *27*, 411–422. [CrossRef]
20. Scholz, M.; Grabowiecki, P. Review of permeable pavement systems. *Build. Environ.* **2006**, *42*, 3830–3836. [CrossRef]
21. Turco, M.; Brunetti, G.; Palermo, S.A.; Capano, G.; Grossi, G.; Maiolo, M.; Piro, P. On the environmental benefits of a permeable pavement: Metals potential removal efficiency and Life Cycle Assessment. *Urban Water J.* **2020**, *17*, 619–627. [CrossRef]
22. Decree 195/1963 of 17 January, Approving "MV101-62. Acciones en la Edificación". Available online: https://www.boe.es/diario_boe/txt.php?id=BOE-A-1963-4613 (accessed on 2 January 2024).
23. Royal Decree 1370/1988 of 25 July 1988, Approving the Basic Building Standard "NBE-AE/88. Acciones en la Edificación". Available online: https://www.boe.es/eli/es/rd/1988/11/11/1370 (accessed on 2 January 2024).
24. Hamdan, M.; Mirzaei, P.; Gillott, M. Life Cycle Cost Assessment and Retrofit in Community Scale: A Case Study of Jordan. *E3S Web Conf.* **2023**, *396*, 0401211. [CrossRef]
25. Borràs, J.G.; Lerma, C.; Mas, A.; Vercher, J.; Gil, E. Contribution of green roofs to energy savings in building renovations. *Energy Sustain. Dev.* **2022**, *71*, 212–221. [CrossRef]
26. Valencian Building Institute. Data Base for Construction. Available online: https://bdc.f-ive.es/BDC23/1 (accessed on 18 January 2024).
27. Fayos, H. Comparativa ambiental y económica del ciclo de vida de cubiertas planas. Master's Thesis, Universitat Jaume I, Castellón, Spain, 2013.
28. Ramos-Martín, M. Metodología integrada para la evaluación de la sostenibilidad de actuaciones regenerativas sobre la envolvente de células urbanas en climas templados. Ph.D. Thesis, Universidad de Sevilla, Seville, Spain, 2020.
29. Pitarch, A.; Ruá, M.J.; Reig, L.; Arín, I. Contribution of Roof Refurbishment to Urban Sustainability. *Sustainability* **2020**, *12*, 8111. [CrossRef]
30. Mutani, G.; Alehasin, M.; Yang, H.; Zhang, X.; Felmer, G. Urban Building Energy Modeling to Support Climate-Sensitive Planning in the Suburban Areas of Santiago de Chile. *Buildings* **2024**, *14*, 185. [CrossRef]
31. Martín-Consuegra, F.; de Frutos, F.; Oteiza, I.; Hernández-Aja, A. Use of cadastral data to assess urban scale building energy loss. Application to a deprived quarter in Madrid. *Energy Build.* **2018**, *171*, 50–63. [CrossRef]
32. Abdeen, A.; Mushtaha, E.; Hussien, A.; Ghenai, C.; Maksoud, A.; Belpoliti, V. Simulation-based multi-objective genetic optimization for promoting energy efficiency and thermal comfort in existing buildings of hot climate. *Results Eng.* **2024**, *21*, 101815. [CrossRef]
33. Aguacil, S.; Lufkin, S.; Rey, E.; Cuchí, A. Application of the cost-optimal methodology to urban renewal projects at the territorial scale based on statistical data—A case study in Spain. *Energy Build.* **2017**, *144*, 42–60. [CrossRef]
34. Gangolells, M.; Casals, M.; Ferré-Bigorra, J.; Forcada, N.; Macarulla, M.; Kàtia, G.; Tejedor, B. Office representatives for cost-optimal energy retrofitting analysis: A novel approach using cluster analysis of energy performance certificate databases. *Energy Build.* **2019**, *206*, 109557. [CrossRef]
35. Ferreira, A.; Duarte Pinheiro, M.; de Brito, J.; Mateus, R.; Mendonça, R. Wall and roof solutions for a retail building considering cost investment and life cycle approach: A case study in Portugal. *J. Clean. Prod.* **2023**, *383*, 135314. [CrossRef]

36. Jaber, J.O. Assessment of Retrofitting Old Residential Buildings in Urban Districts: Expected Performance of Selected Energy Efficiency Measures. *Jordan J. Mech. Ind. Eng.* **2023**, *17*, 555–570. [CrossRef]
37. Madushika, U.G.D.; Ramachandra, T.; Karunasena, G.; Udakara, P.A.D.S. Energy Retrofitting Technologies of Buildings: A Review-Based Assessment. *Energies* **2023**, *16*, 4924. [CrossRef]

sustainability

MDPI

Article

Towards Sustainable Material: Optimizing Geopolymer Mortar Formulations for 3D Printing: A Life Cycle Assessment Approach

Charlotte Roux [1,*], Julien Archez [2], Corentin Le Gall [2], Myriam Saadé [2], Adélaïde Féraille [2] and Jean-François Caron [2]

[1] MINES Paris—PSL Research University, CEEP (Centre Energie Environnement Procédés), 60 Boulevard Saint Michel, 75006 Paris, France

[2] Navier Laboratory, Ecole des Ponts ParisTech, Gustave Eiffel University CNRS, 77454 Champs sur Marne, France

* Correspondence: charlotte.roux@minesparis.psl.eu

Abstract: Geopolymer-based concretes have been elaborated among others for their potential to lower the environmental impact of the construction sector. The rheology and workability of fresh geopolymers make them suitable for new applications such as 3D printing. In this paper, we aim to develop a potassium silicate- and metakaolin-based geopolymer mortar with sand and local earth additions suited for 3D printing and an environmental assessment framework for this material. The methodology aims at the optimization of both the granular skeleton and the geopolymer matrix for the development of a low-environmental-impact material suited for 3D printing. Using this approach, various metakaolin/earth geopolymer mortars are explored from a mechanical and environmental point of view. The environmental assessment of the lab-scale process shows an improvement for the climate change category but a degradation of other indicators, compared to Portland-cement-based concrete. Several promising options exist to further optimize the process and decrease its environmental impacts. This constitutes the main research perspective of this work.

Keywords: sustainable material; geopolymers; 3D printing; material characterization; environmental optimization; life cycle assessment

Citation: Roux, C.; Archez, J.; Le Gall, C.; Saadé, M.; Féraille, A.; Caron, J.-F. Towards Sustainable Material: Optimizing Geopolymer Mortar Formulations for 3D Printing: A Life Cycle Assessment Approach. *Sustainability* **2024**, *16*, 3328. https://doi.org/10.3390/su16083328

Academic Editors: Uroš Klanšek and Tomaž Žula

Received: 13 March 2024
Revised: 9 April 2024
Accepted: 12 April 2024
Published: 16 April 2024

1. Introduction

Concrete is responsible for 8% of global greenhouse gas (GHG) emissions and 5.2% of particulate matter (PM_{10}) emissions [1]. With the gradual awareness of the urgency of decarbonizing the concrete and cement industries, alternatives have been developed, particularly to replace Portland cement, which largely contributes to the carbon footprint of concrete [2,3]. Some alternatives rely on the replacement of cement using by-products of high-emitting industries such as coal and steel production, providing fly ashes and slags, respectively. Other emerging technologies aim at reducing emissions and energy use in cement production [4,5]. The wide adoption of such technologies, however, faces multiple barriers, ranging from regulatory issues to supply, product confidence, and technical obstacles [6].

In addition to being a major contributor to climate change, concrete production also consumes notable amounts of natural resources such as aggregates and sand with a certain quality. The world consumption of sand and aggregates is estimated to reach about 41 billion tons per year and is expected to increase soon [7]. The sand supply has long been taken for granted but nowadays faces issues of local resource depletion, impacts on the ecosystem, and climatic instabilities [8].

In this context, geopolymers recently benefited from renewed interest. Geopolymers (GPs) are inorganic, amorphous three-dimensional alumino-silicate materials synthesized at an ambient temperature through the activation of an aluminosilicate source (i.e., metakaolin) by an alkaline solution [9]. The polycondensation reaction between alumina and silicates occurring under basic alkali activation results in a geopolymer network. Such geopolymer provides resistance to high temperatures, as well as high mechanical or chemical resistance due to GP covalent bonds [10,11]. Sodium and potassium alkaline solution are largely used as activators but another synthesis path involving phosphoric acid is also studied, to a lesser extent [12].

Geopolymers are seen as an alternative to Ordinary Portland Cement (OPC), with expected lower CO_2 emissions [9]. Portland cement requires heating up to 1450 °C for several hours which causes the calcination of the limestone (clinkerization process). The combustion of fuels to reach such a temperature and the decarbonization of limestone both emit a high quantity of CO_2—about one ton of CO_2 per ton of cement [13]. In comparison, geopolymer production requires the use of a chemical solution of alkali silicates and the heating of kaolin clay at a lower temperature (around 850 °C), without the calcination of limestone. The synthesis of raw material is, therefore, less emitting than the Portland cement production process, when considering the CO_2 emitted directly during the production process [9]. However, its environmental advantage has to be rigorously confirmed in a life cycle and multi-criteria perspective [14].

Life Cycle Assessment (LCA) is a method currently widely used to assess the environmental impacts of products and services [15,16]. It is a multi-step and multi-criteria approach developed to avoid impacts shifting along the value chain and among impact categories. Multiple Life Cycle Assessment studies have been applied to geopolymer (GP) matrices, concretes, and mortars, with the latter being constituted of a geopolymer matrix and a granular skeleton [17–22]. So far, no scientific consensus has been reached about their environmental performances, mainly because the results highly depend on the formulation and raw materials used for their synthesis. Some authors show a significant reduction of environmental impacts when using GP concrete based on slag, fly ash, and alkali-silicate systems [18,23], while others confirm a slight reduction of GHG emissions using fly ash and blast-furnace geopolymers [19]. They also highlight a trade-off between impacts on climate change and other environmental categories such as abiotic resources, eutrophication, and acidification, for which GP concrete presents higher impacts. GP concrete could even have a higher carbon footprint than conventional concrete depending on upscaling commercial scenarios for 3D printing concrete [24].

Most studies agree on the importance of the contribution to the impacts of alkali activator production. To reduce the contribution of the geopolymer matrix, fillers such as sand can be added to form the geopolymer mortar. Another way is to use raw earth instead of sand [25] as GP chemistry makes it more prone to interact with the earth, especially clays, than ordinary concrete [26,27]. In this way, from a circular economy perspective, geopolymer synthesis could help recover excavated earth.

In addition to their potential intrinsic environmental performances, GP mortars could be suited for 3D printing. Applied to the construction sector, this technique, currently under development, is considered as a way to reduce the amount of construction materials used [28]. Geopolymers present a totally different physic than 3D-printed rock analogs [29] and a different setting than cement mortars—they are closer to polymeric glues than hydrated mortars. They also may be highly completed with different fillers. For these reasons, they become good challengers for this new way of building. Their interesting durability properties would also be relevant in a growing number of applications (structural materials, heat-resistant pavement, sewer pipes, sub-aqueous seawater, etc.) [30]. Improved knowledge of geopolymers is thus broadly recommended to understand their potential to mitigate carbon emissions in the construction sector [31–33] and precisely evaluate the domain of environmental relevance for this technology. Specific work on geopolymer mortars for 3D printing shows low embodied carbon per m^3 compared to cementitious material [34]. However, the contribution of transport and the curing and mixing process are not clearly presented. Moreover, most formulations reviewed are based on used fly ash and blast-furnace slags (Table 1), which are by-products of polluting industries and are only available in limited quantities. Even if the use of fly ash and blast-furnace slags could be envisioned in countries where electricity is still mainly coal-based, they remain by-products. They might not be a long-term option for replacing cementitious blends on a large scale. The development of a potassium silicate- and metakaolin-based geopolymer mortar with low environmental impact, sufficient mechanical properties, and suitability for 3D printing could therefore be beneficial for the construction sector. Its environmental impact must, however, be thoroughly studied using a comprehensive Life Cycle Assessment.

Table 1. Comparison of embodied GHG emissions for GP concrete or mortar and Ordinary Portland Concrete found in the literature.

Geopolymer (GP) Description	GHG Emissions of GP ($kgCO_2eq/m^3$)	GHG Emissions of Ref. OPC ($kgCO_2eq/m^3$)	Ref	Comments
Binder, suitable for 3D printing—FA50% + GGBS50%	107	556	[34]	Carbon accounting and not full LCA (f.e. water carbon footprint is set to zero). Reference binder is 80% OPC and 20% FA.
Binder, suitable for 3D printing—FA78.5% + GGBS13.8% + SF7.7%	134	556	[34]	Carbon accounting and not full LCA (f.e. water carbon footprint is set to zero). Reference binder is 80% OPC and 20% FA.
Slag and FA GP binder for 3D printing	677	493	[24]	Full, complete LCA, including sensitivity analysis on allocations.
"Standard" FA GP concrete	320	354	[35]	Carbon accounting and not full LCA.
FA GP concrete	169	306	[19]	Also investigating MK-GP, but clear figures are not available. MK-GP impacts are higher than FA-GP impacts.
FA and slag GP cement	267	895	[18]	Indian context, cement and not concrete.

FA: Fly ash, GGBS: granulated blast-furnace slag.

The objective of this paper is to propose a formulation for a low-environmental-impact geopolymer mortar to be used for 3D-printing applications exploiting locally available materials. The research explores a lab-scale process developed in a French context. After a brief description of the studied class of geopolymer, the article describes the LCA-based approach used to optimize geopolymer formulation, relying on its environmental performance. An initial formulation of a geopolymer mortar is tested, and its environmental impacts are evaluated to identify the life cycle stages with the main impact. The article then presents the results of formulation optimization using LCA and examines the influence of LCA parameters such as transport. Based on a comparison of three 3D-printing materials, it discusses potential improvements of the process to further decrease the environmental impacts of metakaolin geopolymers, highlighting the need to improve the accuracy of LCA data.

2. Materials and Methods

2.1. Methodological Approach

Our methodological approach is summed up in Figure 1. To understand the environmental impacts of a printable geopolymer formulation, a first environmental assessment was led using a GP matrix previously developed in the NAVIER laboratory and suitable for 3D printing, as described in [36]. This first matrix serves as a reference for further optimization. It is composed of 46%w metakaolin, 37%w potassium alkaline solution, 15%w wollastonite, and 2%w glass fibers and will be called GP–GfW in this study.

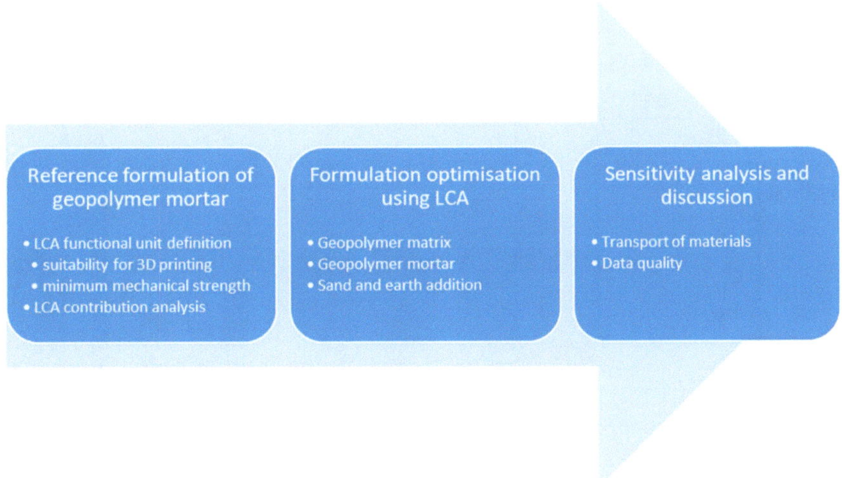

Figure 1. Methodological approach.

Based on the environmental impact of a geopolymer-based 3D-printing material, an optimization of the matrix was undertaken to lower the environmental impacts while maintaining sufficient mechanical property. Then, geopolymer formulations (matrix + granular skeleton) were developed to decrease GHG emissions while maintaining printability and a minimum mechanical strength of 32.5 MPa (NF-EN 197-1 standard [37]).

The results were further analyzed to understand the influence of certain parameters on the environmental impacts of the formulation. The studied process is a lab process and is not optimized as the processes currently in place within the cement and concrete sector are. Transport distances and modes were investigated in a sensitivity analysis to understand how impacts could be decreased by a scale effect. The poor quality of some data, related to low temporal representativity, for instance, is also discussed.

2.2. *Raw Materials and Sample Preparation*

2.2.1. Raw Materials

Geopolymer formulations were synthesized at the NAVIER laboratory with potassium silicate solution (Geosil14515: [K] = 7.0 M, SiO_2 = 19%w, K_2O = 22%w and H_2O = 59%w) and metakaolin aluminosilicate source (M1000: SiO_2: 55%w Al_2O_3: 40%w, D50 = 10 μm) [38]. The developed formulations differ from most GPs presented in the literature, which are sodium silicate solution and industrial waste (fly ashes, ground granulated blast slag)-based GPs [39]. Given their limited availability on the French territory and the uncertainties of their future supply, industrial wastes were not included in the tested formulations.

A granular skeleton constituted partly of masonry sand (0–4 mm diameter) and partly of raw earth was then added to the geopolymer matrix to form a geopolymer mortar. Raw earth was supplied locally from the excavation works of the "Grand Paris" project. The earth was dried in an oven (24 h at 100 °C) and then sieved consecutively into four particle sizes (1.6–2.5 mm, 0.8–1.6 mm, 0.4–0.8 mm, and <0.4 mm). Qualitative tests regarding earth addition in geopolymers showed water absorption issues. The earth has a higher water demand than sand due to the presence of clays (like illite or smectite for the earth used in this study), especially for the low granulometry (<0.8 mm). To quantify the water absorption, the water demand of earth particles and sand was quantified by measuring the quantity of water (in increments of 10 μL) to be added to 10 g of material to reach a moist state. The moist state corresponds to a visual criterion where the sand or earth is fully wet. Their values are reported in Table 2. Given that the earth absorbs most of the water from the mix, the mortar becomes dry too quickly and the earth addition becomes impossible. The absorption of water by the earth modifies the ratio of chemical components and consequently the polycondensation reaction of the geopolymer.

Table 2. Granulometry and water demand of sand and earth used in this study.

Particle Type	Sand	Earth			
Grain size (mm)	0–4	1.6–2.5	0.8–1.6	0.4–0.8	<0.4
Water absorption ± 0.02 (mL/g)	0.13	0.06	0.28	0.58	0.75
Picture	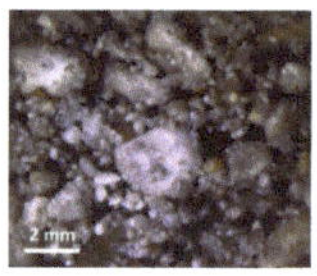				

To optimize the earth quantity, the granular skeleton of the mix was improved following the Fuller-Thompson method [36], considering Equation (1).

$$pi = (d\_i/D)^{0.45} \quad (1)$$

where pi is the percent passing ith sieve, di (mm) is the opening size of the ith sieve, and D (mm) is the maximum particle size.

An optimized granularity was designed to compensate for the high water absorption of the small earth fractions (diameter under 0.8 mm). The final composition consisted of sand for aggregates with a diameter lower than 0.8 mm and earth for coarse aggregates, with a size grain higher than 0.8 mm. The optimized grain size distribution is exposed in Figure 2.

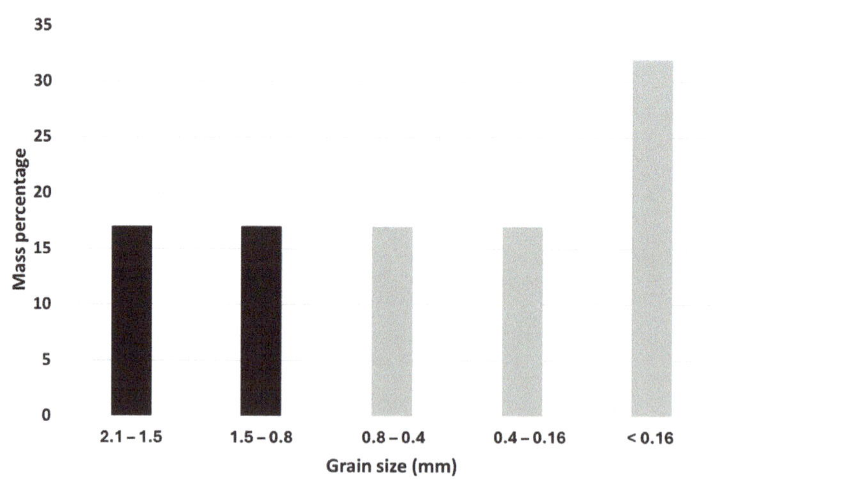

Figure 2. Grain size distribution of ■ earth and ■ sand in the optimized granularity.

2.2.2. Sample Preparation

The nomenclature used in this work is $M_aS_bSa_yE_z$, where a and b characterize the ratio between potassium silicate solution (S) and metakaolin (M), so a + b = 100%. Moreover, "y" and "z" represent, respectively, the quantity in grams of sand (Sa) and earth (E) added to form the geopolymer mortar for 100 g of geopolymer paste (a + b). The sample preparation consisted of adding metakaolin into a silicate solution progressively while mixing with a planetary mixer. After 3 min of mixing, sand and/or earth were added progressively to the geopolymer paste and mixed for 5 min. The sample was then cast in a 40 × 40 × 160 mm closed mold and the bubbles were removed with a vibrating needle (50 Hz, 1 min). The samples were demolded after 24 h and stored in a sealed plastic bag. The printability of the sample was quantified by overlaying manually the material with a syringe with a 15 mm diameter. The normal compressive strengths were evaluated after 7 days on 12 (half 40-40-160 mm) samples using an MTS with a 100 kN load cell at 0.5 mm/min constant speed.

2.3. Environmental Characterization

2.3.1. Environmental Assessment: Methodological Choices and Perimeter of the LCA for 3D Printing Geopolymer Mortars

Definition of the functional unit: The functional unit is defined as: "producing 1 m^3 of geopolymer mortar suitable for 3D printing". The impact of the formulation or 3D-printing process on the mechanical properties of these geopolymers has already been studied in the past [40], and according to the NF EN 197-1 standard, the compressive strength of the mortar should reach at least 32.5 MPa to ensure its suitability for 3D printing. The material printability also needs to be assured. Such printability was qualified by manually overlaying the material with a syringe, as a preliminary test [38].

System boundaries: The flowchart of the system is presented in Figures 3 and 4. The study is a cradle-to-gate assessment, and we included in the system the production of GP mortar as well as upstream activities and processes. GP applications are not considered, and consequently, neither are transportation, use, and end-of-life. The equipment used to process the materials (e.g., mixing unit, oven, molds) and the land occupation generated by the lab were also excluded from the system processes, as no data were accessible. Moreover, no material loss during the process was accounted for in this study. To consider the localized characteristic of this supply, LCA data were as much as possible adapted to the French context.

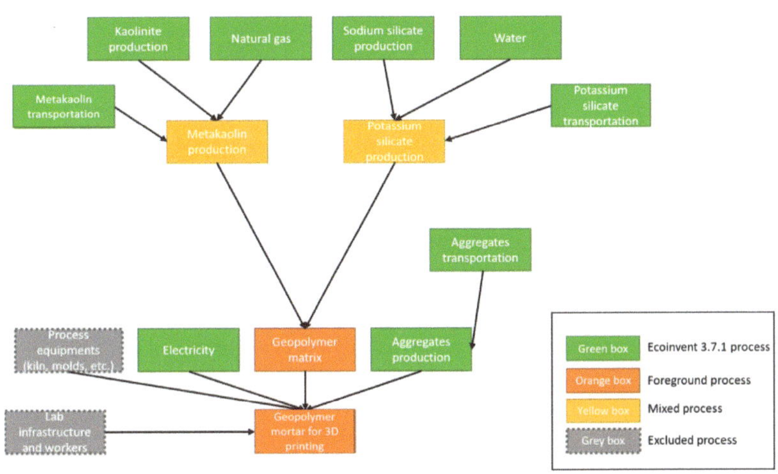

Figure 3. Granularity flowchart of the GP studied system. Mixed processes are ecoinvent processes adapted or modified for the study. Foreground processes are processes defined by the authors.

The study was led using the open-source framework Brightway 2 and its graphic interface Activity-Browser [41,42], relying on the database Ecoinvent 3.7 Cutoff [43,44].

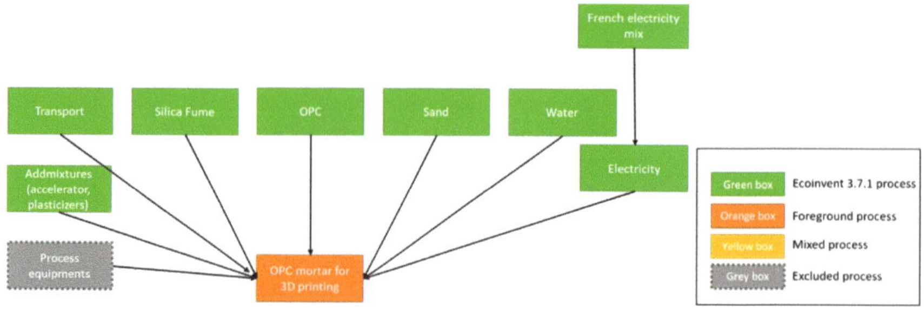

Figure 4. Flowchart of the OPC studied system. Mixed processes are ecoinvent processes adapted or modified for the study. Foreground processes are processes defined by the authors.

2.3.2. Life Cycle Inventory of Individual Processes

The assumptions made to derive the process inventory for each component of the studied system are exposed below. Exhaustive information on the system modeling is given in Supplementary Materials, Section S1.

Metakaolin (MK): The considered metakaolin (MK) was provided by the French company Ceradel under the label METAKAOLIN ARGICAL M1000 (Ceradel, Clérac, France). According to commercial communication by Imerys, it was assumed that the MK was formed from a process involving high-purity kaolinite heated at 750 °C in a rotary kiln. The modeling process considers 1.16 kg of kaolin production (ecoinvent process, contextualized using a French electricity mix) and 2.5 MJ of natural gas heating to obtain 1 kg of metakaolin [22].

Potassium silicate solution (PSS): The potassium silicate solution (PSS) used in this process is a commercial solution distributed by the German chemical company Woellner (Ludwigshafen, Germany), under the name Geosil 14515. An existing process in the ecoinvent database for sodium silicate was selected based on the same synthesis process (hydrothermal). A molar equivalent was applied in the ongoing flows, to replace sodium with potassium [45]. This process had a close solid ratio with the product Geosil 14515 used in this study (48% mass of dried content compared to 45% for this study). Finally, water flow was added to the process to match the dried content of the Geosil 14515. This way of designing a process for PSS was confirmed by the industrial producer.

Glass fiber: Previously introduced in a formulation developed in the NAVIER laboratory and suitable for 3D printing [36], glass fibers were used in our reference formulation. The glass fiber came from a French producer located in the south of France (700 km from the lab), and the global market in the ecoinvent database was adjusted accordingly. A French electricity mix was considered for glass fiber production.

Wollastonite: Wollastonite production is not represented in the ecoinvent database. It is usually a mined stone, although it can be artificially produced. Here, only open mining was considered, using the asbestos chrysotile global production process as a reference. Transport distances were also adjusted, as the lab providers are located in Mexico.

Other raw materials: The other constituents were directly taken from the ecoinvent 3.7 databases. If possible, data for the French context were taken. When not available, Swiss (CH) or European (RER) data were considered.

Transportation of the matrix components: The transportation process was directly extracted from the Ecoinvent 3.7 library database without changes. We used the EURO4 class for all transportation vehicles (trucks). For PSS and MK, distances between the actual production sites and the NAVIER laboratory were calculated. The following driving distances were estimated using Google Maps:

- MK: Clérac—Champs sur Marne: 535 km.
- PSS: Ludwigshafen—Champs sur Marne: 504 km.

For the supply of cement, sand, and gravel, distances were estimated. A distance of 50 km was taken to account for displacement from the cement factory to the lab. This estimated distance is about 30 km for the aggregates (sand and gravel).

As the GP mortar was synthesized at the laboratory scale, and given the small number of required materials, it was considered that the vehicles used belonged to the 3.5–7.5 T category, for MK, G, and aggregates. As the process for Ordinary Portland Concrete (OPC) is more usual, 16–62 T lorries were considered for its transportation.

Electricity: The electricity consumption of the process was evaluated at 4 kWh per m^3 of produced concrete or mortar. This value is an expert-based estimation for regular concrete provided by the French National Project RECYBETON (https://www.pnrecybeton.fr/ accessed on 11 April 2024). It should be studied in more depth and adapted to GPs in further research. The French low-voltage market for electricity included in Ecoinvent v3.7.1 was chosen for the inventory.

2.3.3. Environmental Indicators

The main environmental issue usually related to traditional concrete and mortar is its impact on climate change (CC). This environmental category is thus a major focus of this study. Nevertheless, other impact categories (listed in Table 3) were examined to provide an overview of the environmental profile of the GP technology and avoid impact shifting.

Table 3. Abbreviations and units of the environmental indicator used in this study.

Impact Category	Abbreviation	Unit
Climate Change Total	CC	kg CO_2-eq
Freshwater and Terrestrial Acidification	FTA	mol H^+-eq
Freshwater Ecotoxicity	Fex	CTUe
Freshwater Eutrophication	Feu	kg P-eq
Marine Eutrophication	Meu	kg N-eq
Terrestrial Eutrophication	Teu	mol N-eq
Carcinogenic Effects	CE	CTUh
Ionizing Radiation	IR	kg Bq U^{235}
Non-Carcinogenic Effects	nCE	CTUh
Ozone Layer Depletion	OD	kg CFC-11-eq
Photochemical Ozone Creation	POCP	kg NMVOC eq
Respiratory Effects, Inorganics	RE	disease incidences
Resources, Dissipated Water	DW	m^3 water deprived
Resources, Fossils	RF	Megajoule
Resources, Land Use	RLU	soil quality index—dimensionless
Resources, Minerals, and Metals	RMM	kg Sb-Eq
Cumulative Energy Demand	CED	MJ-Eq

The LCA was first performed using the European consensus set of environmental indicators reached around the International Life Cycle Data initiative (ILCD). The midpoint set of indicators from the methodology ILCD 2.0 2018 was used [46,47]. To give an overview of the impact at the damage level to raise potential impact shifting among categories, a second assessment was performed using the ReCiPe2016, hierarchist method [48].

Although the cumulative energy demand is sometimes depreciated by LCA experts and seen as additional information more than a full LCA indicator [49], it was still added to the midpoint indicators because of its very frequent use in the construction sector. This set of impact categories aims to provide a comprehensive overview to avoid or at least quantify the phenomena of pollution transfers and impact shifts.

3. Results

3.1. Environmental Impact Assessment of Elementary Processes of Geopolymer Mortars and Reference Situation

In order to identify the parameters to be further optimized, a first environmental assessment of the reference formulation (GP–GfW) was performed. The impact factors of the elementary processes of the geopolymer production system used for the calculation are available in Supplementary Materials, Section S2. The contribution analysis of this formulation is exposed in Figure 5. Although the additions (wollastonite and glass fiber) are not negligible, the main driver for environmental impact in most categories is the potassium silicate solution (PSS), followed by the metakaolin and transport. The electricity contribution appears to be negligible in this study.

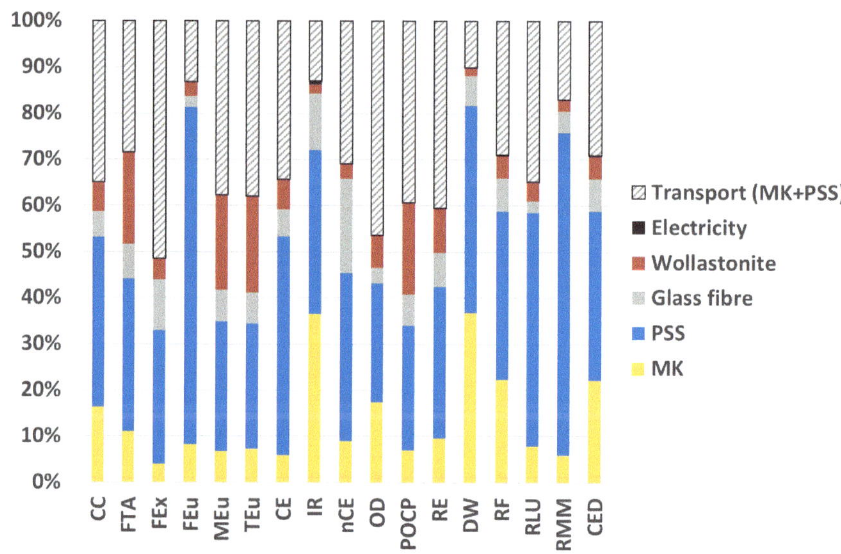

Figure 5. Reference situation for 3D-printing geopolymer formulation. CC = climate change, FTA = freshwater and terrestrial acidification, Fex = freshwater ecotoxicity, Feu = freshwater eutrophication, Meu = marine eutrophication, Teu = terrestrial eutrophication, CE = carcinogenic effect, IR = ionizing radiation, nCE = non-carcinogenic effects, OD = ozone depletion, POCP = photochemical ozone creation, RE = respiratory effects, DW = water depletion, RF = fossil resources, RMM = minerals and metals resources, CED = cumulative energy demand.

3.2. *Optimization of the Geopolymer Formulation to Lower GHG Emissions*

Based on the preliminary LCA results, the optimization of the matrix is undertaken to lower the quantity of PSS while maintaining sufficient mechanical properties. A formulation that lowers the matrix quantity by integrating a maximum of sand and earth additions and is suitable for the 3D-printing process was then developed and characterized mechanically.

3.2.1. The Geopolymer Matrix

In order to obtain a geopolymer formulation adapted for 3D printing with sufficient compressive strength and low GHG emissions, the geopolymer matrix was first optimized. Different ratios of metakaolin/potassium silicate solution were investigated with a metakaolin mass percentage ranging from 40 to 60%. The mechanical curves obtained for a $M_{50}S_{50}$ geopolymer are displayed in Figure 6a. This formulation exhibits a 60 $\pm$ 3 Mpa compressive strength and a brittle failure. The compressive strength and the climate change impact, expressed in $kgCO_2eq/m^3$ of the different geopolymer matrices are presented in Figure 6b. The compressive strength presents small variations but stays in the same order of magnitude (from 51 to 60 Mpa). When the metakaolin content increases from 40 to 50%, the compressive strength slightly increases due to enhanced polycondensation reaction [50]. It then slightly decreases with a further increase in metakaolin content, from 50 to 60%, probably because of the presence of unreactive particles and a decrease in the paste workability [51]. The climate change impact per m^3 increases slightly with increased

PSS content. Indeed, the PSS has a bigger impact per mass unit than metakaolin (0.63 and 0.58 $kgCO_2eq/kg$, respectively, for the $M_{40}S_{60}$ and $M_{60}S_{40}$ formulations). However, the density of the matrix increases with the decreased proportion of PSS which leads to an increase in impact per volume unit. Since the values of climate change impact are of the same order of magnitude, the $M_{50}S_{50}$ matrix presenting better mechanical properties seems optimal. In terms of buildability, printing tests show that every tested geopolymer matrix is not adapted for 3D printing because the paste flows when several layers are stacked during printing. Their yield stress has then to be increased to improve the buildability, which means to be able to carry the weight of the layers [52]. For that purpose, it is possible to add reinforcement elements such as glass fibers or wollastonite [38] or to change the volume fraction by adding, for instance, sand or earth [53].

(**a**) (**b**)

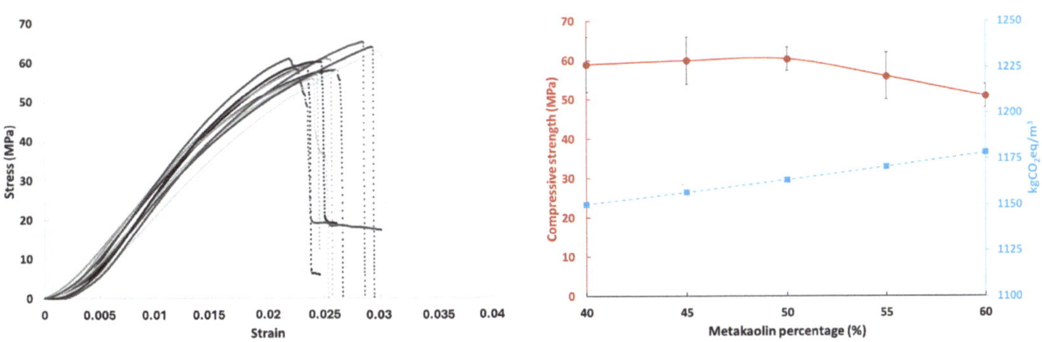

Figure 6. (**a**) Compressive curves of a $M_{50}S_{50}$ geopolymer and (**b**) • compressive strength and ■ climate change in $kgCO_2eq/m^3$ as a function of metakaolin percentage in the geopolymer matrix.

3.2.2. The Geopolymer Mortar

In order to decrease the matrix quantity and increase the buildability, sand was added to the different geopolymer matrices. To define the maximum quantity of sand to be added, two thresholds were determined. The first threshold corresponds to the quantity of sand for which the mortar begins to shear under mixing. After reaching this threshold, sand was continuously added to a point where a vibrating needle could not fluidize the mix anymore. The admissible range of sand quantity suitable for 3D printing lies between these two thresholds; before the first, the mortar is not stackable, and after the second, the mortar is not pumpable. These two thresholds are presented in Figure 7a for the different geopolymer binders. The thresholds are almost similar for a quantity of metakaolin ranging from 40 to 50% of the binder. With higher proportions of metakaolin, a saturation level is reached, and the value of the second threshold decreases. In that range, the water coming from the silicate solution is not sufficient in the geopolymer matrices, and the amount of sand it is possible to add decreases.

To ensure that the geopolymer mortar formulations are mechanically reliable, the compression strength of the $M_{60}S_{40}$ geopolymer binder was measured. This is reported in Figure 7b as a function of the quantity of added sand. The mechanical strength decreases slightly from 51 ± 3 MPa without sand to 44 ± 2 MPa with the addition of sand ($M_{60}S_{40}Sa_{60}$). Consequently, the addition of sand to form a printable geopolymer mortar has no significant effects on the properties of the material.

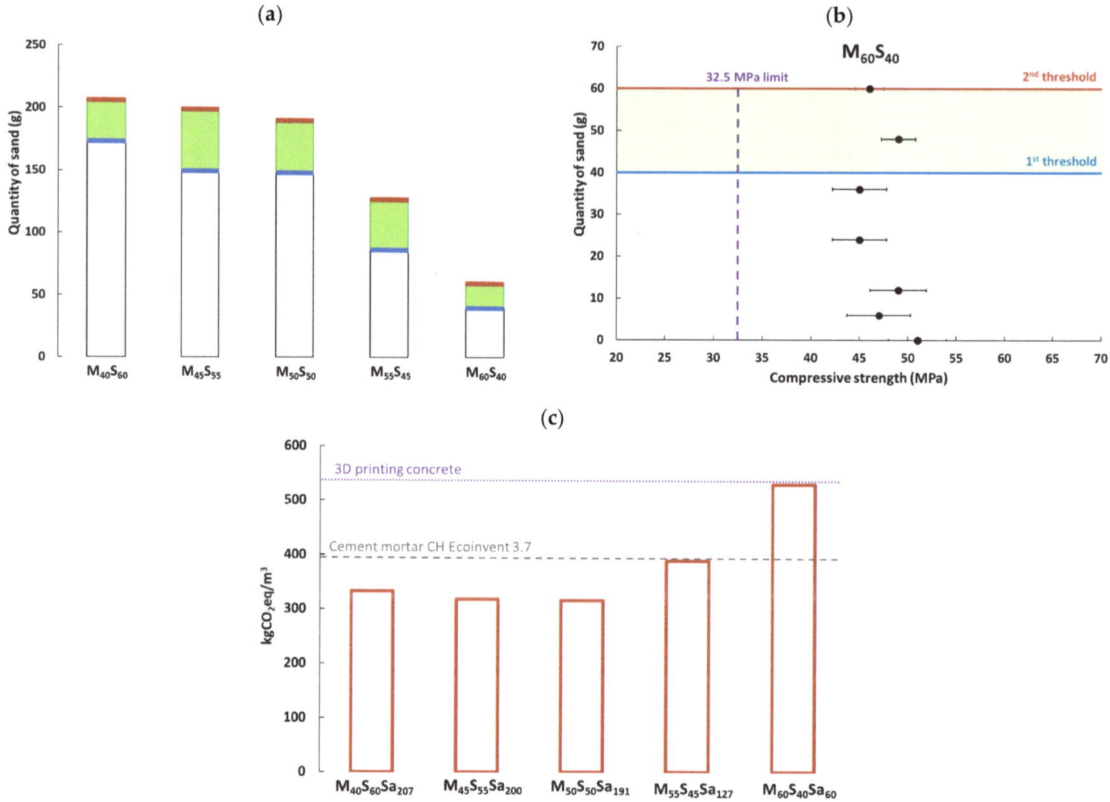

Figure 7. (**a**) quantity of sand added in different geopolymer matrices with the (■) printable domain between the ▪ first and ▪ second thresholds (**b**) quantity of sand as a function of compressive strength for an $M_{60}S_{40}$ formulation and (**c**) quantity of $kgCO_2/m^3$ for the second threshold mortar formulations.

The value of climate change impact per volume unit ($kgCO_2eq/m^3$) was then calculated for mortar formulations at the second threshold and is reported in Figure 7c. The climate change impact decreases slightly with a decrease in silicate solution until it reaches an optimum at 316 $kgCO_2eq/m^3$ for the $M_{50}S_{50}Sa_{191}$ formulation. It increases afterward. This optimum can be explained by two phenomena: (i) the silicate solution has a big impact on climate change and (ii) the formulation that contains a high quantity of silicate solution allows us to add more sand, which decreases the final impact. The impact of the $M_{50}S_{50}Sa_{191}$ formulation is 3.7 times lower than the ($M_{50}S_{50}$) matrix impact. The addition of sand in the geopolymer formulation then drastically decreases the climate change impact of geopolymers per unit of volume. Nevertheless, their impact depends on the quantity it is possible to add to the GP matrix. Moreover, the geopolymer formulations with high silicate content ($M_{40}S_{60}Sax$, $M_{45}S_{55}Sax$, $M_{40}S_{60}Sax$) have climate change impacts lower than 3D-printed concrete (529 $kgCO_2eq/m^3$) or generic mortar (393 $kgCO_2eq/m^3$) while keeping sufficient mechanical properties (above 40 MPa). These formulations can then be selected to progress toward a sustainable material for 3D printing.

3.2.3. Optimization with Earth Addition

After optimizing the geopolymer formulation through the addition of sand, tests were conducted using the mix of earth and sand determined in Figure 1. This mix was added in a 1:1 mass proportion with an $M_{50}S_{50}$ geopolymer matrix to obtain a printable material ($M_{50}S_{50}Sa_{66}E_{34}$). The printability and buildability of this formulation were tested with a manual extrusion tool, as shown in Figure 8a. As shown in previous work [38], this preliminary test provides an indication of the buildability of the material.

Figure 8. Photos of (**a**) printed layers of the earth–sand–M50S50Sa66E34 mix extruded with a manual tool and (**b**) a 40 × 40 × 160 mm sample.

The tested formulation presents a compressive strength equal to 31 ± 4 MPa. This is lower than the geopolymer matrix due to the earth's inclusion (Figure 8b) and a modification of the polycondensation reaction. However, it is still acceptable for construction applications. These results show that the optimization of the granular skeleton allows us to add local earth into the formulation to improve the circular economy while keeping sufficient mechanical properties and obtaining a mix adapted for 3D printing.

3.3. Comparison of the Optimized Geopolymer Formulation with Other Printing Materials

The environmental performance of the optimized formulations as printing materials was assessed in a comparative approach at a material scale. The robotic process necessary to print the material [54–56] was excluded from the scope of the study. The formulation with the addition of sand ($M_{50}S_{50}Sa_{191}$—GP-S) and the formulation with the addition of earth and sand ($M_{50}S_{50}Sa_{66}E_{34}$)—GP-SE) were compared to two other 3D-printing formulations—one based on Portland cement [54], named 3DCM, and one based on a geopolymer [36], with glass fibers and wollastonite additions (GP–GfW). The four formulations are summarized in Table 4.

Table 4. Formulation of the 3D-printing mortars.

Quantity (in kg/m^3)	3D Cement Mortar—«3DCM» [54]	Quantity (in kg/m^3)	GP Mortar «GP–GfW» [36]	GP Mortar «GP–S» (Section 3.2.2)	GP Mortar «GP–SE» (Section 3.2.3)
OPC	540	MK	915.4	327.4	505
Silica Fume	480	PSS	736.3	327.4	505
Sand	1033	Sand	0	1248.8	666.6
Water	212	Wollastonite	298.5	0	0
Accelerator	6	Glass fibers	39.8	0	0
Plasticizers	8.8	Steamed earth	0	0	333.3

Considering the Portland cement-based mortar 3DCM, most upstream requirements are accounted for in the market processes of the ecoinvent databases. Electricity consumption for the 3D cement mortar was added considering 4 kWh/m^3 and transport is also accounted for, using the process "market for concrete, 50 MPa, global" contextualized to Europe. The 4 scenarios are considered to have equivalent properties in terms of mechanical resistance, pumpability and extrudability, making them suitable for 3D printing in the perspective of structural uses. Relative results are provided for a clearer interpretation of the results, 3D cement mortar (3DCM) being the reference/denominator (Figure 9).

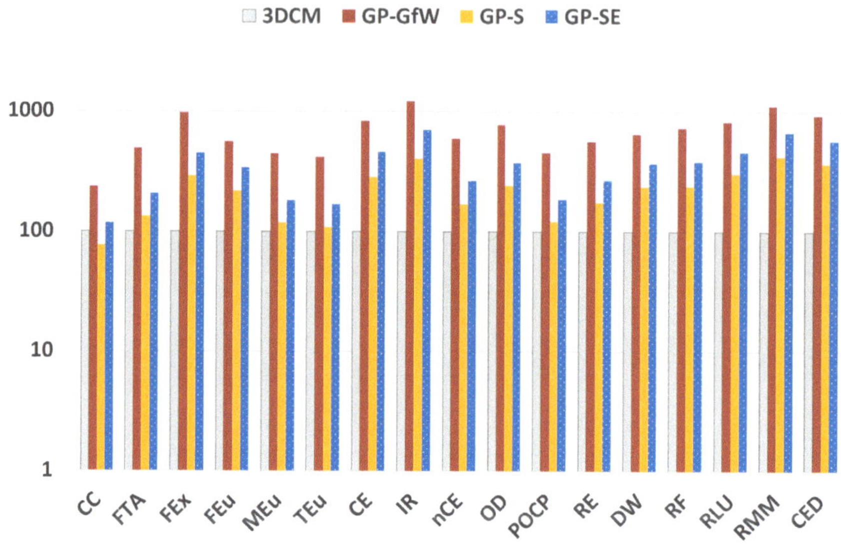

Figure 9. LCA of GP-S and GP-SE mortar, compared to 3DCM (used as reference, fixed at 100%) and GP–GfW (log scale).

With the currently tested formulations and in a French context, the results show that the GP formulations present impacts equivalent to or higher than 3DCM, except for climate change, for which GP-S has a lower impact (−23%). Secondly, the GP-S and GP-SE perform significantly better than the previous formulation with a higher share of the matrix (GP–GfW). However, GP-SE does not perform better than GP-S due to a higher matrix proportion. The climate change impact of geopolymer mortar highly depends on its formulation—whether by the nature or quantity of its raw material or by the optimization of the granular skeleton.

Figure 10 shows the important contribution of transportation of the material for GP mortar: from 7 to 47% according to certain categories (28% on average). The same figures for 3DCM vary from 4 to 42% (16% on average). This is related to the lab-scale and poor development of production sites for kaolinite. Another source of improvement clearly stands in the material production itself: PSS and MK productions are significant contributors to most categories, and further research should investigate the improvement of the production process, especially on toxicity-related impacts for PSS. Further effort to decrease the needed matrix quantity would also be very efficient as it will decrease both impacts from production and transport.

Figure 11 shows the contribution of different environmental issues to damage to human health, biodiversity, and resources, using the ReCiPe2016 method endpoint level. If climate change clearly dominates human health, other categories are not negligible, such as toxicity (cancer and non-cancer) and particulate matter (which can be associated with "respiratory effects"). On the biodiversity side, global warming dominates, followed by terrestrial acidification, photochemical ozone formation, and land use. A multi-criteria assessment is therefore essential to clearly prove the environmental relevance of the technology, restricting the evaluation to carbon footprint is not sufficient and could lead to impact shifting.

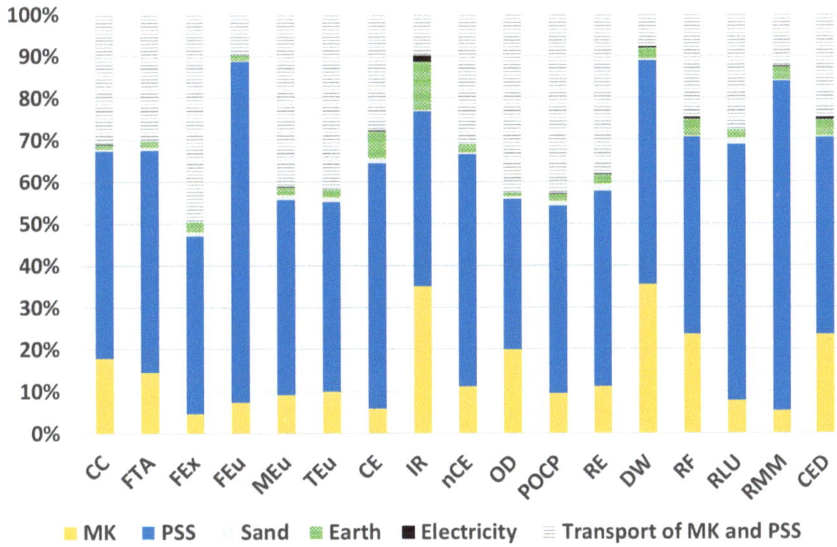

Figure 10. Contribution analysis of GP-SE mortar.

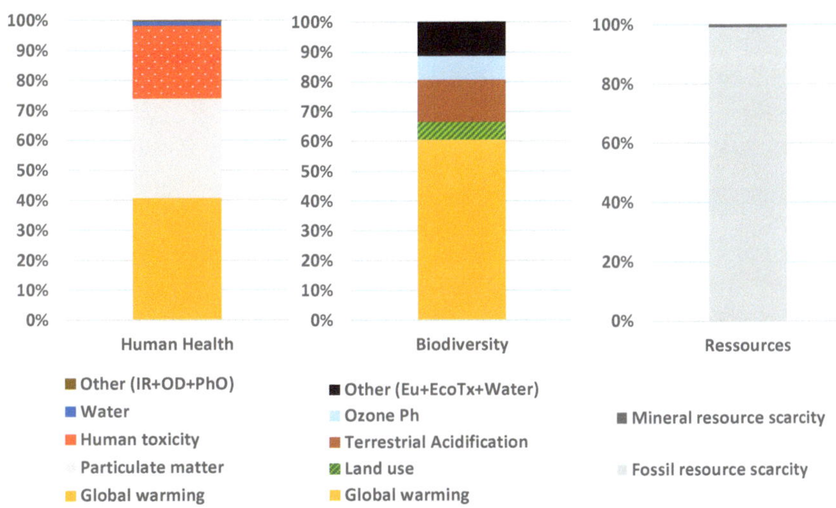

Figure 11. Midpoint contribution to Endpoint for GP-S concrete, Human Health (**left**), Biodiversity (**center**), and resource (**right**) using ReCiPe2016, hierarchist method.

Results for the GP materials are largely hampered by scale factors, as we compared a lab-scale technology to a fully deployed technology that has benefited from many years of innovation and improvements, and its small-scale production units are close to final users. Consequently, these direct results are to be taken with caution. A scale-up of the GP technology will help assess its long-term potential. It should cover both the possible optimization of the geopolymer elaboration process (equipment efficiency, raw materials) and the overall value chain (optimization of transport type and distance). This is a major perspective of this work and is further detailed in the Section 4.

4. Discussion

4.1. Value Chain Optimization

The developed process is currently an under-optimized laboratory process compared to well-established conventional materials such as concrete mortars. This explains why the impact of transport is so high compared to 3DCM. To account for this bias, a sensitivity analysis was performed on transport type and transport distances; first, we improved transport to 16–32 T EURO6 lorries (scenario T++) and secondly, we decreased transport distances from 500 km to 50 km for MK and PSS (scenario T++D−). The results are presented in Figure 12 below, which provides a selection of seven indicators among the most important ones according to the endpoint assessment.

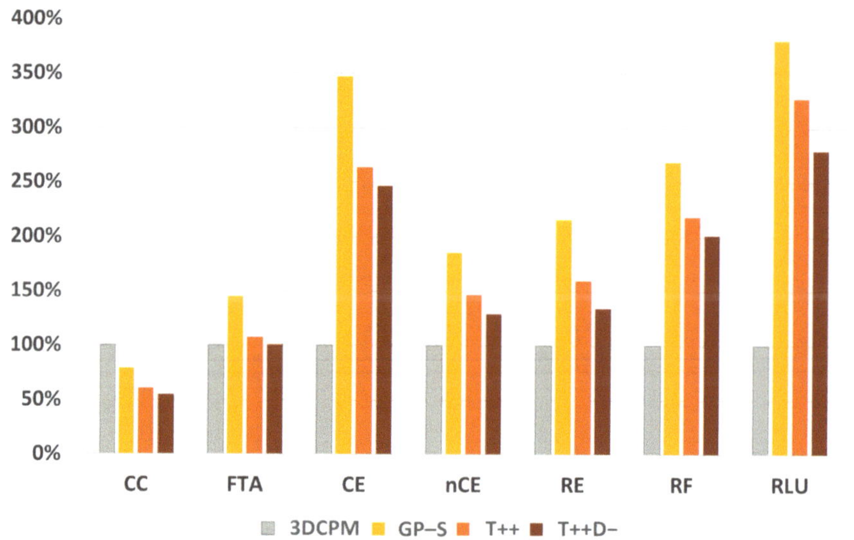

Figure 12. Assessment of the reference lab process and scale-up scenarios, midpoint level, considering 3DCM mortar as a baseline scenario.

There is great room for improvement, which could be made thanks to the scale effect. This limited analysis based on only two parameters should be further investigated to produce full scale-up scenarios, focusing both on process optimization and distribution options. Process improvements can be obtained thanks to bigger and more efficient equipment, such as MK flash calcination [57,58], a change in energy suppliers, or a change in PSS suppliers. Distribution options could involve the use of alternative raw materials such as low-kaolinite metakaolin that could be produced from clays with a high content of kaolinite [59]—an abundant resource often available locally in France [60]. Freight by train or by barge could also be envisioned to further decrease transport impacts. This represents an important research perspective of this work.

4.2. Matrix Optimization

The geopolymer matrix formulation was selected based on its mechanical performance and its climate change impact. This procedure could be further deployed to evaluate candidates in all environmental impact categories. The authors want to highlight the methodology set up to develop the geopolymer, which is simplified and restrained to one single indicator in the first step for readability. They acknowledge the importance of the multi-criteria approach in LCA.

The LCA results also showed that the formulation with sand (GP-S) exhibits better results than the formulation with earth (GP-SE). The use of locally excavated earth could, however, be beneficial as a strategy to reduce impacts related to the extraction of sand in the context of aggregate scarcity and a lack of environmental regulation [7,8].

Moreover, the use of sodium instead of potassium in the alkaline solution could also produce geopolymers suited for structural applications while reducing both the cost and the environmental burden [9]. Other synthesis processes could be investigated in a future study to use broadly available potash salts instead of potassium hydroxide in the confection of potassium silicate solution (waterglass).

Finally, the printability of the formulation has been measured in a simple way, even if better methods exist, like the squeeze test [61] or modified Vicat [62], for instance. In [63], we developed an original and efficient in-line test for 3D concrete, the slug test, which is able to deduce the yield stress from the weight of drops falling from the nozzle, which directly relies on the printability. In [36], we successfully adopted this method for the GP–GfW geopolymer used in this study for comparisons, and future tests will be conducted for the formulations integrating earth, to confirm their printability.

4.3. Uncertainties

Uncertainty values are high, as the data suffers from some severe flaws, especially in foreground processes. The estimation of this uncertainty is a critical issue for comprehensive LCA results [64]. For PSS, the data on potassium are inexistent, following an existing trend in the chemical industry [65]. The values for alkali silicates come from a 25-year-old study from Fawer and colleagues [45], as is the case in the vast majority of other LCA studies such as [19,22,66]. It strongly affects the temporal and technological correlation parameters and consequently gives high uncertainty to the related process. For MK, the material comes from widely different processes, and a high variability is observed in the literature [19,22,66]. Consequently, the technological correlation is low. Furthermore, an uncertainty assessment is not explained in this article, although it is considered to be a major perspective of this work.

The uncertainty related to the freshwater ecotoxicity indicator will hopefully soon be reduced with ongoing work on improving the characterization factor of metals. In the meantime, this result has to be taken with caution. For instance, the most important contributing elementary flows differ between the ILCD and Impact World+ indicators [67], with questions on both results and the overall contribution of this category in the environmental assessment of GP mortar.

4.4. Damage Assessment Method

As endpoint characterization methods are subject to higher uncertainties than the midpoint, a sensitivity analysis was performed using another method: ImpactWorld+ [67]. The results significantly differ, especially for biodiversity assessment and the contribution of long-term effects regarding freshwater ecotoxicity, which clearly dominates the ImpactWorld+ results. This impact category is subject to very high uncertainties, and this result is the consequence of only a few compounds highly persistent in the environment. Here, these are aluminum, copper, and iron, which come from upstream secondary processes related to sulfidic tailing from copper production. There is an important debate on ecotoxicity characterization factors for metals, which could eventually lead to a decrease in their contribution by at least two orders of magnitude [68,69]. There is also a current debate on the possible overestimation of tailing releases to the environment [70]. Excluding long-term effects to avoid this metal issue, the dominant category is climate change, closely followed by land use (related to transport) and acidification. Details on stressor contribution to freshwater ecotoxicity impacts are available in Supplementary Materials, Section S3.

The ReCiPe2016 method, a hierarchist scenario, used in this work does not consider the long-term effects of ecotoxicity (over 100 years), which could potentially lead to an underestimation of the impacts of very persistent compounds. However, stressor contribution results were consistent with the ILCD method used for midpoint assessment. Uncertain impacts must be assessed even if uncertain and ongoing important research efforts might soon lead to a better comprehension of occurring phenomena and a reduction of related uncertainties and variations among models.

4.5. Comparative LCA with 3D Printing Mortar Based on OPC

For simplicity reasons, the choice of a conventional equivalent was based on concrete with 100% OPC. Although 100% OPC is still widely used, a wide range of alternative clinkers have been developed [71]. It would be highly consistent to further include the proposed geopolymer in a wider perspective, comparing it to a wide scope of 3D-printing materials.

5. Conclusions and Perspective

The research presented in this paper focuses on geopolymer compositions iteratively developed for structural applications, with the goal of lowering the environmental impact. An LCA model for MK- and PSS-based geopolymer mortars was built. This model was used to optimize the GP matrix and granular skeleton starting from an existing formulation, through the reduction of GHG emissions. Two optimized formulations of GP mortars suitable for 3D printing were then compared with a 3D-printed mortar based on Portland cement. The results were completed by a sensitivity analysis of the transport of raw materials. The conclusions of this LCA show that 3D-printed GP formulations are not yet a mature technology, and short-term applications of this technology are not necessarily environmentally beneficial. Nevertheless, GPs hold great potential as they can divide GHG emissions by a factor close to two, with conceivable innovations. In the meantime, human and environmental toxicities, mineral depletion, and fossil resources are increased by the GP mortars—a trade-off already underlined by previous studies. The designed material shows good extrudability and buildability and could be used in high-performance applications thanks to its high compressive strength. Despite its low proportion of aggregates, this material displays significant environmental assets as it contains an important share of earth that is widely available and often treated as waste.

Some other prospective works lie ahead and are worth exploring. First, there is a need for a prospective resource-availability assessment for the GP technology (alkali, kaolin clays) to ensure that the required resources are widely available or to define locations that could most benefit from the technology. Secondly, an environmental assessment of the large-scale use of 3D printing in building applications must be conducted, as the literature is still scarce. A robust uncertainty assessment should be undertaken with rigorous sampling for processes specific to geopolymers. Furthermore, the social impact of such a new technique is also worth exploring before its wider application.

Supplementary Materials: The following supporting information can be downloaded at: https://www.mdpi.com/article/10.3390/su16083328/s1. Section S1: Process of Detailed Modelization; Section S2: Impact Categories; Section S3: LCA Results: Stressor Contribution to Ecotoxicity.

Author Contributions: C.R.: Conceptualization, Writing—Original draft preparation, visualization, validation—environmental assessment part; J.A.: Conceptualization, Writing—Original draft preparation, Visualization, validation—mechanical and workability characterization part; C.L.G.: Conceptualization, Methodology, Investigation, Software; M.S.: Writing—Original draft preparation, Critical Review and Editing, A.F.: Writing—Review and Editing, supervision; J.-F.C.: Writing—Review and Editing, supervision, Project administration, funding acquisition. All authors have read and agreed to the published version of the manuscript.

Funding: This work was performed using the funds of the NAVIER research lab.

Informed Consent Statement: Not applicable.

Data Availability Statement: Data is contained within the article or Supplementary Materials. A dataset for more detailed information is available on request from the authors.

Conflicts of Interest: The authors declare no conflicts of interest.

References

1. Miller, S.A.; Moore, F.C. Climate and health damages from global concrete production. *Nat. Clim. Chang.* **2020**, *10*, 439–443. [CrossRef]
2. Shi, C.; Jiménez, A.F.; Palomo, A. New cements for the 21st century: The pursuit of an alternative to Portland cement. *Cem. Concr. Res.* **2011**, *41*, 750–763. [CrossRef]
3. Habert, G.; Miller, S.A.; John, V.M.; Provis, J.L.; Favier, A.; Horvath, A.; Scrivener, K.L. Environmental impacts and decarbonization strategies in the cement and concrete industries. *Nat. Rev. Earth Environ.* **2020**, *1*, 559–573. [CrossRef]
4. Makul, N. Modern sustainable cement and concrete composites: Review of current status, challenges and guidelines. *Sustain. Mater. Technol.* **2020**, *25*, e00155. [CrossRef]
5. Sivakrishna, A.; Adesina, A.; Awoyera, P.O.; Rajesh Kumar, K. Green concrete: A review of recent developments. *Mater. Today Proc.* **2020**, *27*, 54–58. [CrossRef]
6. Van Deventer, J.S.J.; Provis, J.L.; Duxson, P. Technical and commercial progress in the adoption of geopolymer cement. *Miner. Eng.* **2012**, *29*, 89–104. [CrossRef]
7. UNEP. *Sand and Sustainability: Finding New Solutions for Environmental Governance of Global Sand Resources*; GRID-Geneva, United Nations Environment Programm: Geneva, Switzerland, 2019; Available online: https://unepgrid.ch/storage/app/media/documents/Sand_and_sustainability_UNEP_2019.pdf (accessed on 26 March 2021).
8. Bendixen, M.; Best, J.; Hackney, C.; Iversen, L.L. Time is running out for sand. *Nature* **2019**, *571*, 29–31. [CrossRef] [PubMed]
9. Davidovits, J. *Geopolymer Chemistry and Applications*; Geopolymer Institute: Saint-Quentin, France, 2008; ISBN 978-2-9514820-1-2.
10. Zhang, P.; Zheng, Y.; Wang, K.; Zhang, J. A review on properties of fresh and hardened geopolymer mortar. *Compos. Part B Eng.* **2018**, *152*, 79–95. [CrossRef]
11. Amran, Y.H.M.; Alyousef, R.; Alabduljabbar, H.; El-Zeadani, M. Clean production and properties of geopolymer concrete; A review. *J. Clean. Prod.* **2020**, *251*, 119679. [CrossRef]
12. Pradhan, P.; Dwibedy, S.; Pradhan, M.; Panda, S.; Panigrahi, S.K. Durability characteristics of geopolymer concrete—Progress and perspectives. *J. Build. Eng.* **2022**, *59*, 105100. [CrossRef]
13. Monteiro, P.J.M.; Miller, S.A.; Horvath, A. Towards sustainable concrete. *Nat. Mater.* **2017**, *16*, 698–699. [CrossRef] [PubMed]
14. Provis, J.L. Green concrete or red herring?—Future of alkali-activated materials. *Adv. Appl. Ceram.* **2014**, *113*, 472–477. [CrossRef]
15. Curran, M.A. *Life Cycle Assessment Handbook: A Guide for Environmentally Sustainable Products*; John Wiley & Sons: Hoboken, NJ, USA, 2012; ISBN 978-1-118-52841-9.
16. Guinée, J.B.; Lindeijer, E. *Handbook on Life Cycle Assessment: Operational Guide to the ISO Standards*; Springer Science & Business Media: Berlin/Heidelberg, Germany, 2002; ISBN 978-1-4020-0228-1.
17. Ghadir, P.; Zamanian, M.; Mahbubi-Motlagh, N.; Saberian, M.; Li, J.; Ranjbar, N. Shear strength and life cycle assessment of volcanic ash-based geopolymer and cement stabilized soil: A comparative study. *Transp. Geotech.* **2021**, *31*, 100639. [CrossRef]
18. Meshram, R.B.; Kumar, S. Comparative life cycle assessment (LCA) of geopolymer cement manufacturing with Portland cement in Indian context. *Int. J. Environ. Sci. Technol.* **2022**, *19*, 4791–4802. [CrossRef]
19. Habert, G.; d'Espinose de Lacaillerie, J.B.; Roussel, N. An environmental evaluation of geopolymer based concrete production: Reviewing current research trends. *J. Clean. Prod.* **2011**, *19*, 1229–1238. [CrossRef]
20. Weil, M.; Dombrowski, K.; Buchwald, A. 10—Life-cycle analysis of geopolymers. In *Geopolymers*; Provis, J.L., van Deventer, J.S.J., Eds.; Woodhead Publishing: Sawston, UK, 2009; pp. 194–210. ISBN 978-1-84569-449-4.
21. Matheu, P.S.; Ellis, K.; Varela, B. Comparing the Environmental Impacts of Alkali Activated Mortar and Traditional Portland Cement Mortar using Life Cycle Assessment. *IOP Conf. Ser. Mater. Sci. Eng.* **2015**, *96*, 012080. [CrossRef]
22. Heath, A.; Paine, K.; McManus, M. Minimising the global warming potential of clay based geopolymers. *J. Clean. Prod.* **2014**, *78*, 75–83. [CrossRef]
23. Davidovits, J. False Values on CO2 Emission for Geopolymer Cement/Concrete, Scientific Papers, Technical Paper #24, Geopolymer Institute Library. 2015. Available online: http://www.geopolymer.org/wp-content/uploads/False-CO2-values.pdf (accessed on 2 September 2021).
24. Yao, Y.; Hu, M.; Maio, F.D.; Cucurachi, S. Life cycle assessment of 3D printing geo-polymer concrete: An ex-ante study. *J. Ind. Ecol.* **2020**, *24*, 116–127. [CrossRef]
25. Morel, J.-C.; Charef, R.; Hamard, E.; Fabbri, A.; Beckett, C.; Bui, Q.-B. Earth as construction material in the circular economy context: Practitioner perspectives on barriers to overcome. *Philos. Trans. R. Soc. B Biol. Sci.* **2021**, *376*, 20200182. [CrossRef]
26. Luzu, B.; Duc, M.; Djerbi, A.; Gautron, L. High Performance Illitic Clay-Based Geopolymer: Influence of the Mechanochemical Activation Duration on the Strength Development. In *Calcined Clays for Sustainable Concrete*; Bishnoi, S., Ed.; Springer: Singapore, 2020; pp. 363–373.
27. Essaidi, N.; Samet, B.; Baklouti, S.; Rossignol, S. Feasibility of producing geopolymers from two different Tunisian clays before and after calcination at various temperatures. *Appl. Clay Sci.* **2014**, *88–89*, 221–227. [CrossRef]
28. De Schutter, G.; Lesage, K.; Mechtcherine, V.; Nerella, V.N.; Habert, G.; Agusti-Juan, I. Vision of 3D printing with concrete—Technical, economic and environmental potentials. *Cem. Concr. Res.* **2018**, *112*, 25–36. [CrossRef]
29. Song, R.; Wang, Y.; Ishutov, S.; Zambrano-Narvaez, G.; Hodder, K.J.; Chalaturnyk, R.J.; Sun, S.; Liu, J.; Gamage, R.P. A Comprehensive Experimental Study on Mechanical Behavior, Microstructure and Transport Properties of 3D-printed Rock Analogs. *Rock. Mech. Rock. Eng.* **2020**, *53*, 5745–5765. [CrossRef]

30. Hassan, A.; Arif, M.; Shariq, M. Use of geopolymer concrete for a cleaner and sustainable environment—A review of mechanical properties and microstructure. *J. Clean. Prod.* **2019**, *223*, 704–728. [CrossRef]
31. Singh, B.; Ishwarya, G.; Gupta, M.; Bhattacharyya, S.K. Geopolymer concrete: A review of some recent developments. *Constr. Build. Mater.* **2015**, *85*, 78–90. [CrossRef]
32. Zakka, W.P.; Abdul Shukor Lim, N.H.; Chau Khun, M. A scientometric review of geopolymer concrete. *J. Clean. Prod.* **2021**, *280*, 124353. [CrossRef]
33. Komnitsas, K. Potential of geopolymer technology towards green buildings and sustainable cities. *Procedia Eng.* **2011**, *21*, 1023–1032. [CrossRef]
34. Zhong, H.; Zhang, M. 3D printing geopolymers: A review. *Cem. Concr. Compos.* **2022**, *128*, 104455. [CrossRef]
35. Turner, L.K.; Collins, F.G. Carbon dioxide equivalent (CO_2-e) emissions: A comparison between geopolymer and OPC cement concrete. *Constr. Build. Mater.* **2013**, *43*, 125–130. [CrossRef]
36. Archez, J.; Maitenaz, S.; Demont, L.; Charrier, M.; Mesnil, R.; Texier-Mandoki, N.; Bourbon, X.; Rossignol, S.; Caron, J.F. Strategy to shape, on a half-meter scale, a geopolymer composite structure by additive manufacturing. *Open Ceram.* **2021**, *5*, 100071. [CrossRef]
37. *NF-EN 197-1*; Cement—Part 1: Composition, Specifications and Conformity Criteria for Common Cements. NFE: Paris, France, 2001.
38. Archez, J.; Texier-Mandoki, N.; Bourbon, X.; Caron, J.F.; Rossignol, S. Adaptation of the geopolymer composite formulation binder to the shaping process. *Mater. Today Commun.* **2020**, *25*, 101501. [CrossRef]
39. Gökçe, H.S.; Tuyan, M.; Nehdi, M.L. Alkali-activated and geopolymer materials developed using innovative manufacturing techniques: A critical review. *Constr. Build. Mater.* **2021**, *303*, 124483. [CrossRef]
40. Archez, J.; Texier-Mandoki, N.; Bourbon, X.; Caron, J.F.; Rossignol, S. Shaping of geopolymer composites by 3D printing. *J. Build. Eng.* **2021**, *34*, 101894. [CrossRef]
41. Mutel, C. Brightway: An open source framework for Life Cycle Assessment. *JOSS* **2017**, *2*, 236. [CrossRef]
42. Steubing, B.; de Koning, D.; Haas, A.; Mutel, C.L. The Activity Browser—An open source LCA software building on top of the brightway framework. *Softw. Impacts* **2020**, *3*, 100012. [CrossRef]
43. Frischknecht, R.; Jungbluth, N.; Althaus, H.-J.; Doka, G.; Dones, R.; Heck, T.; Hellweg, S.; Hischier, R.; Nemecek, T.; Rebitzer, G.; et al. The ecoinvent Database: Overview and Methodological Framework (7 pp). *Int. J. Life Cycle Assess.* **2005**, *10*, 3–9. [CrossRef]
44. Wernet, G.; Bauer, C.; Steubing, B.; Reinhard, J.; Moreno-Ruiz, E.; Weidema, B. The ecoinvent database version 3 (part I): Overview and methodology. *Int. J. Life Cycle Assess.* **2016**, *21*, 1218–1230. [CrossRef]
45. Fawer, M.; Concannon, M.; Rieber, W. Life cycle inventories for the production of sodium silicates. *Int. J. LCA* **1999**, *4*, 207. [CrossRef]
46. Chomkhamsri, K.; Wolf, M.-A.; Pant, R. International Reference Life Cycle Data System (ILCD) Handbook: Review Schemes for Life Cycle Assessment. In *Towards Life Cycle Sustainability Management*; Finkbeiner, M., Ed.; Springer: Amsterdam, The Netherlands, 2011; pp. 107–117. ISBN 978-94-007-1898-2.
47. Saouter, E.; Biganzoli, F.; Ceriani, L.; Versteeg, D.; Crenna, E.; Zampori, L.; Sala, S.; Pant, R. *Environmental Footprint: Update of Life Cycle Impact Assessment Methods: Ecotoxicity Freshwater, Human Toxicity Cancer, and Non Cancer*; Publications Office of the European Union: Luxembourg, 2020; ISBN 978-92-76-17143-0.
48. Huijbregts, M.A.J.; Steinmann, Z.J.N.; Elshout, P.M.F.; Stam, G.; Verones, F.; Vieira, M.; Zijp, M.; Hollander, A.; van Zelm, R. ReCiPe2016: A harmonised life cycle impact assessment method at midpoint and endpoint level. *Int. J. Life Cycle Assess.* **2017**, *22*, 138–147. [CrossRef]
49. Huijbregts, M.A.J.; Rombouts, L.J.A.; Hellweg, S.; Frischknecht, R.; Hendriks, A.J.; van de Meent, D.; Ragas, A.M.J.; Reijnders, L.; Struijs, J. Is Cumulative Fossil Energy Demand a Useful Indicator for the Environmental Performance of Products? *Environ. Sci. Technol.* **2006**, *40*, 641–648. [CrossRef]
50. Gharzouni, A.; Sobrados, I.; Balouti, S.; Joussein, E.; Rossignol, S. Control of polycondensation reaction generated from different metakaolins and alkaline solutions. *J. Ceram. Sci. Technol.* **2017**, *8*, 365–376.
51. Jaya, N.A.; Liew, Y.M.; Heah, C.Y.; Abdullah, M.M.A.B. Effect of solid-to-liquid ratios on metakaolin geopolymers. *AIP Conf. Proc.* **2018**, *2045*, 020099. [CrossRef]
52. Roussel, N. Rheological requirements for printable concretes. *Cem. Concr. Res.* **2018**, *112*, 76–85. [CrossRef]
53. Romagnoli, M.; Leonelli, C.; Kamse, E.; Lassinantti Gualtieri, M. Rheology of geopolymer by DOE approach. *Constr. Build. Mater.* **2012**, *36*, 251–258. [CrossRef]
54. Kuzmenko, K.; Roux, C.; Feraille, A.; Baverel, O. Assessing environmental impact of digital fabrication and reuse of constructive systems. *Structures* **2021**, *31*, 1300–1310. [CrossRef]
55. Agusti-Juan, I.; Habert, G. An Environmental Perspective on Digital Fabrication in Architecture and Construction. In Proceedings of the 21st International Conference on Computer-Aided Architectural Design Research in Asia (Caadria 2016), Melbourne, Australia, 30 March–2 April 2016; pp. 797–806. Available online: http://papers.cumincad.org/data/works/att/caadria2016_797.pdf (accessed on 7 May 2021).
56. Agustí-Juan, I.; Müller, F.; Hack, N.; Wangler, T.; Habert, G. Potential benefits of digital fabrication for complex structures: Environmental assessment of a robotically fabricated concrete wall. *J. Clean. Prod.* **2017**, *154*, 330–340. [CrossRef]
57. San Nicolas, R.; Cyr, M.; Escadeillas, G. Characteristics and applications of flash metakaolins. *Appl. Clay Sci.* **2013**, *83–84*, 253–262. [CrossRef]

58. Deteuf, C. Imerys Metakaolin Production.pdf. 2016. Available online: https://geopolymer.org/fichiers/gpcamp-2016/Deteuf%20-%20Imerys%20metakaolin%20production.pdf (accessed on 23 September 2021).
59. Batis, G.; Pantazopoulou, P.; Tsivilis, S.; Badogiannis, E. The effect of metakaolin on the corrosion behavior of cement mortars. *Cem. Concr. Compos.* **2005**, *27*, 125–130. [CrossRef]
60. Handbook on Kaolin and Kaolinic Clays | BRGM. Available online: https://www.brgm.fr/en/reference-completed-project/handbook-kaolin-kaolinic-clays (accessed on 7 April 2023).
61. Toutou, Z.; Roussel, N.; Lanos, C. The squeezing test: A tool to identify firm cement-based material's rheological behaviour and evaluate their extrusion ability. *Cem. Concr. Res.* **2005**, *35*, 1891–1899. [CrossRef]
62. Sleiman, H.; Perrot, A.; Amziane, S. A new look at the measurement of cementitious paste setting by Vicat test. *Cem. Concr. Res.* **2010**, *40*, 681–686. [CrossRef]
63. Ducoulombier, N.; Mesnil, R.; Carneau, P.; Demont, L.; Bessaies-Bey, H.; Caron, J.-F.; Roussel, N. The "Slugs-test" for extrusion-based additive manufacturing: Protocol, analysis and practical limits. *Cem. Concr. Compos.* **2021**, *121*, 104074. [CrossRef]
64. Heijungs, R.; Huijbregts, M.A. A Review of Approaches to Treat Uncertainty in LCA. 2004. Available online: http://scholarsarchive.byu.edu/iemssconference/2004/all/197/ (accessed on 7 September 2017).
65. Oberschelp, C.; Hellweg, S.; Bradford, E.; Pfister, S.; Huo, J.; Wang, Z. Poor Data and Outdated Methods Sabotage the Decarbonization Efforts of the Chemical Industry. 2023. Available online: https://doi.org/10.26434/chemrxiv-2023-8c86t (accessed on 25 May 2023).
66. Shobeiri, V.; Bennett, B.; Xie, T.; Visintin, P. A comprehensive assessment of the global warming potential of geopolymer concrete. *J. Clean. Prod.* **2021**, *297*, 126669. [CrossRef]
67. Bulle, C.; Margni, M.; Patouillard, L.; Boulay, A.-M.; Bourgault, G.; De Bruille, V.; Cao, V.; Hauschild, M.; Henderson, A.; Humbert, S.; et al. IMPACT World+: A globally regionalized life cycle impact assessment method. *Int. J. Life Cycle Assess.* **2019**, *24*, 1653–1674. [CrossRef]
68. Hedberg, J.; Fransson, K.; Prideaux, S.; Roos, S.; Jönsson, C.; Odnevall Wallinder, I. Improving the Life Cycle Impact Assessment of Metal Ecotoxicity: Importance of Chromium Speciation, Water Chemistry, and Metal Release. *Sustainability* **2019**, *11*, 1655. [CrossRef]
69. Gandhi, N.; Diamond, M.L. Freshwater ecotoxicity characterization factors for aluminum. *Int. J. Life Cycle Assess.* **2018**, *23*, 2137–2149. [CrossRef]
70. Muller, S.; Lassin, A.; Lai, F.; Thiéry, D.; Guignot, S. Modelling releases from tailings in life cycle assessments of the mining sector: From generic models to reactive transport modelling. *Miner. Eng.* **2022**, *180*, 107481. [CrossRef]
71. Gartner, E.; Sui, T. Alternative cement clinkers. *Cem. Concr. Res.* **2018**, *114*, 27–39. [CrossRef]

sustainability

MDPI

Article

Enhancing Sustainable Construction: Optimization Tool for Glulam Roof Structures According to Eurocode 5

María Simón-Portela [1], José Ramón Villar-García [2], Pablo Vidal-López [1,*] and Desirée Rodríguez-Robles [1]

[1] Mechanical and Fluid Engineering Research Group, Department of Forest and Agricultural Engineering, School of Agricultural Engineering, University of Extremadura, Av. Adolfo Suarez s/n, 06071 Badajoz, Spain; msimonu@alumnos.unex.es (M.S.-P.); desireerodriguez@unex.es (D.R.-R.)

[2] Forest Research Group, Department of Forest and Agricultural Engineering, Universitary Center of Plasencia, University of Extremadura, Av. Virgen del Puerto 2, 10600 Plasencia, Spain; jrvillar@unex.es

* Correspondence: pvidal@unex.es

Abstract: The construction industry has a notably negative impact on the environment; thus, the promotion of the use of timber structures is an alternative to mitigate its effects. This research develops an artificial intelligence-based decision approach in the calculation of timber structures focused on the enhancement of the sustainability of roof structures. Based on the optimization carried out through genetic algorithms and the framework established in Eurocode 5, a general set of equations has been proposed for a laminated timber roof structure. The tool, which determines the most suitable roof structure for each strength class of laminated timber, allows for the determination of the dimensions of beams and purlins and their respective separations in order to minimize wood consumption. The ultimate goal is to offer multiple solutions regarding strength classes and structural designs in order to foster sustainability-informed choices that promote efficient use of resources in construction.

Keywords: artificial intelligence optimization; genetic algorithm; timber; strength class; roof structure; Eurocode 5

check for updates

Citation: Simón-Portela, M.; Villar-García, J.R.; Vidal-López, P.; Rodríguez-Robles, D. Enhancing Sustainable Construction: Optimization Tool for Glulam Roof Structures According to Eurocode 5. *Sustainability* **2024**, *16*, 3514. https://doi.org/10.3390/su16093514

Academic Editors: Uroš Klanšek and Tomaž Žula

Received: 8 March 2024
Revised: 18 April 2024
Accepted: 19 April 2024
Published: 23 April 2024

1. Introduction

Its economic and social relevance notwithstanding, the construction sector in the European Union (EU) consumes a great amount of resources, with figures up to 50% of all extracted materials [1], and produces vast quantities of waste, accounting for 37.5% of the total [2]. Moreover, the built environment is responsible for 40% of the total energy consumption and 36% of all CO_2 emissions in the EU [3].

Among some of the most commonly employed construction materials are concrete and steel. Understanding the properties and constraints of materials is essential to both the design and future construction and use; however, evaluating the environmental impact has become equally critical as climate change and sustainability concerns increase. Thus, the use of life cycle assessment (LCA) databases allows for the assessment and comparison of construction materials' environmental performance through the use of indicators such as embodied energy or CO_2 emissions. For instance, the Ecoinvent database [4] indicates that steel has an embodied energy of 27.90 MJ/kg and emissions of 1.71 kg CO_2/kg. Similarly, values of 0.618 MJ/kg and 0.112 kg CO_2/kg are reported for concrete [4].

Nowadays, similarly to other economic sectors, construction is veering towards practices within material selection, construction methods, operational efficiency, and end-of-life strategies that regard social and environmental well-being to further sustainability. In this regard, the European Green Deal [5] promotes the use of timber as a means to reduce the environmental pressure of the construction industry.

Although the use of timber is not new, approaches to alleviate some of the inherent limitations of wood (i.e., dimensions, strength, instability, fire resistance, susceptibility to biodegradation, etc.) through the development of wood-engineered products [6,7] and various technological advances (e.g., chemical, thermal, or mechanical treatments) [6,8,9] are propelling its current popularity. Moreover, another factor driving the increase in use is the comparatively better environmental profile with respect to other construction materials [10]. As a renewable resource, timber can be continually sourced from sustainably managed forests, which contributes to the maintenance and expansion of forested areas, thereby mitigating deforestation and promoting ecological quality [11]. Since it originates from a photosynthetic organism, uptake of carbon occurs during its growth due to biomass conversion, with approximately 1.5 t CO_2/m^3 of wood [12]. Thus, sequestration or carbon storage could be considered in the manufacture of timber, provided that the harvested tree is used in long-life products such as construction materials. Moreover, the use of timber in substitution of other more environmentally damaging materials constitutes an additional benefit in CO_2 mitigation, since the manufacturing emissions of the replaced construction material are avoided. After sawing, lumber is naturally or force-dried to achieve dimensional stability and planed according to use or processing into other wood-engineered products. Undoubtedly, these operations pose a negative impact on the environment. Although precise figures vary for the different wood products, both carbon emissions and embodied energy are significantly lower in timber construction [10,13–16].

Among the wood-engineered materials, this research focuses on glued laminated timber (also known as glulam). This construction material is composed of layers of sawn lumber bonded together, with the grain running parallel to the length of the structural element. This manufacturing allows for large spans and variable cross sections, as well as high strength-to-weight ratios. Nevertheless, the different strength classes (e.g., from GL 20h to GL 32h for homogeneous glulam [17]) exhibit specific mechanical performances relating to the wood species. For instance, lower strength classes could be associated with softwoods (*Thuja plicata*, *Picea sitchensis*, *Abies magnifica*, *Abies grandis*, *Abies concolor*, *Abies procera*, *Abies amabilis*, etc.), whereas greater strength classes are connected to *Pinus sylvestris*, *Larix decidua*, *Pseudotsuga menziesii*, etc. [18].

In this regard, the scientific community has predominantly centered its attention on the study of a single strength class, and the interest has not been uniformly distributed among them. Conversely, it is worth mentioning the research by Baranski et al. [19], who addressed the optimization of different beam geometries as well as different qualities of glulam (i.e., GL 22h, GL 24h, GL 26h, GL 28h, GL 30h, and GL 32h).

From the literature analysis, GL 24h stands out as the most evaluated class, with a greater incidence of studies on bending elements such as beams. In the context of structure optimization, two investigations can be highlighted. Firstly, Jelušič and Kravanka [20] assessed the optimization design of timber floor joists for a given imposed load and span of the structure. Then, De Vito et al. [21], who developed a topology optimization of Douglas fir GL 24h beams, considered the orthotropic nature of wood and the layering manufacture. Kilincarslan and Turker [22] considered the strengthening of the GL 24h spruce timber column–beam connection with carbon fiber-reinforced polymer through experimental evaluation. Wang et al. [23] evaluated the stiffness of GL 24h timber from Scotch pine beam–column joints with bolted connections and introduced a stiffness prediction method. Moreover, in the current state of transition to more sustainable materials, there are also some examples of the hybrid use of GL 24h with inert materials. Fu et al. [24] studied the optimization of the bonding performance between prefabricated concrete and GL 24h timber from spruce. Similarly, Giv et al. [25] studied the effect of adhesive type on the bending behavior of the GL 24h timber–concrete composite panel. Ferrara et al. [26] conducted real-scale experiments on the mechanical performance of a GL 24h timber–concrete composite floor supported at two edges. Gomez-Ceballos et al. [27] presented a numerical method for the analysis of the mechanical behavior of GL 24h Douglas beams reinforced with ultrahigh-performance fiber-reinforced concrete. For the timber–steel conjunction, Ching et al. [28] developed a topology optimization framework for trusses made of GL 24h timber and steel aimed at reduce global warming potential.

Several authors have also assessed the GL 28h class, mostly in the context of reinforced elements such as tendons or steel bars. For instance, De Luca and Marano [29] tested the failure of prestressed GL 28h spruce timber reinforced with steel bars. Also for European spruce, McConnell et al. [30] tested GL 28h timber with steel tendons in post-tensioning conditions. In regard to optimization, Mam et al. [31] focused on GL 28h bracing structures. For a timber–timber composite with glulam ribs and cross-laminated timber flanges, Suárez-Riestra et al. [32,33] tested the GL 28h from *Picea abies* and proposed an estimation model for the long-term behavior of the composite.

Amid the less studied strength classes, an investigation carried out by Jelušič [34], who proposed an optimization approach to variable cross-section beams of GL 30h timber should be mentioned, as well as the research performed on GL 32h timber by Šilih et al. [35] for timber trusses and Simón-Portela [36] for an entire timber roof structure.

Despite the current clear focus on specific strength classes, the complete consideration of the strength class range within the material selection could benefit the optimization of the timber volume required for a specific structure, which is one of the objectives of this research. In this regard, there is extensive literature on reducing material consumption in the design of steel and concrete structures; some examples can be found in [37–39]. Albeit more limited, optimization approaches have also been made for timber construction, such as beams [19,34,36] and trusses [20,35,40,41].

Additionally, the consideration of the use of the strength classes could also result in advantages from a sustainability standpoint, since a greater variety of wood species would be considered for construction purposes. It should be noted that besides their performance and classification within a strength class, the selection of available tree species would also be motivated by their intrinsic characteristics (e.g., climatic adaptation, growing rate [34], etc.) as well as the final acquisition cost of the material. In this regard, research shows that tree species richness can enhance wood productivity while maximizing ecosystem functioning [42–44]. Notwithstanding, to adequately supply the increasing demand for wood as a construction material, the source material must originate from responsibly managed forests.

Therefore, the present investigation assesses the influence of glulam strength classes in timber construction, specifically, on the design of a timber roof structure consisting of timber double-tapered beams and purlins according to Eurocode 5 [45] requirements. To that end, an optimization tool based on genetic algorithms has been developed, and based on the results, several equations have been proposed to determine the optimal geometry (width and height) of the structural elements (beams and purlins) as well as their spatial configuration given the roof length, span, snow load, and strength class. Similarly, a general equation has been derived to predict the optimum volume required for the roof structure considering the different strength classes, which would promote efficient use of resources and economic advantages while complying with structural and safety requirements.

2. Materials and Methods

This section presents the materials and methods employed throughout the research. Firstly, to establish a comprehensive framework of construction solutions relevant to the European context, the range of study for the different design parameters (material properties, geometric dimensions, and load scenarios) is considered. Subsequently, the specifics regarding the development of the optimization tool are detailed. On the one hand, the Eurocode 5 [45] equations utilized to evaluate the potential of each proposed design solution are mentioned. On the other hand, the concrete parameters within the genetic algorithm, as well as the objective function and the associated penalization criteria implemented, are defined. Finally, the statistical analysis conducted on the optimal solutions to derive the optimization equations intended to serve as a decision tool is described.

2.1. Design Parameters: Material, Dimensions and Loads

A roof structure comprised of timber double-tapered beams and purlins has been examined in this research. For the assessment, six homogeneous glued laminated (glulam) timber strength classes, as defined in EN 14080 [17], have been considered: GL 20h, GL 22h, GL 24h, GL 26h, GL 28h, GL 30h and GL 32h (Table 1). This broad selection of strength classes would allow for consideration beyond the traditionally used wood species and thus the promotion of previously unexploited or underutilized species in timber construction.

Table 1. Characteristic properties for each homogeneous glulam strength class [17].

Property	GL 20h	GL 22h	GL 24h	GL 26h	GL 28h	GL 30h	GL 32h
Bending strength (N/mm^2)	20	22	24	26	28	30	32
Tensile strength: parallel to the grain (N/mm^2)	16	17.6	19.2	20.8	22.3	24	25.6
Tensile strength: perpendicular to the grain (N/mm^2)				0.5			
Compression strength: parallel to the grain (N/mm^2)	20	22	24	26	28	30	32
Compression strength: perpendicular to the grain (N/mm^2)				2.5			
Modulus of elasticity: parallel to the grain (N/mm^2)	8400	10,500	11,500	12,100	12,600	13,600	14,200
Modulus of elasticity: perpendicular to the grain (N/mm^2)				300			
Density (kg/m^3)	340	370	385	405	425	430	440

The optimization was carried out for different timber roof dimensions. The roof length was studied at 30 m, 45 m, 60 m and 75 m; the assessed span varied from 15 m to 30 m in 1.25 m increments; and inclination angles between 5° and 10° were examined for the tapered beam. The geometry of the structural elements was also considered a variable within the experimental program. The heights of beams (Hb) and purlins (Hp) ranged from 120 to 1200 mm, with 40 mm intervals reflecting the laminate thickness. Similarly, the widths for beams (Wb) and purlins (Wp) spanned from 90 to 220 mm, adjusted in 10 mm increments. Finally, regarding the arrangement of the structural elements within the roof structure, the spacing between beams and purlins was also studied as a variable. Beam spacing (Sb) was considered within a range of 3 m to 7 m, while purlin spacing (Sp) varied from 0.625 to 1.25 m, which was the maximum allowed separation.

Similarly, specific limits were established regarding the range of potential load conditions that the roof structure is designed to withstand. All contemplated loads were combined as established in Eurocode 1 [45] to assess the worst-case scenario in the structural integrity verification. As the roof elements fall within service class 1, a 1.25 safety factor was considered for the glulam material combined with a modification factor (k_{mod}) of 0.9. For permanent surface loads, in addition to the self-weight of the beams and purlins that was automatically calculated during the optimization process, a dead load of 0.45 kN/m^2 was also included. For variable loads, snow loads ranging from 0.4 to 3 kN/m^2 were considered to accommodate typical values in different locations across the European Union [46]. It should be noted that in accordance with the Spanish Technical Building Code (also known as CTE) [47], the combination of snow and wind loads with the maintenance load is avoided due to the assumption that during extreme conditions, no one would be present on the roof, thus resulting in an overestimation of the total load on the structure. Conversely, snow load values below 0.4 kN/m^2 were excluded to ensure that in scenarios with negligible snow loads, the maintenance load is applied. Regarding the wind load, it was considered for a gable roof without openings in a building with a height of 5 to 7 m [48], and since it has been proven not to affect the optimization results [36], a fixed value of 0.07 kN/m^2 was employed.

2.2. Optimization

The optimization goal is to identify the most effective design for a roof structure, taking into account six different glulam strength classes that reduce timber material consumption and comply with the requirements of strength, stability, and stiffness set out in Eurocode 5 [45]. As such, initially, this method aims to balance structural performance (i.e., function and safety) with material conservation. However, it is worth mentioning that the interpretation of the outcomes across the different strength classes should also be regarded as a sustainability equalization, e.g., the selection of an optimal result from a different strength class that could be represented by less exploited wood materials, which alleviates the demand for certain species while promoting biodiversity.

Genetic algorithms grounded in biological evolution principles were employed to execute the optimization process (Figure 1).

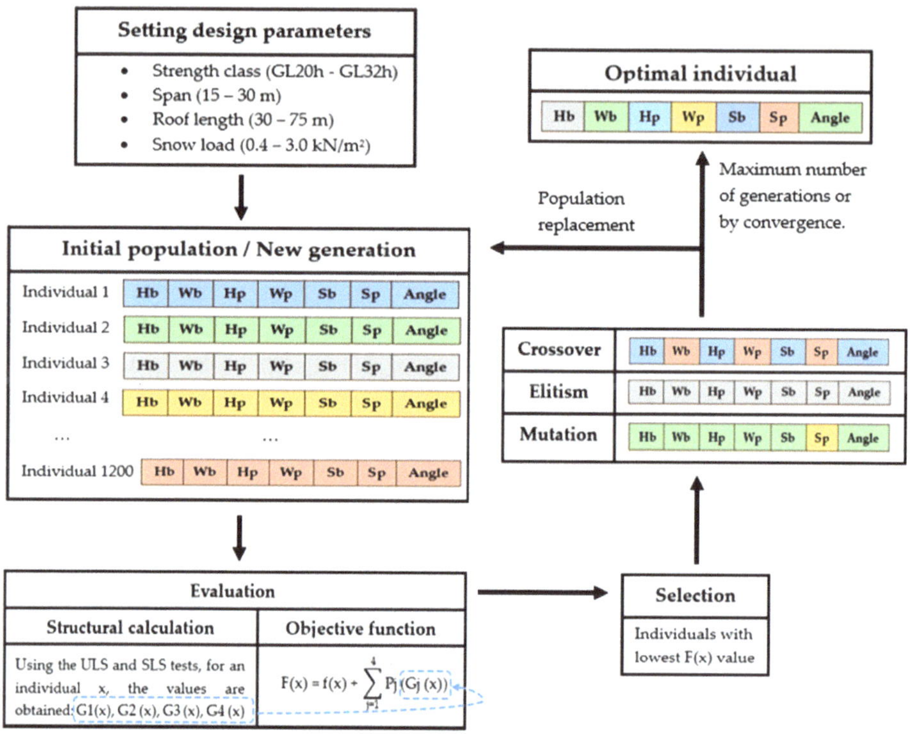

Figure 1. Genetic algorithm optimization process.

To represent different potential designs for the roof structure, an initial population of 1200 random individuals was generated. The selected population size is a compromise between the computational cost and the probability of identifying the optimal solution. The possible dimensions (i.e., Hb, Hp, Wb, Wp, and angle) and arrangements (i.e., Sb, Sp) of beams and purlins were considered through specific values in their chromosomes. Each individual within the population corresponds to a potential structural solution, encoded in a chromosome comprising seven genes: Hb, Hp, Wb, Wp, Sb, Sp, and angle. Then, the evaluation of each design solution depended on the utilization ratios determined within the structural calculation program developed in Matlab to verify the Eurocode 5 [45] criteria, which was previously collected in [36], as follows.

1. Ultimate limit state (ULS) tests:
   - Beam verification for shear strength ($I_v \leq 1$).
   - Beam verification for bending strength ($I_m \leq 1$).
   - Beam verification in the apex zone ($I_{t,90} \leq 1$).
   - Purlin verification for combined bending and shear strength.

2. Serviceability limit state (SLS) tests:
   - A limit value of l/400 for the instantaneous deflection (w_{inst}).
   - A limit value of l/300 for the final deflection (w_{fin}).
   - A limit value of l/225 for the net final deflection (w_{netfin}).

Then, the modified objective function (F(x) in Equation (1)) measures the fitness of individuals in terms of volume, in m^3, and applies a penalty to those that do not meet the structural safety criteria and according to the aforementioned utilization rates:

$$F(x) = f(x) + \sum_{j=1}^{4} Pj(Gj(x)) \tag{1}$$

where x denotes an individual within the study population, f(x) is the objective function in terms of volume (m^3), j is the number of variables under examination, Pj is the penalization term conforming to the restrictions imposed on each structural element, and Gj is the penalty parameter, with the same order of magnitude as the objective function, based on the maximum utilization ratio observed in each structural component, with j assuming values from 1 to 4, as follows.

- G1(x) is the highest ultimate limit state utilization ratio of the beam:
  $0 > G1(x) > 1$ then, $P1(G1(x)) = 3 \times 10^{(1-G1(x))}$.
  $G1(x) = 1$ then $P1(G1(x)) = 0$.
  $G1(x) = 0$ then $P1(G1(x)) = 40$.
  $G1(x) > 1$ then $P1(G1(x)) = G1(x) \times 400$.
- G2(x) is the highest serviceability limit state utilization ratio of the beam:
  $0 > G2(x) < 1$ then $P2(G2(x)) = 1.8 \times 10^{(1-G2(x))}$.
  $G2(x) = 1$ then $P2(G2(x)) = 0$.
  $G2(x) = 0$ then $P2(G2(x)) = 24$.
  $G2(x) > 1$ then $P2(G2(x)) = 60 \times G2(x)$.
- G3(x) is the highest ultimate limit state utilization ratio of the purlin:
  $0 > G3(x) > 1$ then $P3(G3(x)) = 1.725 \times 10^{(1-G3(x))}$.
  $G3(x) = 1$ then $P3(G3(x)) = 0$.
  $G3(x) = 0$ then $P3(G3(x)) = 23$.
  $G3(x) > 1$ then $P3(G3(x)) = G3(x) \times 11.5$.
- G4(x) is the highest serviceability limit state utilization ratio of the purlin:
  $0 > G4(x) > 1$ then $P4(G4(x)) = 172.5 \times 10^{(1-G4(x))}$.
  $G4(x) = 1$ then $P4(G4(x)) = 0$.
  $G4(x) = 0$ then $P4(G4(x)) = 23$.
  $G4(x) > 1$ then $P4(G4(x)) = G4(x) \times 11.5$.

Then, the individuals with the lowest values are roulette-selected to be subjected to two-point crossover (i.e., the combination of their characteristics), elitism (i.e., the retention of the fittest individuals) and mutation (i.e., the introduction of variability) operations in order to generate new individuals that replace the initial population. Values of 80%, 10% and 1% were employed for these operators, respectively. The cycle is repeated until convergence is reached, or up to a maximum set value of 50 generations. Nonetheless, convergence was always reached after 15 to 20 generations. For instance, Figure 2 showcases the characteristic evolution of the modified objective function across successive generations.

2.3. Statistical Analysis

A total of 1792 optimal individuals were generated based on combinations of specified values for span, depth, snow load, and strength class. A comprehensive statistical analysis was performed on the attributes of these configurations to assess for each glulam strength class, the optimal geometry (width and height) of the structural elements (beams and purlins) comprising the roof structure, and their spatial configuration. Additionally, the relationships among the different considered variables (i.e., snow load, span, depth, beam height and width, purlin height and width, beam spacing, and purlin spacing) as a function of the glulam strength class were examined through multiple linear regression analysis. The analysis also focused on the comparison of the optimum timber volume across different strength classes, offering a basis for making informed decisions based on both the material consumption (i.e., type and amount) and the cost-effectiveness. Therefore, several equations have been formulated to enable reliable predictions regarding the optimal volume and geometry parameters for a given span and roof length, snow load, and timber strength class. The validity of each predictive equation was assessed through the normality and homoscedasticity of the residuals as well as the correlation value.

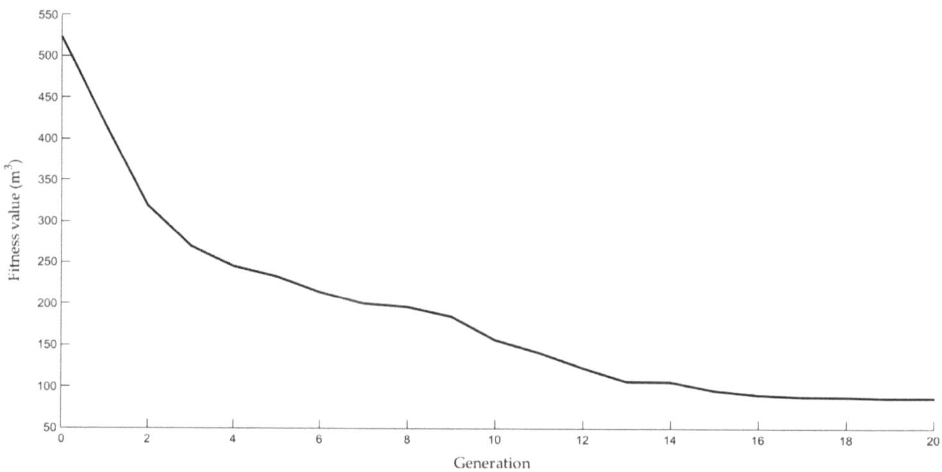

Figure 2. Illustrative example of the evolution of the modified objective function for a GL 32h roof structure of 22.5 × 45 m subjected to a snow load of 1 kN/m^2.

3. Results and Discussion

The optimization outcomes for the timber roof structures studied are presented as predictive equations according to the design parameters and glulam strength classes. Initially, the analysis focuses on the optimal geometrical values of beams (i.e., inclination angle, width and height) and purlins (i.e., width and height), as well as their respective spatial arrangements in the roof structure. Subsequently, the exploration encompasses the optimization of the overall timber volume across the different strength classes. Moreover, examples for the application of the proposed predictive equations are included for each of the approaches mentioned.

3.1. Optimal Geometry of Structural Elements and Their Spatial Configuration

3.1.1. Inclination Angle of the Double-Tapered Beam

Consistently, the optimal inclination angle was 5°, the minimum for the studied range, regardless of span, snow load, and strength class. Nonetheless, it was initially hypothesized that the optimal angle might exceed the 5° considered minimum. It was assumed that an increased angle and the resulting increase in the central cross-section of the tapered beam could reduce the dimensions of the initial cross-section of the beam or allow for wider beam spacing, leading to a decrease in overall timber volume. However, a direct comparison with the findings reported by Simón-Portela [36] at a 10° inclination reveals that the reduction achieved due to the increased spacing between beams (i.e., halving the number of beams) does not compensate for the added volume due to the steeper angle. In fact, for identical loading conditions and roof dimensions, the volume at a 10° inclination is 30% greater than that at a 5° inclination. Minimal angle inclinations (<5°) were also reported in the optimizations carried out by Baranski et al. [19] and Jelušič [34].

3.1.2. Width, Height and Spacing of Beams

A multiple linear regression analysis was carried out to study the influence of the span and snow load variables on the dimensions of beams across various strength classes and snow load intervals. It should be noted that the data did not exhibit heteroscedasticity, and there were no significant deviations from the normal distribution (Levene's test: $p > 0.05$; Shapiro–Wilk test: $p > 0.05$). The findings revealed that:

- The span, both as a standalone factor and in combination with the snow load, significantly influences (p-value < 0.001) the width of the beam.
- Both the span and snow load, whether considered separately or together, have a significant effect (p-value < 0.001) on the height of the beam.

As the correlation coefficients observed were greater than 0.95, reliable prediction equations of the width and height of beams could be achieved for a given span and snow load values. In this regard, Equations (2) and (3) were proposed based on the correlation coefficients from the multiple linear regression.

The optimal beam width (Wb), in meters, could be determined using Equation (2):

$$\mathrm{Wb} = 90 + 10 \times (\mathrm{A} + \mathrm{B} \times \mathrm{Span} + \mathrm{C} \times \mathrm{Snow\ load} \times \mathrm{Span}) \tag{2}$$

where span is expressed in meters and the snow load is expressed in kN/m^2. Coefficients A, B, and C are detailed in Table 2 categorized according to the strength class and snow load.

Table 2. Coefficients for the determination of the beam width as a function of the span, snow load, and strength class according to Equation (2).

Strength Class	Snow Load Range (kN/m^2)	A	B	C
GL 20h	[0.4, 1.0]	−2.62363	0.08474	0.09558
	(1.0, 3.0]	−5.54945	0.23634	0.06286
GL 22h	[0.4, 1.0]	−1.92033	0.07653	0.03823
	(1.0, 1.8]	−3.41484	0.03779	0.12499
	(1.8, 3.0]	−6.76484	0.38397	0.02239
GL 24h	[0.4, 2.4]	−3.20080	0.14541	0.02972
	(2.4, 3.0]	−8.5714	0.1154	0.1461
GL 26h	[0.4, 2.0]	−2.49573	0.12007	0.01960
	(2.0, 3.0]	−7.0249	0.000	0.1611
GL 28h	[0.4, 2.0]	−2.53236	0.11526	0.02117
	(2.0, 3.0]	−6.4864	0.00000	0.1511
GL 30h	[0.4, 1.6]	−1.946625	0.098350	0.009656
	(1.6, 3.0]	−6.19152	0.13680	0.09266
GL 32h	[0.4, 0.6]	−1.61538	0.07035	0.01870
	(1.4, 3.0]	−6.15110	0.17779	0.07631

It is worth noting that the outcomes from Equation (2) must be rounded to the nearest tenth to further be employed in the height beam determination (Equation (3)). Thus, the resulting width of the beams has to be rounded to the closest value in 10 mm increments, with a minimum value set at 90 mm.

For the different strength classes, Figure 3 illustrates the optimal values of the beam width as a colored area within the snow load and span plot. It can be inferred that the optimal model frequently employs the minimum value as the optimal beam width across many span–snow load scenarios, which becomes more prevalent as the strength class increases. In terms of maximum width, it is typically set at values 66.66% greater than the minimum width. Additionally, the graphical representation highlights a marked disparity in the coverage area between widths of 130 mm to 170 mm, which is substantially smaller than the area spanned by widths from 90 mm to 130 mm. A noteworthy observation is the similarity across the graphical representations for the GL 28h to GL 32h strength classes, indicating consistent behavior in beam width selection within this range.

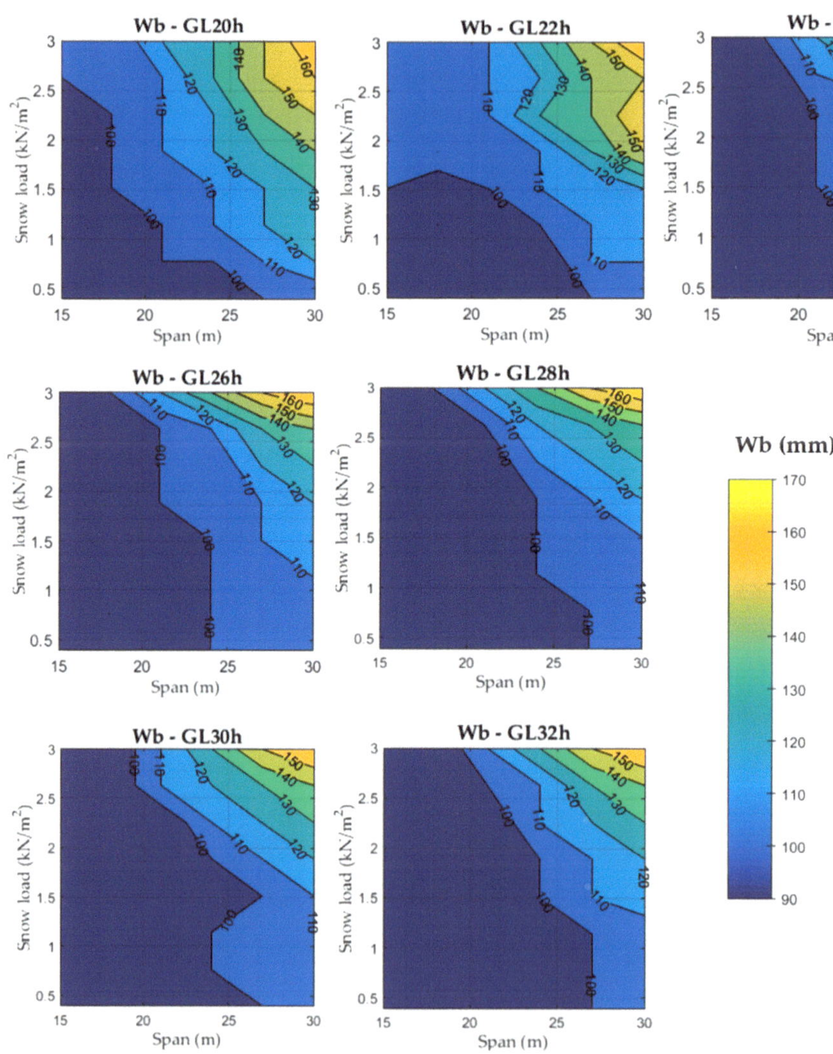

Figure 3. Optimal width of the beam, in mm, for the different timber strength classes.

Although a direct comparison to other optimization research [19,34] is not possible due to different optimization thresholds as well as material and load conditions, some commentary can be made regarding the observed patterns within the optimization results. For instance, Jelušič [34] reported that the maximum beam height-to-width ratio always ranged between 7 and 8. Conversely, for comparable strength (GL 30h) and separation, the optimization proposed in this research resulted in ratios ranging from 13.4 to 24.3, which justifies the lower beam widths arising from this investigation. Contrarily, Baranski et al. [19] limited the optimization to a maximum beam height-to-width ratio of 10, whereas in the present research, such a ratio is not restricted in order to study the complete 90–200 mm range with the sole limitations imposed by the structural standards. It is also worth mentioning that Baranski et al. [19] found the same width dimensions of beams to be valid for the complete strength range.

The optimal beam height (Hb), in meters, can be determined using Equation (3):

$$\mathrm{Hb} = \frac{(\mathrm{A}+\mathrm{B} \times \mathrm{Span} + \mathrm{C} \times \mathrm{Snow\ load} + \mathrm{D} \times \mathrm{Snow\ load} \times \mathrm{Span})^{\frac{1}{\mathrm{E}}} \times 100}{\mathrm{Wb}} \quad (3)$$

where span and beam width (Wb) are expressed in meters and the snow load is expressed in $\mathrm{kN/m^2}$. Coefficients A, B, C, D, and E are detailed in Table 3, categorized according to the strength class and snow load.

Table 3. Coefficients for the determination of the beam height as a function of the span, snow load, beam width, and strength class according to Equation (3).

Strength Class	Snow Load Range ($\mathrm{kN/m^2}$)	A	B	C	D	E
GL 20h	[0.4, 0.8]	−2437.7	364.9	−387.0	0.0	1.3
	(0.8, 3.0]	43.6916	7.7027	−0.6686	1.8943	0.8
GL 22h	[0.4, 1.6]	−107.79	32.44	0.000	10.64	1.0
	(1.6, 3.0]	44.525	8.276	0.000	1.516	0.8
GL 24h	[0.4, 2.4]	−82.268	30.655	0.000	9.972	1.0
	(2.4, 3.0]	9.7869	0.4573	2.1313	0.1799	0.5
GL 26h	[0.4, 1.8]	−1195.93	149.32	0.000	47.62	1.2
	(2.2, 3.0]	35.029	0.000	0.000	1.920	0.7
GL 28h	[0.4, 2.6]	−1899.6	255.0	−931.2	152.6	1.3
	(2.6, 3.0]	−67.787	2.356	31.563	0.000	0.6
GL 30h	[0.4, 2.0]	−790.05	121.95	−260.80	66.66	1.2
	(2.0, 3.0]	480.66	0.000	−233.57	23.77	1.0
GL 32h	[0.4, 0.6]	227.14	61.13	−1924.63	111.49	1.2
	(0.6, 1.4]	−143.50	18.99	138.30	0.000	0.9
	(1.4, 2.4]	−565.42	55.17	220.20	0.000	1.0
	(2.4, 3.0]	−1288.73	66.47	396.61	0.000	1.0

In this case, the height of the beam must be rounded to the nearest multiple of 40, reflecting the laminate thickness, up to a maximum value of 1200 mm.

Figure 4 shows the optimal beam height for the different span–snow load scenarios across the different timber structural strength classes. The analysis revealed greater variability in beam height compared to width, since the genetic algorithm predominantly optimizes the structure by adjusting the height due to the influence of this variable on the moment of inertia of the cross section. This behavior is particularly noticeable for the lower-strength classes that ultimately reach an optimum at the maximum available height value for some scenarios. Similarly to the optimal width, the distribution of the beam optimal height across the GL 28h to GL 32h strength classes shows comparable patterns.

No proportional relationship was observed between the strength class and the beam height-to-span ratio, since the model adjusts the height of the beam based on all geometric variables. For GL 30h, this lack of proportionality could also be observed in Jelušič [34]. In the investigation carried out by Baranski et al. [19], who studied a roof structure for a load of 0.9 $\mathrm{kN/m^2}$, a threshold of span/20 to span/40 was employed in the optimization. The aforementioned range is close to that resulting from this investigation, with values of beam height oscillating between span/20 and span/32. To assess the influence of the span, roof length and snow load on the optimal number of beams, a multiple linear regression analysis was conducted. No evidence of heteroscedasticity or significant deviations from a normal distribution of data was observed (Levene's test: $p > 0.05$; Shapiro–Wilk test: $p > 0.05$). The span was found to be not significant, whereas the roof length, both independently and in combination with the snow load, significantly influences (p-value < 0.001) the number of beams. Equation (4) was derived from the correlation coefficients of the multiple regression model ($R^2 > 0.95$).

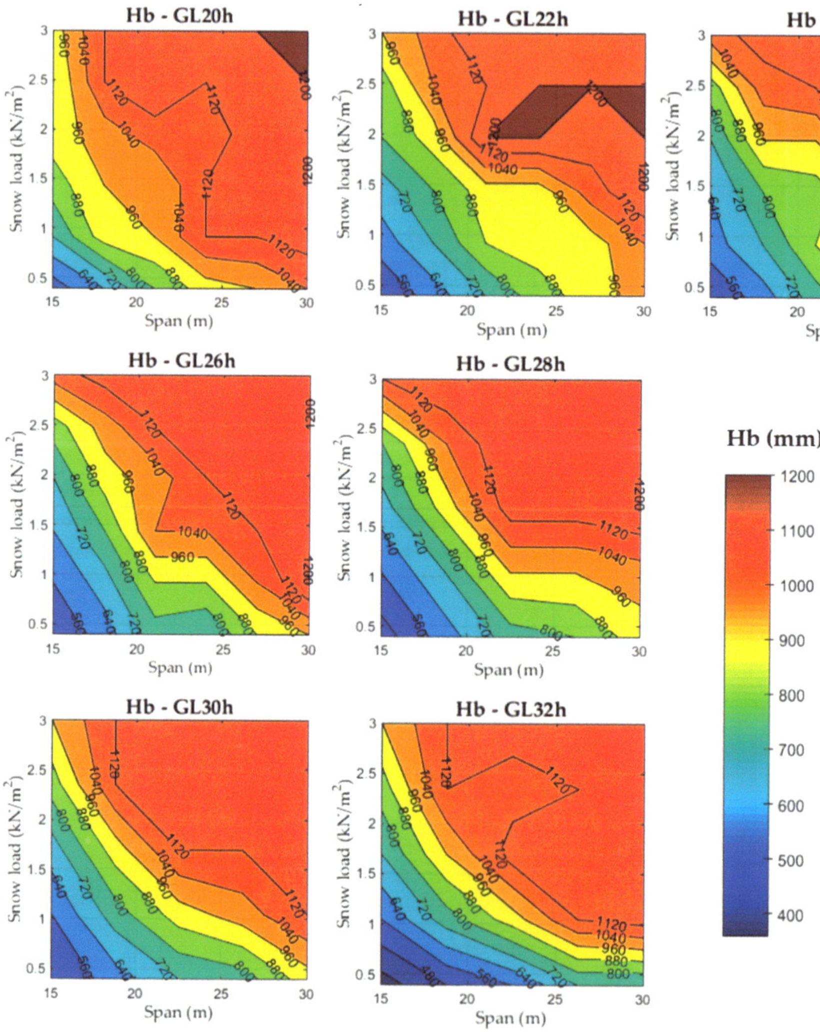

Figure 4. Optimal height of the beam, in mm, for the different timber strength classes.

The optimal values for the number of beams (Nb) and their spacing (Sb), in meters, can be determined using Equations (4) and (5), respectively:

$$\text{Nb} = (\text{A} + \text{B} \times \text{Roof length} + \text{C} \times \text{Roof length} \times \text{Snow load})^{\frac{1}{\text{D}}} \quad (4)$$

$$\text{Sb} = \frac{\text{Roof length}}{(\text{Nb} - 1)} \quad (5)$$

where roof length is expressed in meters and the snow load is expressed in kN/m^2. Coefficients A, B, C, and D are detailed in Table 4, categorized according to the strength class and snow load. It should be noted that the optimal number of beams must be rounded to the next integer.

Table 4. Coefficients for the determination of the number of beams as a function of the span, snow load and strength class according to Equation (4).

Strength Class	Snow Load Range (kN/m^2)	A	B	C	D
GL 20h	[0.4, 1.0)	1.714653	0.029203	0.003352	0.5
	[1.0, 1.4)	1.886763			
	[1.4, 1.8)	2.01307			
	[1.8, 2.6)	2.079799			
	[2.6, 3.0]	2.190146			
GL 22h	[0.4, 0.6)	1.387500	0.179969	0.005556	1.0
	[0.6, 1.0)		0.227129		
	[1.0, 2.2)		0.246142		
	[2.2, 2.8]		0.279475		
	(2.8, 3.0]		0.309969		
GL 24h	[0.4, 0.6)	2.183385	0.035969	0.003365	0.6
	[0.6, 1.0)		0.044658		
	[1.0, 1.4)		0.050095		
	[1.4, 1.8)		0.053231		
	[1.8, 2.5)		0.057995		
	[2.5, 3.0)		0.051065		
GL 26h	[0.4, 0.6)	−13.01695	0.79546	−0.04728	1.4
	[0.6, 1.0)		0.9864		
	[1.0, 1.4)		1.13949		
	[1.4, 1.8)		1.31396		
	[1.8, 3.0]		1.49259		
GL 28h	[0.4, 0.6)	1.416667	0.175884	0.002162	1.0
	[0.6, 1.0)		0.219464		
	[1.0, 1.4)		0.245759		
	[1.4, 1.8)		0.279462		
	(1.8, 3.0]		0.312182		
GL 30h	[0.4, 0.6)	−0.23722	0.24909	0.01508	1.1
	[0.6, 1.0)		0.29747		
	[1.0, 1.4)		0.34151		
	[1.4, 1.8)		0.39152		
	(1.8, 3.0]		0.43494		
GL 32h	[0.4, 1.0)	2.232859	0.041501	0.005099	0.6
	[1.0, 1.8)		0.045316		
	[1.8, 3.0]		0.050558		

In other optimization research [19,34], the separation between beams is not subjected to optimization, but has a fixed value of 4 m. Nonetheless, the results arising from this optimization point to 4 to 6 m separation for the lower snow load values and between 3 and 3.5 m for greater snow load values, both regardless of the considered strength class.

3.1.3. Width, Height and Spacing of Purlins

It was found that the width of the purlin remains unaffected by the span and snow load, as in [36], or by the strength classes. Therefore, it could be consistently set at the optimal minimum of 90 mm. In contrast, the snow load and strength class impact the optimal height value of the purlins. Thus, for the GL 32h class, the ideal height falls between 120 and 160 mm. For the remaining classes, the optimal height lies between 160 and

200 mm, as indicated in Table 5. As for the spacing between purlins, the optimal distance is consistently the maximum allowable, set at 1.25 m.

Table 5. Optimal purlin height values as a function of the snow load and strength class.

Strength Class	Snow Load Range (kN/m^2)	Purlin Height (mm)
GL 20h	[0.4, 0.6] (0.6, 3.0]	160 200
GL 22h	[0.4, 1.6] (1.6, 3.0]	160 200
GL 24h	[0.4, 2.4] (2.4, 3.0]	160 200
GL 26h	[0.4, 2.6] (2.6, 3.0]	160 200
GL 28h	[0.4, 2.8] (2.8, 3.0]	160 200
GL 30h	[0.4, 3.0]	160
GL 32h	[0.4, 0.6] (0.6, 3.0]	120 160

Example 1. *Case study of a roof structure located in an area subjected to a snow load of 1 kN/m^2 with a span of 20 m and a depth of 50 m consisting of tapered beams and purlins of glued laminated timber pertaining to GL 24h strength class.*

From Equation (2):

$$\text{Wb} = 90 + 10 \times (-3.20080 + 0.14541 \times 20 + 0.02972 \times 1 \times 20) = 93.018 \cong 90 \text{ mm}$$

From Equation (3):

$$\text{Hb} = \frac{(-82.268 + 30.655 \times 20 + 0 \times 1 + 9.972 \times 20 \times 1)^{\frac{1}{1}} \times 100}{90} = 811.41 \cong 840 \text{ mm}$$

From Equation (4):

$$\text{Nb} = (2.183385 + 0.050095 \times 50 + 0.003365 \times 50 \times 1)^{\frac{1}{0.6}} = 13.93 \cong 14 \text{ beams}$$

From Equation (5):

$$\text{Sb} = \frac{50}{(14 - 1)} = 3.85 \text{ m}$$

Hence, through the application of the proposed equations (Equations (1)–(3)), the optimal solution involves the use of 14 beams with a 5° inclination angle and cross-sectional dimensions of 90 × 840 mm spaced at intervals of 3.85 m. The optimal dimensions for purlins are 90 mm width and 160 mm height at a separation of 1.25 m.

3.2. Optimum Timber Volume as a Function of Glulam Strength Class

Determining the optimal volume required for the construction of a timber roof structure offers relevant information regarding material usage and cost estimations. Nonetheless, it could also be used as a comparison parameter between timber materials pertaining to different strength classes, since it facilitates an assessment of which strength class provides the best value in terms of material volume or cost while complying with the structural and safety requirements.

A multiple linear regression analysis was performed to explore the influence of the span, roof length and snow load on the timber volume required for the roof structure, taking into account different strength classes and ranges of span and snow load. The data exhibited no heteroscedasticity or significant deviations from a normal distribution, as confirmed by Levene's test ($p > 0.05$) and Shapiro–Wilk test ($p > 0.05$). It was noticed that the span in combination with roof length, the snow load in combination with roof length, as well as the interaction between all three variables significantly influenced the timber volume ($p < 0.001$). Correlation values greater than 0.95 support the reliability of Equation (6), which was derived from the correlation coefficients arising from the multiple linear regression model.

Thus, for a roof structure consisting of double-tapered beams and purlins, the optimal timber volume (V), in m^3, can be determined using Equation (6):

$$V = A + \text{Roof length} \times (B + C \times \text{Snow load} + D \times \text{Span} + E \times \text{Snow load} \times \text{Span}) \quad (6)$$

where roof length and span are expressed in meters and the snow load is expressed in kN/m^2. Coefficients A, B, C, D, and E are detailed in Table 6. categorized according to the strength class, snow load and span.

Table 6. Coefficients for the determination of the volume of timber as a function of the roof length, snow load and span according to Equation (6).

Strength Class	Snow Load Range (kN/m^2)	Span Range (m)	A	B	C	D	E
GL 20h	[0.4, 3.0]	[15, 24]	2.49589	−0.39406	−0.33466	0.04909	0.03310
		(24, 30]	7.38722	−2.81552	0.0000	0.12969	0.0275
GL 22h	[0.4, 3.0]	[15, 23]	1.06634	−0.30129	−0.27391	0.04309	0.02919
		(23, 30]	7.26380	−0.65811	−0.98976	0.05306	0.06075
GL 24h	[0.4, 3.0]	[15, 23]	0.92186	−0.26346	−0.29997	0.03967	0.03067
		(23, 30]	6.42904	−0.61534	−0.95922	0.04981	0.05996
GL 26h	[0.4, 3.0]	[15, 22]	0.07444	0.00000	−0.39930	0.02437	0.03664
		(22, 29]	5.13592	0.00000	−1.12414	0.02561	0.06663
		(29, 30]	6.0173	0.8492	0.8507	0.0000	0.0000
GL 28h	[0.4, 2.5]	[15, 22]	0.14786	0.0000	−0.4101	0.02465	0.03664
		(22,30]	4.40794	0.0000	−1.05264	0.02806	0.06222
	(2.5, 3.0]	[15, 22]	−1.19404	0.0000	−0.35815	0.01146	0.03943
		(22,30]	8.53219	0.0000	−1.12923	0.0000	0.07523
GL 30h	[0.4, 3.0]	[15, 23]	1.25132	−0.18641	−0.35579	0.03370	0.03375
		(23, 30]	5.65268	−0.57068	−0.97329	0.04615	0.06066
GL 32h	[0.4, 3.0]	[15, 22]	0.91510	−0.20407	−0.30354	0.03422	0.03088
		(22, 30]	4.25579	−0.50446	−0.87409	0.04427	0.05749

Figure 5 illustrates the linear correlation between the timber volumes predicted by the proposed equations and those arising from the genetic optimization. The regression line ($R^2 = 0.9977$) virtually overlaps the one-to-one ratio relationship indicated by the dashed line. This finding highlights a near-perfect equivalence between the predicted and optimized volumes that points to an exceptionally low prediction error.

To provide a comprehensive analysis of the differences in timber volume across the examined strength classes, Figure 6 showcases the percentage of variation between each strength class relative to the GL 32h reference class. This comparison is calculated based on the impact of the snow load and the span on the roof structure. The overview intentionally omits the factor of roof depth, as it has been determined to have negligible impact on the percentage difference between any given class and the benchmark GL 32h class.

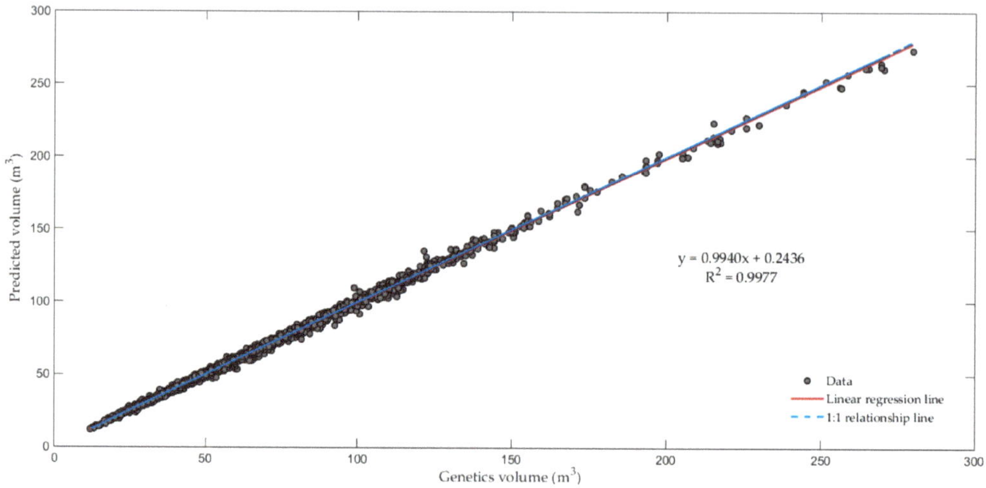

Figure 5. Linear regression between the genetics and predicted timber volumes, in m^3. The dashed line corresponds to the 1:1 relationship.

Therefore, from a close examination of Figure 5 and in direct comparison with the reference class GL 32h, the following insights emerge.

- The selection of the GL 20h strength class roughly requires 17–20% additional volume for snow loads ranging from 0.4 to 1.0 kN/m^2. For higher snow loads of 1 to 2 kN/m^2 and 2 to 3 kN/m^2, the increase in timber volume needed is about 12–15% and 10–12%, respectively.
- For the GL 22h class, an increase in timber volume of approximately 11–13% is needed for snow loads ranging from 0.4 to 0.7 kN/m^2. The requirement decreases to 8–11% for loads between 0.7 and 1.8 kN/m^2, 6–8% for 1.8 and 3 kN/m^2 with spans of 15 to 18 m, and 3–6% for the same snow load with spans of 18 to 30 m.
- The GL 24h class demands an additional 6–8% timber volume for snow loads from 0.4 to 0.8 kN/m^2, 4–6% for 0.8 to 1.8 kN/m^2 with spans of 15 m to 20 m, 5–7% for 0.8 to 1.6 kN/m^2 with spans of 20 to 30 m, 3–5% for 1.8 to 3.0 kN/m^2 with spans of 15 to 18 m, and a minimal increase up to 3% for 1.4 to 3.0 kN/m^2 with spans of 18 to 30 m.
- The GL 26h class roughly requires 6–8% additional timber volume for snow loads between 0.4 and 0.8 kN/m^2, 4–6% for 0.8 and 1.2 kN/m^2, 2–4% for 1.2 and 2.0 kN/m^2, and a slight increase to 2% for snow loads of 2.0 and 3.0 kN/m^2.
- For the GL 28h class, a 5–7% timber volume increase is necessary for snow loads of 0.4 to 0.8 kN/m^2 across spans of 15 to 27 m. A lower volume increase of 3–5% occurs for three scenarios: snow loads of 0.8 to 2 kN/m^2 and spans of 15 to 25 m, 0.4 to 1.3 kN/m^2 and spans of 27 to 30 m as well as snow loads between 2.6 and 3.0 kN/m^2. Moreover, a -1% timber volume variation for snow loads of 2.0 to 2.6 kN/m^2 and spans of 15 to 25 m, as well as for snow loads between 1.3 and 2.0 kN/m^2 with spans of 25–30 m should be noted. The difference reaches a -2% value for snow loads of 1.3 to 2.6 kN/m^2 and spans of 25 to 30 m.
- The selection of the GL 30h class requires a 4–6% increase in volume for snow loads of 0.4 to 0.6 kN/m^2 and spans of 15 to 20 m, with a marginal 0–2% increase for similar snow loads across spans of 20 to 30 m as well as for snow loads of 1.0 to 2.0 kN/m^2 for spans of 15 to 20 m. For the remaining loads and spans, a decrease of up to 2% in timber volume could be achieved.

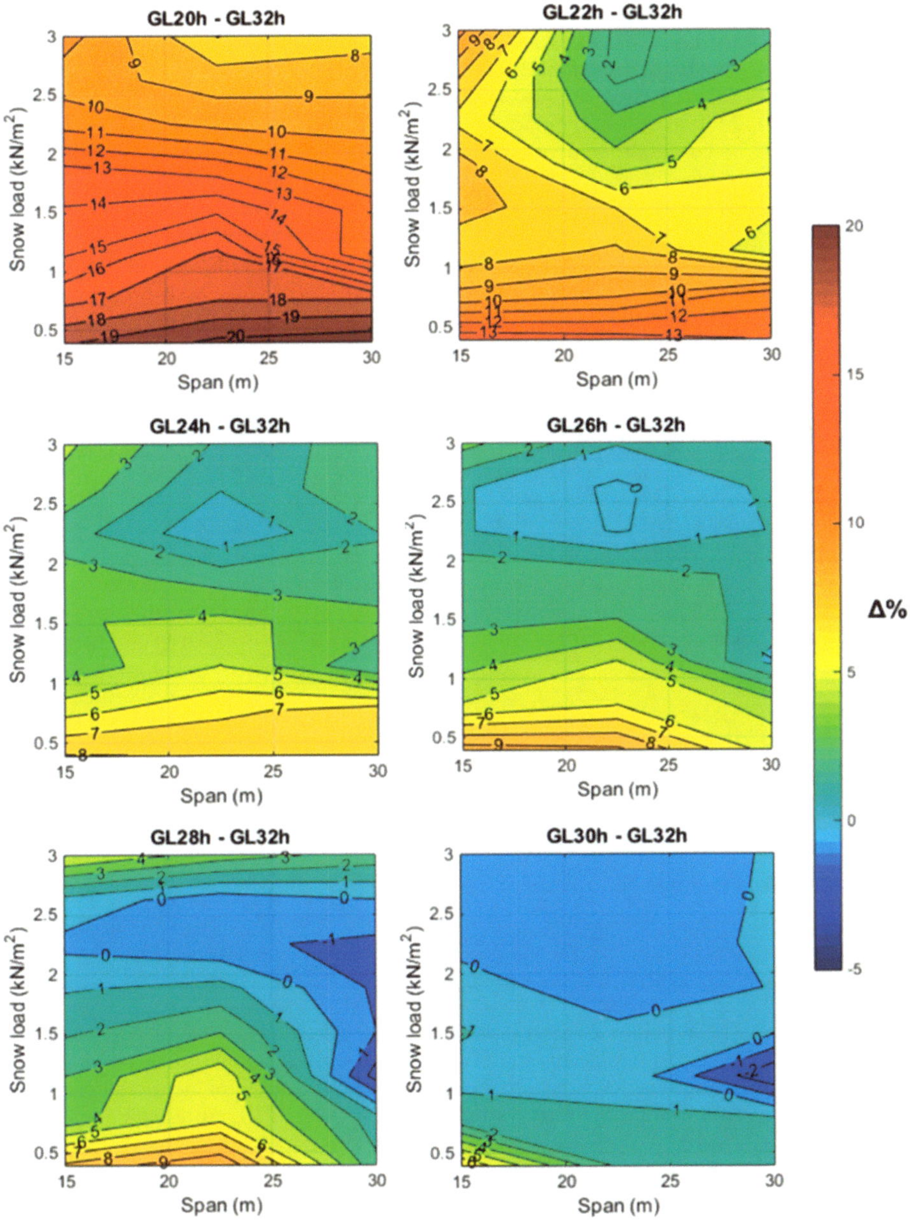

Figure 6. Timber volume variation, in percentage, between the selected strength class and the reference GL 32h class as a function of the snow load and span.

In order to analyze the aforementioned cost approach, a Spanish construction cost database [49] was employed to assess the cost of glulam timber pertaining to different strength classes. For instance, the price for 1 m^3 of GL 24h timber was 1080 EUR, whereas 1 m^3 of GL 32h timber reached 1140 EUR. Since GL 24h resulted in a 5.26% lower cost, it could be inferred that for snow loads in the range of 2 to 3 kN/m^2, using the GL 24h material would be desired, while for snow loads ranging from 0.4 to 2 kN/m^2, the preferred decision lies in the GL 32h class. In this regard, there are numerous cases where, for snow loads greater than 2 kN/m^2, cost considerations would be the sole driver in the decision due to the same timber usage for both GL 30h and GL 32h classes. Although in a smaller capacity, there are also instances following this pattern for GL 26h and GL 28h classes. Moreover, this approach should go beyond strength class and contemplate the difference in cost of different species with the suitable strength properties required for the studied strength class. Through the consideration of wood species, both within the same or a different strength class, the role of the different tree species in the ecosystem is regarded. Then, the sustainability of such derived decisions includes both material reduction and the promotion of less commonly used materials, which further biodiversity and future resilience against climate change.

Example 2. *Two worked examples initially proposed by Argüelles Álvarez and Arriaga Martitegui [50] are compared with the corresponding optimal timber volume derived from Equation (6). Specifically, their Example 8.1 consisting of GL 28h double-tapered beams and a modification of Example 6.1 focusing on purlins were employed. It should be noted that their Example 6.1 was modified to account for the same material and load conditions as in Example 8.1. The final result for a roof structure of 20 × 45 m exposed to a snow load of 0.47 kN/m^2 resulted in a GL 28h timber consumption of 39.75 m^3.*

Table 7 shows the calculated optimal timber volumes (Equation (6)) required for the aforementioned roof structure, while also including the volume incurred for the remaining strength classes. Moreover, the variation in the timber requirements across all strength classes is compared to the reference 39.75 m^3, providing a comprehensive overview of material efficiency.

Table 7. Optimal timber volumes (Equation (5)) for Example 2 and variation of timber consumption compared to the reference [50].

Strength Class	Optimal Volume (m^3)	Variation (%)
GL 20h	35.87	−9.76
GL 22h	32.84	−17.38
GL 24h	31.4	−21.01
GL 26h	29.06	−26.89
GL 28h	29.16	−26.64
GL 30h	29.94	−24.68
GL 32h	29.17	−26.62

From the direct comparison for the same timber material (GL 28h), the optimization leads to a significant reduction of 26.64% in the material volume required for the roof structure. Considering material costs [49], the reference roof incurs an expense of 43,315 EUR, whereas the optimized structure costs 33,243 EUR, representing a 10,000 EUR saving. Hence, the optimization results in more efficient resource use and lower construction costs without compromising structural integrity.

Similarly, the selection of timber pertaining to a different strength class also results in savings while allowing for diversification in the wood species selected or adaptation to locally available materials. Thus, such a strategy could seek economic and material efficiency but also include sustainability considerations in the decision-making process.

Finally, to provide a simplified approach to the comparison between classes, Table 8 shows the percentage reduction between the modulus of elasticity parallel to the grain (Table 1) of a specific strength class and that of the GL 32h reference (RMOD) as well as the percentage reduction between the required timber volume of a specific class compared to the GL 32h reference (RVOL). As proved by the statistical analysis, the snow load range significantly explains the variability of the correlation between RVOL and RMOD

(Equation (7)), with a slope approximately 1.53 times steeper for the 0.4 to 1.5 kN/m^2 snow range.

$$\begin{array}{l}\text{For Snow load range } [0.4,\ 1.5] \rightarrow \text{RVOL} = -2.0375 + 0.4349 \times \text{RMOD} \\ \text{For Snow load range } (1.5,\ 3.0] \rightarrow \text{RVOL} = -2.0375 + 0.2847 \times \text{RMOD}\end{array} \quad (7)$$

Table 8. RMOD and RVOL values for the comparison of a strength class to the GL 32h reference class as a function of the snow load range.

Snow Load Range (kN/m^2)	%	GL 20h	GL 22h	GL 24h	GL 26h	GL 28h	GL 30h
	RMOD	40.8	26.1	19	14.8	11.3	4.2
[0.4, 1.5]	RVOL	15.9	9.8	6	3.2	2.6	1.1
(1.5, 3.0]	RVOL	10.2	5.2	2.7	1.6	0.7	0.1

4. Conclusions

This research focused on the optimization of roof structures comprised of glulam beams and purlins of different strength classes and exposed to varying snow loads. The developed genetic algorithm tool, along with the structural calculation program, was employed to generate 1792 optimal roof structures with dimensions (i.e., roof length and span) and load conditions (i.e., snow loads below 3 kN/m^2) typical of European construction. From a systematic statistical analysis, several predictive equations were proposed as a reliable method for optimizing timber roof design in accordance with Eurocode 5 [45]. Firstly, the predictive model could be used to optimize the beam width and height as well as their spatial arrangement as a function of the roof dimensions, loads, and the desired strength class. Additionally, the developed model also enabled the determination of the overall optimal timber volume required for a roof structure as a function of the strength class. Although this approach seeks material usage reduction and economic savings, it ultimately could include sustainability considerations in the decision-making process. Since an optimal solution could arise from timber pertaining to various strength classes, it would allow for diversification in the wood species selected or adaptation to locally available materials. Thus, such a strategy would assist in the promotion of the use of alternative wood species, which would further biodiversity and future resilience against climate change.

Among the findings, it was found that double-tapered beams exhibited their optimal inclination angle at 5°, leading to a 30% reduction in the total timber volume used for the roof structure compared to an inclination of 10°. The optimal spacing between purlins coincided with the maximum set value, i.e., that allowed by the roofing material. This pointing to consideration of non-traditional roofing materials as a future optimization strategy.

For structures exposed to snow loads greater than 2.5 kN/m^2, a similar timber volume requirement, with at most a 3% difference, was noticed for most strength classes (GL 24h, GL 26h, GL 28h, GL 30h, and GL 32h). Thus, suggesting the existence of a broad range of wood species that effectively could rival the most commonly employed (GL 24h and GL 32h). Similarly, for snow loads greater than 1 kN/m^2, GL 30h and GL 32h classes also resulted in similar consumption of timber volumes (up to 1%). Across all strength classes, the largest differences in volume occur at span values close to 15 m and for lower snow loads, with differences up to 20% for GL 20h compared to GL 32h. In any case, a strong relationship exists between the increase in the timber volume and the reduction in the strength class or the modulus of elasticity.

Author Contributions: Conceptualization, methodology, software, validation, investigation, data curation, writing—original draft preparation, review and editing, visualization, M.S.-P., J.R.V.-G., P.V.-L. and D.R.-R.; supervision, funding acquisition, J.R.V.-G., P.V.-L. and D.R.-R. All authors have read and agreed to the published version of the manuscript.

Funding: This research was funded by Junta de Extremadura and the European Regional Development Fund of the European Union through grants GR21163 and GR21091.

Institutional Review Board Statement: Not applicable.

Informed Consent Statement: Not applicable.

Data Availability Statement: The data presented in this study are available on request from the corresponding author.

Acknowledgments: Administrative and technical support from the Forest Research Group and the Mechanical and Fluid Engineering Research Group of the University of Extremadura is gratefully acknowledged.

Conflicts of Interest: The authors declare no conflicts of interest.

References

1. European Commission. *Communication from the Commission to the European Parliament, the Council, the European Economic and Social Committee and the Committee of the Regions. A New Circular Economy Action Plan For a Cleaner and More Competitive Europe*; COM(2020) 98 Final; Publications Office of the EU: Luxembourg, 2020.
2. Eurostat Waste Statistics. Available online: https://ec.europa.eu/eurostat/statistics-explained/index.php?title=Waste_statistics (accessed on 6 March 2024).
3. European Commission. *Directive (EU) 2023/1791 of the European Parliament and of the Council of 13 September 2023 on Energy Efficiency and Amending Regulation (EU) 2023/955 (Recast) (Text with EEA Relevance)*; Publications Office of the EU: Luxembourg, 2023; Volume 231.
4. Ecoinvent LCA Database. Available online: https://ecoinvent.org/database/ (accessed on 6 March 2024).
5. European Commission. *European Parliament Resolution of 15 January 2020 on the European Green Deal (2019/2956(RSP))*; Publications Office of the EU: Luxembourg, 2020.
6. Ayanleye, S.; Udele, K.; Nasir, V.; Zhang, X.; Militz, H. Durability and Protection of Mass Timber Structures: A Review. *J. Build. Eng.* **2022**, *46*, 103731. [CrossRef]
7. Huang, Y.; Hu, J.; Peng, H.; Chen, J.; Wang, Y.; Zhu, R.; Yu, W.; Zhang, Y. A New Type of Engineered Wood Product: Cross-Laminated-Thick Veneers. *Case Stud. Constr. Mater.* **2024**, *20*, e02753. [CrossRef]
8. Sandberg, D.; Kutnar, A.; Mantanis, G. Wood Modification Technologies—A Review. *iForest—Biogeosci. For.* **2017**, *10*, 895. [CrossRef]
9. Cabral, J.P.; Kafle, B.; Subhani, M.; Reiner, J.; Ashraf, M. Densification of Timber: A Review on the Process, Material Properties, and Application. *J. Wood Sci.* **2022**, *68*, 20. [CrossRef]
10. Puettmann, M.; Pierobon, F.; Ganguly, I.; Gu, H.; Chen, C.; Liang, S.; Jones, S.; Maples, I.; Wishnie, M. Comparative LCAs of Conventional and Mass Timber Buildings in Regions with Potential for Mass Timber Penetration. *Sustainability* **2021**, *13*, 13987. [CrossRef]
11. Condé, T.M.; Tonini, H.; Higuchi, N.; Higuchi, F.G.; Lima, A.J.N.; Barbosa, R.I.; dos Santos Pereira, T.; Haas, M.A. Effects of Sustainable Forest Management on Tree Diversity, Timber Volumes, and Carbon Stocks in an Ecotone Forest in the Northern Brazilian Amazon. *Land Use Policy* **2022**, *119*, 106145. [CrossRef]
12. Arriaga, F.; Wang, X.; Íñiguez-González, G.; Llana, D.F.; Esteban, M.; Niemz, P. Mechanical Properties of Wood: A Review. *Forests* **2023**, *14*, 1202. [CrossRef]
13. Martínez-Alonso, C.; Berdasco, L. Carbon Footprint of Sawn Timber Products of Castanea Sativa Mill. in the North of Spain. *J. Clean. Prod.* **2015**, *102*, 127–135. [CrossRef]
14. Cabral, M.R.; Blanchet, P. A State of the Art of the Overall Energy Efficiency of Wood Buildings—An Overview and Future Possibilities. *Materials* **2021**, *14*, 1848. [CrossRef]
15. Allan, K.; Phillips, A.R. Comparative Cradle-to-Grave Life Cycle Assessment of Low and Mid-Rise Mass Timber Buildings with Equivalent Structural Steel Alternatives. *Sustainability* **2021**, *13*, 3401. [CrossRef]
16. Liang, S.; Gu, H.; Bergman, R.; Kelley, S.S. Comparative Life-Cycle Assessment of a Mass Timber Building and Concrete Alternative. *Wood Fiber Sci.* **2020**, *52*, 217–229. [CrossRef]
17. *EN 14080 Timber Structures—Glued Laminated Timber and Glued Solid Timber—Requirements*; CEN: Belgium, Brussels, 2013.
18. Porteous, J.; Kermani, A. *Structural Timber Design to Eurocode 5*; John Wiley & Sons: Oxford, UK, 2013.
19. Baranski, J.; Szolomicki, J.; Damian, K. Shape Optimization of Glulam Timber Roof Girders. In Proceedings of the World Congress on Engineering and Computer Science, London, UK, 4–6 July 2018; Volume 2.
20. Jelušič, P.; Kravanja, S. Optimal Design and Competitive Spans of Timber Floor Joists Based on Multi-Parametric MINLP Optimization. *Materials* **2022**, *15*, 3217. [CrossRef] [PubMed]

21. De Vito, A.F.; Vicente, W.M.; Xie, Y.M. Topology Optimization Applied to the Core of Structural Engineered Wood Product. *Structures* **2023**, *48*, 1567–1575. [CrossRef]
22. Kilincarslan, S.; Simsek Turker, Y. Experimental Investigation of the Rotational Behaviour of Glulam Column-Beam Joints Reinforced with Fiber Reinforced Polymer Composites. *Compos. Struct.* **2021**, *262*, 113612. [CrossRef]
23. Wang, X.T.; Zhu, E.C.; Niu, S.; Wang, H.J. Analysis and Test of Stiffness of Bolted Connections in Timber Structures. *Constr. Build. Mater.* **2021**, *303*, 124495. [CrossRef]
24. Fu, Q.; Yan, L.; Thielker, N.A.; Kasal, B. Effects of Concrete Type, Concrete Surface Conditions and Wood Species on Interfacial Properties of Adhesively-Bonded Timber—Concrete Composite Joints. *Int. J. Adhes. Adhes.* **2021**, *107*, 102859. [CrossRef]
25. Giv, A.N.; Chen, Z.; Fu, Q.; Leusmann, T.; Yan, L.; Lowke, D.; Kasal, B. Bending Behavior and Bond Analysis on Adhesively Bonded Glulam-Concrete Panels Fabricated with Wet Bonding Technique. *J. Build. Eng.* **2023**, *76*, 107140. [CrossRef]
26. Ferrara, G.; Michel, L.; Ferrier, E. Flexural Behaviour of Timber-Concrete Composite Floor Systems Linearly Supported at Two Edges. *Eng. Struct.* **2023**, *281*, 115782. [CrossRef]
27. Gomez-Ceballos, W.G.; Gamboa-Marrufo, M.; Grondin, F. Multi-Criteria Assessment of a High-Performance Glulam through Numerical Simulation. *Eng. Struct.* **2022**, *256*, 114021. [CrossRef]
28. Ching, E.; Carstensen, J.V. Truss Topology Optimization of Timber–Steel Structures for Reduced Embodied Carbon Design. *Eng. Struct.* **2022**, *252*, 113540. [CrossRef]
29. De Luca, V.; Marano, C. Prestressed Glulam Timbers Reinforced with Steel Bars. *Constr. Build. Mater.* **2012**, *30*, 206–217. [CrossRef]
30. McConnell, E.; McPolin, D.; Taylor, S. Post-Tensioning of Glulam Timber with Steel Tendons. *Constr. Build. Mater.* **2014**, *73*, 426–433. [CrossRef]
31. Mam, K.; Douthe, C.; Le Roy, R.; Consigny, F. Shape Optimization of Braced Frames for Tall Timber Buildings: Influence of Semi-Rigid Connections on Design and Optimization Process. *Eng. Struct.* **2020**, *216*, 110692. [CrossRef]
32. Suárez-Riestra, F.; Estévez-Cimadevila, J.; Martín-Gutiérrez, E.; Otero-Chans, D. Experimental, Analytical and Numerical Vibration Analysis of Long-Span Timber-Timber Composite Floors in Self-Tensioning and Non-Tensioning Configurations. *Constr. Build. Mater.* **2019**, *218*, 341–350. [CrossRef]
33. Suárez-Riestra, F.; Estévez-Cimadevila, J.; Martín-Gutiérrez, E.; Otero-Chans, D. Timber-Timber-Composite (TTC) Beam Long-Term Behaviour. Full Scale Experimental Campaign and Simplified Analytical Model. *Constr. Build. Mater.* **2022**, *361*, 129649. [CrossRef]
34. Jelušič, P. Determining Optimal Designs of Timber Beams with Non-Uniform Cross-Section. In Proceedings of the WIT Transactions on The Built Environment, Ljubljana, Slovenia, 11–13 July 2018; Volume 175, pp. 169–175.
35. Šilih, S.; Kravanja, S.; Premrov, M. Shape and Discrete Sizing Optimization of Timber Trusses by Considering of Joint Flexibility. *Adv. Eng. Softw.* **2010**, *41*, 286–294. [CrossRef]
36. Simón-Portela, M.; Villar-García, J.R.; Rodríguez-Robles, D.; Vidal-López, P. Optimization of Glulam Regular Double-Tapered Beams for Agroforestry Constructions. *Appl. Sci.* **2023**, *13*, 5731. [CrossRef]
37. Negrin, I.; Kripka, M.; Yepes, V. Design Optimization of Welded Steel Plate Girders Configured as a Hybrid Structure. *J. Constr. Steel Res.* **2023**, *211*, 108131. [CrossRef]
38. Negrin, I.; Kripka, M.; Yepes, V. Metamodel-Assisted Meta-Heuristic Design Optimization of Reinforced Concrete Frame Structures Considering Soil-Structure Interaction. *Eng. Struct.* **2023**, *293*, 116657. [CrossRef]
39. Gonzalez-Montellano, C.; Ramirez, A.; Gallego, E.; Ayuga, F. On the Steel Cost of Circular Flat-Bottomed Silos Designed Using the Eurocodes. *Struct. Eng. Mech.* **2009**, *33*, 561–572. [CrossRef]
40. Villar, J.R.; Vidal, P.; Fernández, M.S.; Guaita, M. Genetic Algorithm Optimisation of Heavy Timber Trusses with Dowel Joints According to Eurocode 5. *Biosyst. Eng.* **2016**, *144*, 115–132. [CrossRef]
41. Villar-García, J.R.; Vidal-López, P.; Rodríguez-Robles, D.; Guaita, M. Cost Optimisation of Glued Laminated Timber Roof Structures Using Genetic Algorithms. *Biosyst. Eng.* **2019**, *187*, 258–277. [CrossRef]
42. Baeten, L.; Bruelheide, H.; Van Der Plas, F.; Kambach, S.; Ratcliffe, S.; Jucker, T.; Allan, E.; Ampoorter, E.; Barbaro, L.; Bastias, C.C.; et al. Identifying the Tree Species Compositions That Maximize Ecosystem Functioning in European Forests. *J. Appl. Ecol.* **2019**, *56*, 733–744. [CrossRef]
43. Fichtner, A.; Härdtle, W.; Li, Y.; Bruelheide, H.; Kunz, M.; Von Oheimb, G. From Competition to Facilitation: How Tree Species Respond to Neighbourhood Diversity. *Ecol. Lett.* **2017**, *20*, 892–900. [CrossRef]
44. Vilà, M.; Vayreda, J.; Comas, L.; Ibáñez, J.J.; Mata, T.; Obón, B. Species Richness and Wood Production: A Positive Association in Mediterranean Forests. *Ecol. Lett.* **2007**, *10*, 241–250. [CrossRef]
45. *UNE-EN 1995-1-1*; Eurocódigo 5: Proyecto de Estructuras de Madera. Parte 1-1: Reglas Generales y Reglas Para Edificación. AENOR: Madrid, Spain, 2016.
46. *EN 1991-1-3/AC-A1*; Eurocode 1: Actions on Structures—Part 3: Snow Loads. CEN: Brussels, Belgium, 2015.
47. *Royal Decree 314 Technical Building Code*; Spanish Ministry of Housing: Madrid, Spain, 2006.
48. *EN 1991-1-4/AC-A1*; Eurocode 1: Actions on Structures—Part 4: Wind Actions. CEN: Brussels, Belgium, 2010.

49. Instituto Valenciano de la Edificación Cost Database. Available online: https://bdc.f-ive.es/BDC23/1 (accessed on 5 March 2024).
50. Argüelles Álvarez, R.; Arriaga Martitegui, F. *Timber Structures: Design and Calculation [Estructuras de madera: Diseño y cálculo]*; Editorial AITIM: Madrid, Spain, 1997; ISBN 978-84-87381-09-6.

sustainability

MDPI

Article

Towards High-Efficiency Buildings for Sustainable Energy Transition: Standardized Prefabricated Solutions for Roof Retrofitting

Elisa Pennacchia [1,*], Carlo Romeo [2] and Claudia Zylka [3]

1 Interdepartmental Centre for Territory, Building, Conservation and Environment, Sapienza University of Rome, Via A. Gramsci, 53-00197 Rome, Italy

2 ENEA—Italian National Agency for New Technologies, Energy and Sustainable Economic Development, Via Anguillarese, 301-00123 Rome, Italy; carlo.romeo@enea.it

3 Department of Astronautical, Electrical and Energy Engineering, Sapienza University of Rome, Via Eudossiana, 18-00184 Rome, Italy; claudia.zylka@uniroma1.it

* Correspondence: elisa.pennacchia@uniroma1.it

Abstract: Enhancing energy efficiency in buildings plays a pivotal role in realizing the ambitious objective of achieving carbon neutrality by 2050, as outlined in the European Green Deal. Roofs represent the technical element most affected by energy phenomena related to heat transfer: in winter, roofing can lose up to 35% of heat, and the summer heat flux can even be higher. This paper provides a catalogue of optimized and sustainable solutions, with a specific focus on standardization and prefabrication principles, for enhancing the energy efficiency of the most prevalent types of roofs that characterize the national residential building heritage. The methodological approach that guided the research presented in this article was based on the identification and study of the most common roofings in the diverse national residential building heritage, followed by their classification according to their construction era. In the context of essential energy retrofitting of deteriorated residential building stock, 21 optimized standardized solutions have been identified. The outcome of performance evaluations of the proposed solutions allowed the implementation of a matrix that can be a valuable support for designers in selecting the most efficient precalculated and prefabricated solutions for the national residential building heritage based on energy performance and sustainability criteria.

Keywords: energy requalification; standardized sustainable efficiency solutions; existing building stock; building energy performance; cost analysis; multi-criteria analysis; sustainable building

Citation: Pennacchia, E.; Romeo, C.; Zylka, C. Towards High-Efficiency Buildings for Sustainable Energy Transition: Standardized Prefabricated Solutions for Roof Retrofitting. *Sustainability* **2024**, *16*, 3850. https://doi.org/10.3390/su16093850

Academic Editors: Uroš Klanšek and Tomaž Žula

Received: 25 March 2024
Revised: 22 April 2024
Accepted: 26 April 2024
Published: 3 May 2024

1. Introduction

In recent years, a growing interest has emerged in addressing energy efficiency as a pathway for carbon mitigation, constraining energy usage, and enhancing building energy performance to achieve sustainable buildings [1]. This focus has become a pivotal concern for both developing and developed countries in the twenty-first century, with energy efficiency standing as an integral, if not paramount, component of green and sustainable buildings [2]. Notably, the building sector ranks among the highest in energy consumption [3]. As such, increasing awareness of environmental constraints and available resources necessitates a renewed perspective on sustainable development [4]. Within this evolving paradigm, principles of energy efficiency and sustainability should guide all sectoral policies [5]. Given the close relationship between environmental issues and resource consumption, it is evident that the construction sector significantly impacts the utilization of non-renewable raw materials, land occupation, and energy consumption across the entire lifecycle of buildings—from material production to management and eventual decommissioning.

According to the European Commission, most of the European and national real estate stock is still outdated and inefficient, accounting for approximately 40% of the total energy consumption and 36% of energy-related greenhouse gas emissions [6,7]. Buildings are places where a significant amount of time is spent, and therefore, a renovated and more energy-efficient building stock could significantly contribute to combating climate change and minimizing its effects. In Europe, households consume 80% of the energy for heating, cooling, and domestic hot water. Thirty-five percent of the EU building stock is over 50 years old, and nearly 75% of buildings are considered energy inefficient. Moreover, merely 1% of buildings undergo renovation annually [8]. The need for decarbonization in the European construction sector is evident, requiring the development of innovative sustainable strategies, with the importance of interventions for the renovation of both public and private buildings underscored by data and confirmed by the European Green Deal [9]. This deal is an integral part of the European Commission's strategy for implementing the United Nations' Agenda 2030 and sustainable development goals, which is poised to become one of the primary key actions to pursue. The pivotal initiative for promoting energy efficiency identified in the European Green Deal is the Renovation Wave [10,11]. It is at the core of national programs for economic recovery, aiming to double the rate of building renovations by 2030. The Renovation Wave has identified three areas of interest: tackling energy poverty and buildings with the worst performance, improving public buildings and social infrastructure, and decarbonizing heating and cooling [12]. Such an initiative will not only achieve results in terms of environmental sustainability but also provide a substantial boost to the renovation sector. This sector is characterized by a significant number of local businesses, making it a crucial driver for the economic and employment revival of the entire construction industry.

In December 2021, the European Commission introduced a proposed amendment to the Energy Performance of Buildings Directive (EPBD) within the framework of the "fit for 55" legislative package. This proposed revision seeks to realize a minimum reduction of 55% in European Union greenhouse gas (GHG) emissions by 2030, consistent with the stipulations of the 2021 European Climate Law [13,14]. This revision of the EPBD establishes how the EU can achieve a zero-emission and fully decarbonized building stock by 2050. Specifically, it involves increasing the renovation rate for the least performing buildings in each EU member state. The European Commission's proposal includes a requirement, starting from 2030, for all new buildings in the EU to be zero-emission (by 2027 for all new public buildings). To ensure more harmonized standards among member states, minimum energy performance requirements will be set at the EU level. Residential buildings are expected to achieve Class E by 2030 and Class D by 2033.

At the national level, according to data provided by the 15th census of population and housing in 2011 by ISTAT, the overall national residential building stock amounts to 14,515,795 units, among which 12,187,698 are residential buildings or complexes (approximately 84%). As can be seen from the graph shown in Figure 1, over 70% of Italian buildings are over 45 years old, meaning that they predate the first national energy-saving regulation No. 373 of 1976 [15]. Over 25% of buildings constructed before this regulation exhibit annual consumption ranging from a minimum of 160 kWh/m^2 per year to over 220 kWh/m^2 per year [16]. The national real estate stock predominantly comprises buildings falling within energy classes F and G, accounting for 25% and 37.3%, respectively. These statistics are derived from the Energy Performance Certificate Information System (SIAPE) during the period 2016–2019, based on analyses conducted by the National Agency for New Technologies, Energy, and Sustainable Economic Development (ENEA) [16].

It seems imperative to promote the conversion of existing buildings into high-performance energy structures through the implementation of diverse interventions. These interventions may include enhancements to the building envelope, such as roofs, walls, and transparent closures, as well as upgrades to lighting systems, thermal energy production and distribution systems, and the installation of renewable energy-based production systems [17]. It is known that the overall energy efficiency of a building is predominantly influenced by the effectiveness of its envelope, which serves as the delineation between the indoor and outdoor environments [18]. Specifically, half of a building's total energy consumption for general purposes is expended through its envelope [19]. Roofs represent the technical element most affected by energy phenomena related to heat transfer: in winter, a roof can lose up to 35% of heat, and the summer heat flux can even be higher [20] (these data are indicative as they are strongly influenced by geometric parameters, building stratifications and volumes, as well as climatic conditions).

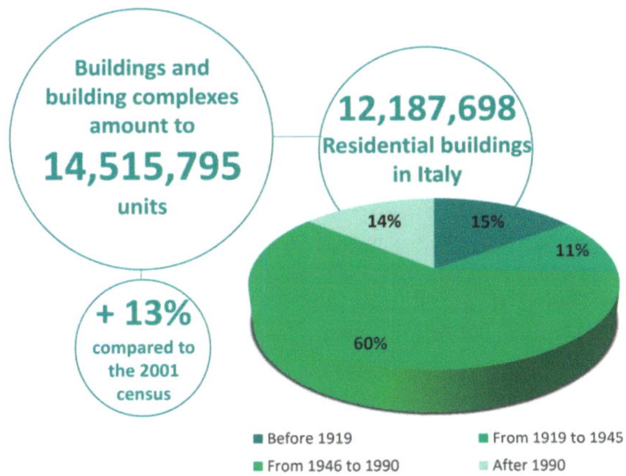

Figure 1. State of the art of the national residential building stock divided by construction era.

Numerous research efforts have focused on evaluating the energy consumption of the existing building stock, and on some factors that can contribute to its reduction, such as the thermal transmittance of both transparent and opaque surfaces, geographical location within specific climatic zones, construction period, orientation, and the characteristics of building elements. Zhenjun Ma et al. [21] asserted that "building retrofit with comprehensive energy simulation and economic analysis is an effective approach to identify the best retrofit solutions, but further research and investigation in this field are required to facilitate economically viable building retrofits." Lizana et al. [22] have stated that selecting the correct method and variables to identify the most effective retrofit solutions still represents a technical challenge. Paraschiv et al. [23] emphasized that a significant portion of the potential for energy savings and, consequently, a plausible reduction in greenhouse gas emissions, resides in the thermal renovation of existing buildings. This requires improvements in thermal efficiency to building energy consumptions. Other investigations typically solely focus on assessing the winter heating demand, concentrating on optimizing the steady-state thermal transmittance of building envelope components [24].

Numerous methodologies and tools have been developed to support the decision-making process towards a conscious and effective promotion of energy efficiency through the renovation of the building envelope such as "façade refurbishment toolbox (FRT) approach" [25], a bottom–up model for analyzing the Italian residential sector [26], IEE-TABULA Project (Typology Approach for Building Stock Energy Assessment) [27], and others [28–31]. Even the most recent Building Information Modeling (BIM) methodology supports energy efficiency enhancement processes. Alhammad et al. [32] asserted that "the integration of Building Information Modeling (BIM) with Building Energy Modeling (BEM) provides an innovative opportunity to leverage the full potential of BIM for optimizing the energy consumption of existing high-energy-efficient Building Automation Systems". However, there are still significant limitations and challenges associated with this method, despite the promised advantages. For instance, transitioning from BIM to BEM through the IFC data model results in the loss of numerous details, such as building envelope stratifications [33–35]. Additionally, there are methods and approaches based on digital-twin technology aimed at developing an intelligent optimization and automation system for energy management, ranging from building scale to neighborhood scale. This is achieved through the use of a three-dimensional data model integrated with Internet of Things (IoT), artificial intelligence (AI), and machine learning [36]. This methodology requires advanced computer skills and significant financial investment for the purchase and installation of sensors. Therefore, various innovative simulation tools are accessible for the energy modeling of new constructions and can also be applied to simulate existing buildings. However, many of them require the user to possess a profound technical understanding of the construction system, the thermal properties of materials, and retrofitting options available on the market, along with their associated costs [37].

These tools do not incorporate standardized insulation solutions, primarily prefabricated, and furthermore, they do not provide an estimate of the costs associated with energy efficiency interventions. It is crucial to also note that these renovation research projects do not encompass an assessment of the environmental sustainability associated with the proposed energy efficiency interventions. In summary, the literature review uncovered that most methodologies and tools associated with building energy efficiency lack the incorporation of verification processes for the mandated energy class upgrades under the new European regulations [38]. It is therefore evidently important that new studies and tools are aimed at promoting and supporting energy efficiency interventions in the building stock to achieve ambitious regulatory goals.

Hence, the final aim of this paper is to improve the energy efficiency of the nation's residential building inventory, emphasizing the construction or the renovation of high-efficiency buildings in accordance with upcoming energy transition regulations. This objective is pursued through the development of a standardized catalogue of optimized building and the implementation of retrofit systems applicable to building roofs. The authors of this research are driven by the aim of developing a methodology and a tool that allow for the identification and promotion of sustainable solutions guiding designers in implementing tailor-made energy renovation interventions, capable of adapting to the different scenarios within the national residential building landscape. The energy efficiency solutions are characterized by significant prefabrication and recycled content, aligning with the prevailing minimum environmental criteria (CAM) [39,40].

The proposed methodology was developed by CITERA, the research center of Sapienza University of Rome, as part of the "Research of the Electricity System" program with ENEA and the Ministry of Economic Development concerning "Solutions and tools for the energy efficiency improvement of buildings on a territorial scale".

2. Materials and Methods

This research aims to empower architects, engineers, and planners in optimizing energy efficiency interventions within residential buildings. The professionals will be equipped with a methodology to calculate insulation scenarios and determine optimal thicknesses for different technological solutions, further promoting energy efficiency. This guidance will steer them towards more efficient, prefabricated, and sustainable standardized solutions, contributing to overall energy efficiency improvements. Economic assessments will enable these professionals to evaluate the feasibility of implementing these solutions in the context of energy efficiency. Moreover, architects, engineers, and planners will benefit from an optimized technological solutions diagram, tailored to climatic, geometric, and technical parameters, aiding in informed decision-making and supporting energy efficiency initiatives. With a focus on prefabrication criteria, they will have clear guidance in selecting suitable solutions for energy efficiency interventions. Performance data compiled into accessible sheets will facilitate them in identifying the most suitable options based on climatic zones, ultimately contributing to enhanced energy efficiency outcomes. This comprehensive approach will empower designers to select effective technological solutions tailored to construction era, building type, and climatic zone, thereby promoting energy efficiency across residential buildings in both winter and summer seasons. This is achieved through the development of a catalogue of standard construction configurations to define sustainable retrofit systems applicable to upper closures.

A schematic of the methodology devised to derive optimized standardized solutions for existing building roofs is depicted in Figure 2.

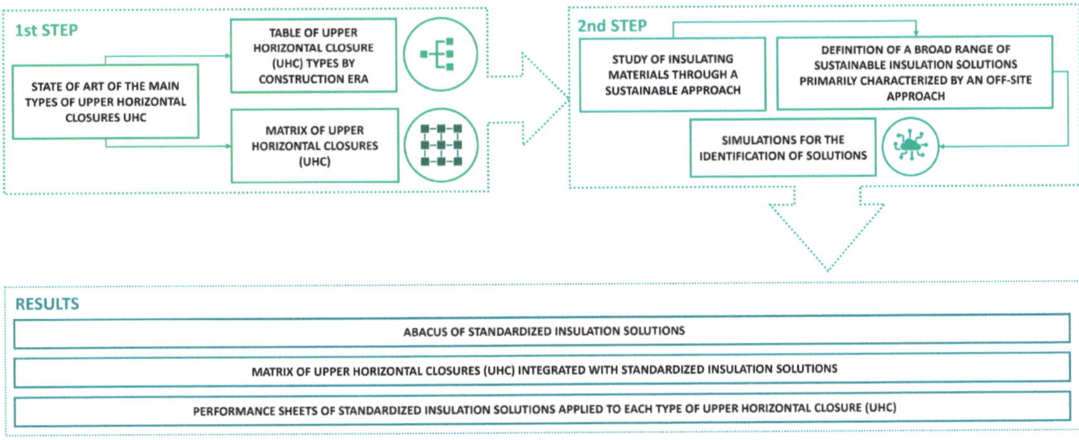

Figure 2. Schematic of the research methodology.

2.1. State of the Art of the Main Types of Upper Horizontal Closures

The subject of this paper is roofing, namely the classes of technical elements that horizontally separate the interior of the building from the exterior. They bear natural loads related to atmospheric agents (such as wind, snow, hail, rain) and those due to the passage of people, animals, and objects. The methodological approach guiding this research involved identifying and studying the most common upper horizontal closures (UHC) in the diverse national residential building stock, classifying them by construction era. The evolution of building envelope typologies in different construction era classes has depended on changes in construction techniques, the materials used, the coupling of building components, and the insulation level. These changes primarily affect the selected insulating materials, the thicknesses and the thermal transmittance values, which constitute essential parameters for selecting the most representative building envelope typologies.

This study identified prevalent construction types and those likely to undergo energy retrofit interventions, predating energy efficiency regulations. The selection of the most representative stratifications and construction era classes for the development of the following abacus was based on the Technical Report elaborated by CTI UNI/TR 11552:2014, which lists all the elements of the building envelope closures, including flat and sloped roofs.

The various types of roof coverings that characterize the national residential building heritage have been classified based on slope, type of resistant layer, accessibility, and construction era. Based on the slope, the following were identified:

- 14 sloped roofs;
- 16 flat roofs.

According to the type of resistant layer, the following were identified:

- 6 roofs with a wooden structure;
- 24 roofs made of reinforced concrete.

Based on the level of accessibility, the following were identified:

- 8 accessible roofs;
- 22 inaccessible roofs.

The construction era classes were identified in relation to specific historical events and the enactment of energy-related regulations that influenced building types and construction techniques, as follows:

- Up to 1950;
- From 1930 to 1975;
- From 1976 to 1990;
- From 1991 to 2005;
- Since 2006.

Figure 3 shows the number of roofs identified and presented in the matrix based on the construction era class.

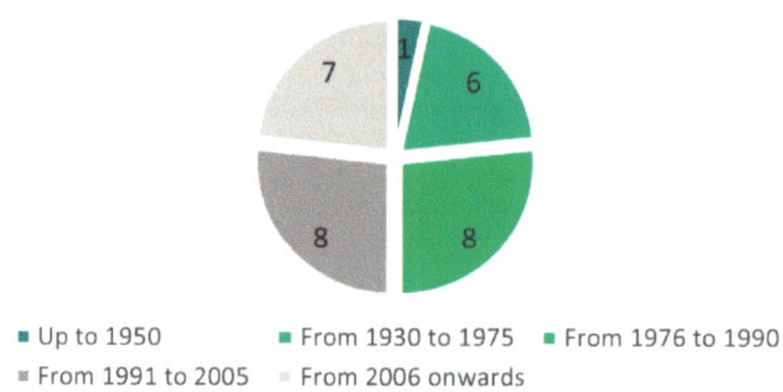

Figure 3. Number of roof types by construction era classes.

Table 1 provides a summary overview of upper horizontal closures (brief description and nomenclature) categorized by construction era class, further explored in detail within Supplement S1, crucial for understanding the evolution of construction systems that characterize the national residential building heritage.

Table 1. Classification of upper horizontal closures by construction era class.

Construction Era Class	Roof Typology
Up to 1950	A pitched roof with a wooden structure and decking (UHC.01)
From 1930 to 1975	Latero-cement pitch roof (UHC.05–UHC.05.1) A non-walkable flat latero-cement roof (UHC.09–UHC.09.1–UHC.09.2–UHC.09.3)
From 1976 to 1990	A pitched roof with a wooden structure and decking, low level of insulation (UHC.02–UHC.02.1) A pitched roof with brick tiles, low level of insulation (UHC.06–UHC.06.1) A non-walkable latero-cement flat roof, low level of insulation (UHC.10–UHC.10.1–UHC.10.2–UHC.10.3)
From 1991 to 2005	A pitched roof with a wooden structure and decking, medium level of insulation (UHC.03–UHC.03.1) A pitched roof with brick tiles, medium level of insulation (UHC.07–UHC.07.1) Non-walkable latero-cement flat roof—medium level of insulation (UHC.11–UHC.11.1–UHC.11.2–UHC.11.3)
Since 2006	A pitched roof with a wooden structure and decking, high level of insulation (UHC.04) A pitched roof with brick tiles, high level of insulation (UHC.08–UHC.08.1) Non-walkable latero-cement flat roof—high level of insulation (UHC.12–UHC.12.1–UHC.12.2–UHC.12.3)

The table below (Table 2) provides an extract from the abacus of various types of upper horizontal closures, including the following information:

- Roof coding;
- Designation;
- An image with highlighted and numbered components;
- The period during which it was most prevalent;
- Description of each numbered component;

- Thickness of each numbered component;
- Energetic performance parameter—steady-state and periodic thermal transmittance.

Table 2. Extract from the abacus of existing upper horizontal closures.

UHC.05	Latero-Cement Pitch Roof
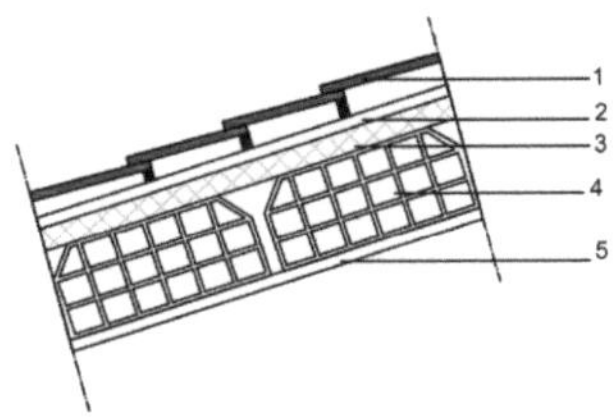	Used from 1930 to 1975 Stratigraphy: 1. Roof tiles/waterproofing element: 1.5 cm 2. Cement mortar: 2 cm 3. Reinforced concrete: 4 cm 4. Latero—cement floor/supporting element: 16 cm 5. Internal plaster: 2 cm U = 1.82 W/m^{2}K YIE = 1.03 W/m^{2}K

As mentioned above, the entire abacus of existing upper horizontal closures has been documented in Supplement S1 through summary tables for each closure.

2.2. Study of Insulating Materials for the Redevelopment of Existing Upper Closure Types

The incorporation of an insulating layer is essential to achieve the thermal transmittance values required by current regulations, ensuring a reduction in heat losses and, consequently, energy consumption. A properly insulated roof allows for the elimination of surface condensation phenomena on the intrados of the technical element, the containment of thermal losses, and the maintenance of comfortable conditions in spaces beneath the upper closure in both summer and winter. Building energy efficiency regulations emphasize the performance attributes of thermal insulation materials, aligning with principles of the circular economy [41]. This has led to the introduction of various high-performance materials to the market. To make an informed and critical choice from the extensive range of insulation materials on the market, which vary in performance, type, and origin, as well as in application and production methods, it is necessary to conduct a comparison of the characteristics of each material [42]. The comparison of insulation materials was carried out to identify the most effective types, particularly from an energy and environmental perspective. To facilitate this comparison, several parameters were selected, not only related to winter and summer insulation capabilities but also considering the environmental impact, usage type, and cost [25]. In particular, in relation to performance indicators, the following parameters were chosen: thermal conductivity (λ), density (ρ), specific heat (c), thermal diffusivity (α), and water vapor diffusion resistance factor (μ) (Figure 4). To estimate the environmental impact, the following aspects have been analyzed: the origin of the raw material (natural organic insulators, synthetic organics, synthetic inorganics), the amount of energy required in the insulation's production phase, linked to the Potential Environmental Impact (PEI) value, expressed in MJ/kg and the possibility of recycling the material at the end of its life. Aspects related to usage have been further investigated since insulators present heterogeneous characteristics regarding both potential use and the format in which they are produced. These indications have been developed based on specific insulator characteristics, such as impact resistance, compactness, or weight. It was deemed important to conduct an economic assessment of insulating materials based on the DEI 2023 price list, as the cost per square meter is considered a significant selection parameter.

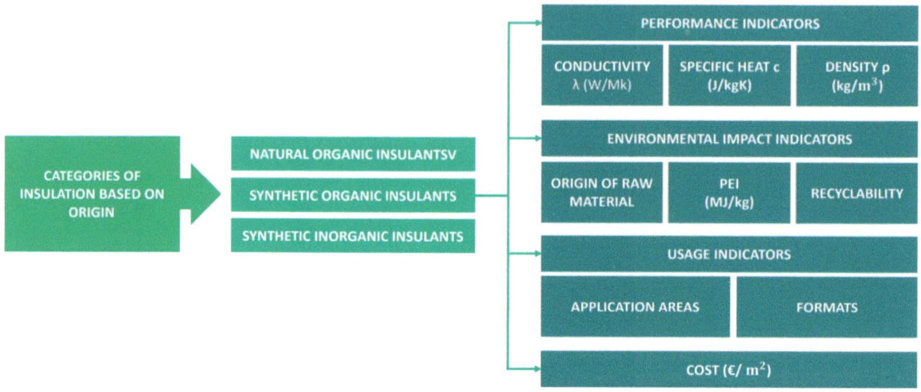

Figure 4. Comparison parameters for insulation materials.

Based on the prices identified in the DEI 2023 price list and through market research, four cost categories were determined:

- Low < 30 EUR/m^2;
- Medium > 30 and ≤70 EUR/m^2;
- Medium-high > 70 and ≤100 EUR/m^2;
- High > 100 EUR /m^2.

To express a comparative evaluation to some characteristic parameters (thermal conductivity, density, specific heat, PEI) of the main selected insulating materials (natural organic, synthetic organic, synthetic inorganic materials), radial diagrams were used. To facilitate the comparison among insulation materials, ranges have been defined for each indicator based on the performance values representative of all insulating materials (Table 3).

Table 3. Range of indicators for insulation materials comparison.

	Insulation Material Indicators			
	Conductivity W/m^{2}K	**Density Kg/m^3**	**Specific Heat J/kgK**	**PEI MJ/kg**
1	$\lambda \leq 0.015$	$\rho > 200$	$c \geq 2000$	PEI ≤ 10
2	$0.015 < \lambda \leq 0.030$	$150 < \rho \leq 200$	$1700 \leq c < 2000$	10 < PEI ≤ 30
3	$0.030 < \lambda \leq 0.040$	$50 < \rho \leq 150$	$1300 \leq c < 1700$	30 < PEI ≤ 60
4	$0.040 < \lambda \leq 0.050$	$20 < \rho \leq 50$	$1000 \leq c < 1300$	60 < PEI ≤ 100
5	$\lambda > 0.050$	$\rho \leq 20$	$c < 1000$	PEI > 100

At the center of the radar, the highest value is assigned, and the values decrease towards the periphery on a scale from 5 to 1, where 1 corresponds to the best performance: the further away from the center, the more relevant the performance is, positively speaking. Each insulation material is represented by a colored trace that allows for the evaluation of similarities and differences in performance; the larger the area delimited by the trace, the better the performance of the insulating material. The comparison graphs clearly show the following:

- Among natural organic insulators, wood fiber insulation performs better across most indicators, particularly in terms of thermal conductivity (Figure 5);
- Among synthetic organic insulators, rigid expanded polyurethane exhibits better performance, especially in terms of thermal conductivity (Figure 6);
- Among synthetic inorganic insulators, aerogel and rock wool demonstrate superior performance, especially in terms of thermal conductivity and density (Figure 7).

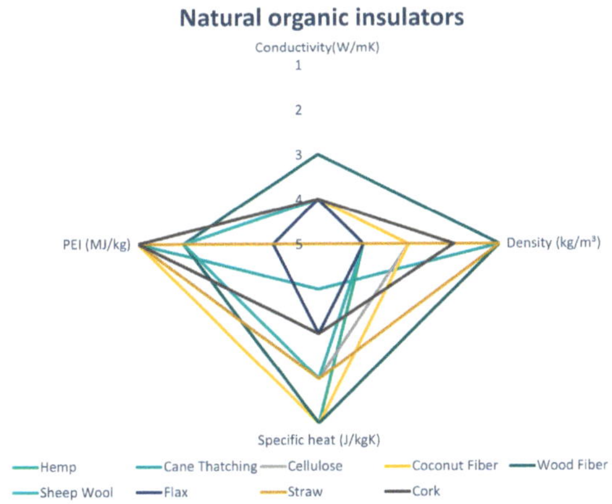

Figure 5. Comparison of natural organic insulation materials.

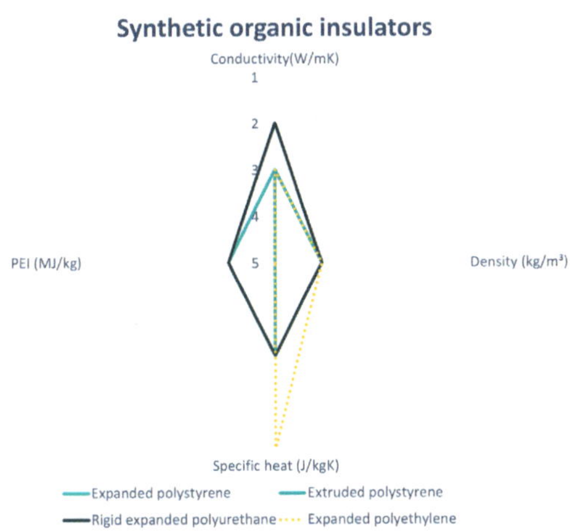

Figure 6. Comparison of synthetic organic insulation materials.

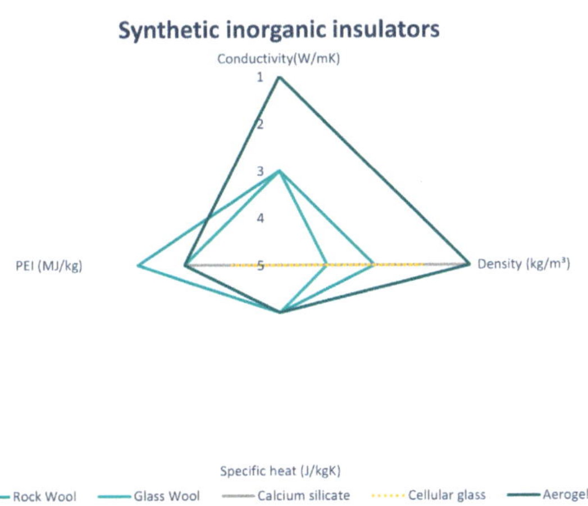

Figure 7. Comparison of synthetic inorganic insulation materials.

The Supplement S2 illustrates the entire process of selecting insulating materials for use in opaque horizontal closure solutions for energy efficiency.

2.3. Definition of a Broad Range of Sustainable Insulation Solutions Primarily Characterized by an Off-Site Approach

The insulation of the roof is a crucial aspect in the energy efficiency enhancement of existing building stock [43]. Insulation solutions for the roofs of existing residential buildings can be applied either on the external side (extrados) or on the internal side (intrados), or on the top floor just beneath the roof. The latter solution was not investigated in the present study. The solutions vary based on the slope of the roof and, consequently, the level of accessibility. In general, three main functional models are distinguished based on the placement of the insulation in relation to the waterproofing layer:

- Warm roof insulation, where the insulation material is positioned below the waterproofing layer;
- Cold (or inverted) roof insulation, where the insulation material is positioned above the waterproofing layer;
- Insulated and ventilated roof, a type of building covering that incorporates an air space between the roof surface and the underlying insulation layer.

To identify insulation solutions for the most common national upper closures, a study of products available on the market was carried out based on the previously described comparison parameters. With the aim of promoting environmental sustainability in the construction sector and minimizing impacts during construction, reducing the time for the implementation of redevelopment interventions was considered. The installation methods of each solution were evaluated, favoring mostly pre-assembled construction systems that significantly reduce execution times and related environmental impacts. By selecting mostly prefabricated technological solutions, the following were identified:

- 7 solutions applicable to non-ventilated or micro-ventilated roofs;
- 8 solutions for insulated and ventilated roofs;
- 4 solutions applicable to the intrados.

The most commonly used insulating materials for thermal insulation of horizontal closures are as follows: expanded polyisocyanurate foam, expanded polystyrene (EPS) with graphite (Neopor), wood fiber, blonde cork, closed-cell polyurethane foam, rock wool, aerogel, and rigid closed-cell polyisocyanurate foam (Figure 8). The identified solutions were studied and compared with a performance, environmental sustainability and economic perspective. Each identified solution has been associated with an acronym accompanied by a progressive number:

- IE—insulation on the extrados;
- VEI—ventilated external insulation;
- II—insulation on the intrados.

To compare the various identified solutions and identify the most efficient ones, the following parameters were taken into consideration:

- From an energy performance perspective, thermal conductivity was primarily assessed;
- From an environmental standpoint, PEI and the percentage of prefabrication were considered;
- From an economic perspective, the cost range and applicability to the main types of roofing existing nationally were evaluated.

Table 4 presents the legend of the comparison parameters used to select the most efficient insulation solutions in various aspects. The colors indicate the ranges identified for each parameter: green represents the best values, ochre-yellow represents average values, and red indicates the least performing values.

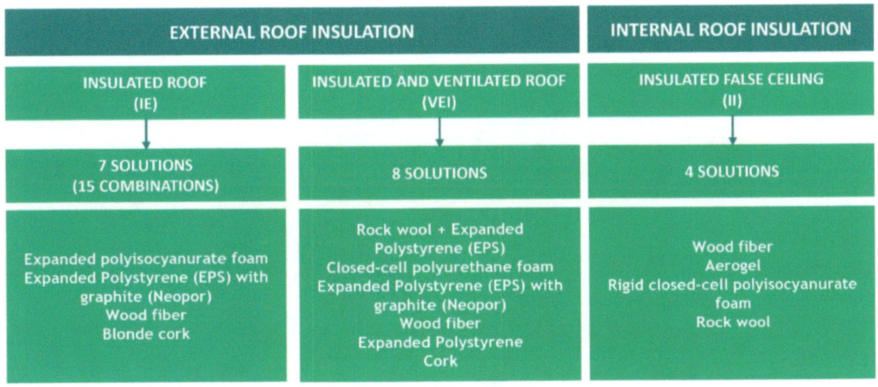

Figure 8. Prefabricated insulation solutions available on the market.

Table 4. Comparison parameters for insulation solutions.

	(green)	(ochre-yellow)	(red)
Cost range (EUR/m^2)	≤100.00	>100.00 and ≤200.00	>200.00
Conductivity (W/m^{2}K)	≤0.025	>0.025 and ≤0.040	>0.040
PEI (MJ/kg)	≤30	>30 and ≤60	>60
Prefabrication (%)	>70	>30 and ≤70	≤30
Applicability (n. of roofs)	>20	>10 and ≤20	≤10

Note: The colors indicate the ranges identified for each parameter: green represents the best values, ochre-yellow represents average values, and red indicates the least performing values.

The following images (Figures 9–11) depict the comparisons made among the various insulation solutions identified in the market. The selected solutions are highlighted in green.

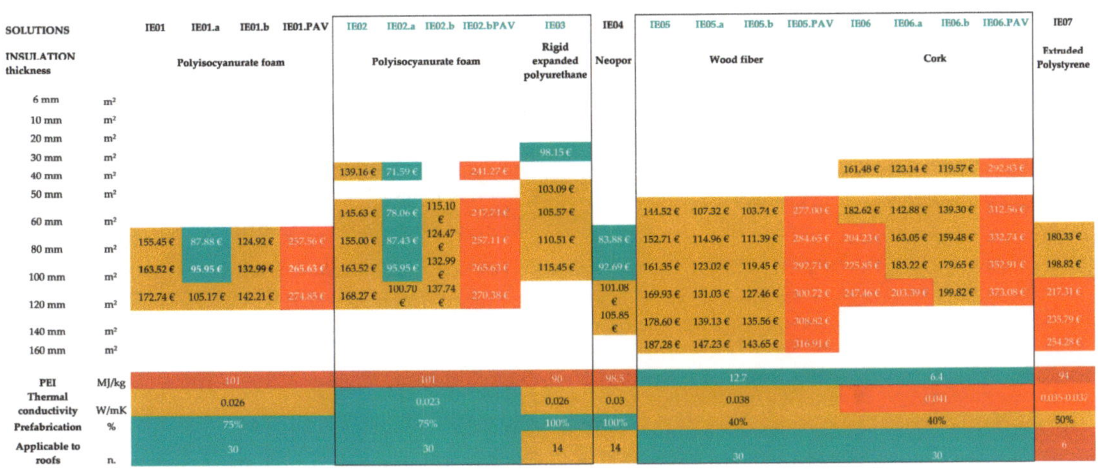

Figure 9. Comparison of selection parameters for unventilated or micro-ventilated roof insulation solutions. (The colors indicate the ranges identified for each parameter: green represents the best values, ochre-yellow represents average values, and red indicates the least performing values).

SOLUTIONS		VEI01	VEI02	VEI03	VEI04	VEI05	VEI06	VEI07	VEI08
INSULATION thickness		Rock wool + Expanded extruded polystyrene foam	Rigid closed-cell polyurethane foam	Neopor	Wood fiber	Neopor	Expanded extruded polystyrene foam	Expanded extruded polystyrene foam with graphite	Cork
60 mm	m²	178.55 €	134.14 €	155.41 €	145.19 €		122.64 €		182.46 €
80 mm	m²	192.68 €	143.90 €	160.44 €	159.05 €	148.29 €	130.24 €	105.88 €	205.72 €
100 mm	m²	204.88 €	153.15 €	165.47 €	173.29 €	156.15 €	139.04 €	114.69 €	228.85 €
120 mm	m²		161.68 €	170.50 €	178.08 €	164.01 €	146.54 €	123.08 €	252.10 €
140 mm	m²			175.53 €	190.87 €	171.87 €	154.04 €	127.85 €	
160 mm	m²				203.60 €	179.73 €		136.67 €	
180 mm	m²				216.32 €				
PEI	MJ/kg	22.12 93.6	92	98.5	12.7	98.5	93.6	98.5	6.4
Thermal conductivity	W/mK	0.038 0.034	0.022	0.03	0.012	0.03	0.033	0.031	0.0375
Prefabrication	%	50%	75%	75%	40%	40%	40%	75%	40%
Applicable to roofs	n.	14	14	14	14	14	14	14	14

Figure 10. Comparison of selection parameters for ventilated roof insulation solutions. (The colors indicate the ranges identified for each parameter: green represents the best values, ochre-yellow represents average values, and red indicates the least performing values).

SOLUTIONS		II01	II02	II03	II04
INSULATION thickness		Wood fiber	Aerogel	Polyiso foam	Rock wool
6 mm	m²		144.34 €		
10 mm	m²		176.58 €		
20 mm	m²		301.06 €		
30 mm	m²		424.96 €	60.02 €	
40 mm	m²	58.66 €	549.29 €	63.68 €	
50 mm	m²		673.91 €	67.34 €	61.40 €
60 mm	m²	65.08 €	797.66 €	71.00 €	65.45 €
80 mm	m²	71.50 €		78.32 €	73.55 €
100 mm	m²			85.64 €	81.65 €
120 mm	m²			92.96 €	89.75 €
140 mm	m²			100.28 €	
160 mm	m²				
PEI	MJ/kg	12.7	35.5	101	22.12
Thermal conductivity	W/mK	0.037	0.015	0.022	0.035
Prefabrication	%	70%	70%	70%	70%
Applicable to roofs	n.	30	30	30	30

Figure 11. Comparison of selection parameters for internal roof insulation solutions. (The colors indicate the ranges identified for each parameter: green represents the best values, ochre-yellow represents average values, and red indicates the least performing values).

Verification of Sustainability through Minimum Environmental Criteria: Disassembly and Recovered or Recycled Material

The prevailing national legislation, encompassing the 'Decreto Rilancio' (Relaunch Decree), offers valuable guidance for securing tax deductions associated with specific energy requalification interventions. It stipulates that the insulating materials employed must adhere to minimum environmental criteria, aimed at identifying the optimal solution from an environmental standpoint across the life cycle, considering market availability. To meet the minimum environmental criteria (CAM), with the perspective of promoting "environmentally preferable" technologies and products, common criteria applicable to all building components were assessed. These include the percentage of disassembly

(Criterion 2.4.14 Disassembly and end-of-life) and the percentage of recovered or recycled material (2.5_Technical specifications for construction products).

Each material may contribute with varying impacts within the framework of current regulations, respecting the specified percentages. At these percentages, each material may contribute with varying impacts in accordance with current regulations. Both in the minimum environmental criteria (CAM) and in the economic assessment, for pitched roofs with a brick sealing layer, a minimum recovery of 70% of existing tiles was considered. In the case of flat roofs with a gravel protective layer for the waterproofing membrane, a minimum recovery of 80% of existing pebbles was considered.

Data regarding the remaining materials were derived from product environmental declarations and literature. Below, there is an extract from the calculations performed to determine the quantity of disassembled material and recycled/recovered material for each proposed insulation solution, aiming to assist designers in preparing the CAM report (Figure 12).

						2.4.14_End-of-life disassembly At least 70% by weight of the building components and prefabricated elements used in the project, excluding the systems, should be capable of end-of-life disassembly or selective demolition (deconstruction) for subsequent preparation for reuse, recycling, or other recovery operations **Criterio 2.4.14**		**2.5_Technical specifications for construction products** Percentage value of recycled material or recovered material or by-products content **Criterio 2.5**	
Material	**Surface (m^2)**	**Thickness (m)**	**Volume (m^3)**	**Density (kg/m^3)**	**Weight (kg)**	**Disassemblyability (%)**	**Disassemblyability (kg)**	**Minimum value of recycled material or recovered material or by-products (%)**	**Minimum value of recycled material or recovered material or by-products (kg)**
IE.01.a									
Vapor barrier	1	0.0005	0.0005	289	0.1445	100.00%	0.1445	20.00%	0.0289
Rigid expanded Polyurethane insulation PIR.GI.	1	0.04	0.04	34	1.36	100.00%	1.36	5.00%	0.068
Bituminous felt cardboard	1	0.001	0.001	1100	1.1	100.00%	1.1	5.00%	0.055
Polymer-modified bitumen membrane	1	0.003	0.003	1200	3.6	0.00%	0	5.00%	0.18
Elevated point supports for canal tiles, using spacer feet made of UV-stabilized copolymer polypropylene	1	0.05			0.88	100.00%	0.88	20.00%	0.176
Clay roof tiles	1	0.015	0.015	1800	27	100.00%	27	70.00%	18.9
						% Disassemblyability assessed on the total of the insulation solution elements	**Total weight of disassemblable materials (kg)**	**Minimum percentage of recovered or recycled material assessed on the total of all materials used (%)**	**Total weight of recovered or recycled material assessed on the total of all materials used (kg)**
				Total weight per m^2	**34.0845**	**89.44**	**30.48**	**56.94**	**19.41**

Figure 12. Verification of CAM for unventilated or micro-ventilated extrados insulation solutions—IE.

2.4. Simulation for the Identification of Solutions

The selected insulation solutions have been applied to the types of roofing found in the Abacus of existing upper horizontal closures, which characterize the residential building heritage.

To identify the thickness of the insulation material for each proposed technological solution, simulations were performed using certified BIM energy software (TerMus-BIM v.52.00h (x64). An excerpt of the energy simulations conducted using this software is provided in Figure 13.

This allowed the verification of periodic and steady-state thermal transmittance values to identify applicable solutions for each roofing type and climatic zone.

The limit values for steady-state thermal transmittance (Table 5) were derived from the 6 August 2020 decree, "Technical Requirements for Accessing Tax Deductions for the Energy Requalification of Buildings—so-called Ecobonus" (Official Gazette General Series no. 246 of 5 October 2020) Attachment E—Requirements for Thermal Insulation Interventions [44] Table 1—Maximum Allowed Thermal Transmittance Values for Accessing Deductions.

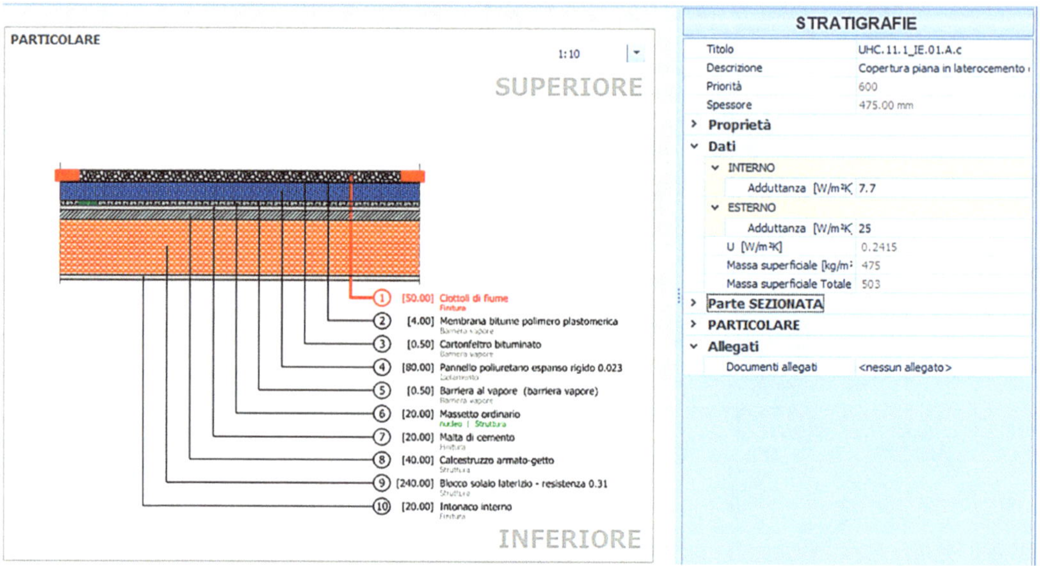

Figure 13. Excerpt of the energy simulations conducted using certified BIM energy software.

Table 5. Steady-state thermal transmittance limit values specified by the Ecobonus Requirements Decree of 05/10/2020 (calculated according to UNI EN ISO 6946) for horizontal opaque structures.

UNI EN ISO 6946	
Climate Zone	**W/m²K**
A	≤0.27
B	≤0.27
C	≤0.27
D	≤0.22
E	≤0.20
F	≤0.19

The periodic thermal transmittance values, attenuation factor, thermal lag, and obtained through the application of various thermal insulation solutions have been compared with the limit values specified by the Minimum Requirements Decree of 26 June 2015 and the UNI EN ISO 13786:2008 standard [45] (Tables 6 and 7).

Table 6. The limit values for periodic thermal transmittance as stipulated by the Minimum Requirements Decree of 26 June 2015.

Periodic thermal transmittance (calculated according to UNI EN ISO 13786:2008 and subsequent updates)	<0.18 W/m²K

To verify the thermal transmittance values of ventilated roofs, the reference standard considered for calculating the thermal resistance of air gaps is UNI EN ISO 6946:2018—Building components and structures—Thermal resistance and thermal transmittance—Calculation methods. This standard identifies three types of air gaps: non-ventilated, weakly ventilated, and strongly ventilated. The following Table 8 outlines the method for determining the thermal resistance of the air gap for each of the three cases.

Table 7. Qualitative parameters—UNI EN ISO 13786:2008.

Lag Time (Hours)	Attenuation	Performance	Performance Quality
S > 12	fa < 0.15	Excellent	I
12 ≥ S > 10	0.15 ≤ fa < 0.30	Good	II
10 ≥ S > 8	0.30 ≤ fa < 0.40	Average	III
8 ≥ S > 6	0.40 ≤ fa < 0.60	Sufficient	IV
6 ≥ S	0.6 0 ≤ fa	Mediocre	V

Table 8. Non-ventilated air gaps thermal resistance values (m^2K/W).

Type of Air Gap	Thickness of the Air Gap (mm)								
	0	5	7	10	15	25	50	100	300
	Thermal Resistance (m^2K/W)								
Horizontal air layer (upward heat flow)	0	0.11	0.13	0.15	0.16	0.16	0.16	0.16	0.16
Vertical air layer (horizontal heat flow)	0	0.11	0.13	0.15	0.17	0.19	0.18	0.18	0.18
Horizontal air layer (downward heat flow)	0	0.11	0.13	0.15	0.17	0.19	0.21	0.22	0.23

To assess the outcomes of the conducted simulations and identify the solutions and corresponding applicable thicknesses for each roof to be included in the matrix, a spreadsheet was created for each horizontal closure. The spreadsheet includes the following columns:

- Indication of the roof type with its corresponding code (UHC and progressive number found in the chart of upper horizontal closures);
- Indication of the solution type with its corresponding code (IE–VEI–II and progressive number);
- Thickness of the simulated solution expressed in millimeters;
- Steady-state thermal transmittance value achieved with the application of the solution, highlighted in light blue to indicate compliance with the limits set by current regulations; in yellow otherwise, and in white when the values specified for the colder climatic zones have already been satisfied by previous thicknesses and, therefore, were not considered in the preparation phase of the performance data sheets;
- Periodic thermal transmittance value achieved with the application of the solution, highlighted in green to indicate compliance with the limits set by current regulations, and in red otherwise.
- Indication of the climatic zones in which solutions are applicable based on the achieved thermal transmittance values;
- The code of the performance sheet given by the combination of the code of the existing roof and the applied and simulated solution, followed by a letter distinguishing them by the thickness of the insulating material;
- Cost of the solution per square meter (the color of the cell indicates the cost range to which it belongs—green if ≤EUR 100; ochre if >EUR 100.00 and ≤EUR 200.00; red > EUR 200.00).

An extract of the spreadsheets for the verification of the analyzed solutions is presented below (Figure 14).

UHC	Solution Encoding	Insulation Thickness [mm]	Steady-state Thermal Transmittance U = W/m2K	Periodic Transmittance YIE < 0.18	CLIMATIC ZONE A-B-C	CLIMATIC ZONE D	CLIMATIC ZONE E	CLIMATIC ZONE F	PERFORMANCE EVALUATING CODING (click and open)	COSTS
UHC.05	IE.01.c	80	0.2432	0.04	A-B-C				UHC.05 IE.01.c	145.63 €
UHC.05	IE.01.d	100	0.2007	0.03	A-B-C	D			UHC.05 IE.01.d	155.00 €
UHC.05	IE.01.e	120	0.1709	0.02	A-B-C	D	E	F	UHC.05 IE.01.e	163.52 €
UHC.05	IE.02.d	80	0.2494	0.04	A-B-C				UHC.05 IE.02.d	110.51 €
UHC.05	IE.02.e	100	0.2049	0.03	A-B-C	D			UHC.05 IE.02.e	115.45 €
UHC.05	IE.03.d	120	0.2596	0.02	A-B-C				UHC.05 IE.03.d	169.93 €
UHC.05	IE.03.f	160	0.2039	0.01	A-B-C	D			UHC.05 IE.03.f	187.28 €
UHC.05	VEI.01.a	60	0.2374	0.03	A-B-C				UHC.05 VEI.01.a	178.55 €
UHC.05	VEI.01.b	80	0.2111	0.03	A-B-C	D			UHC.05 VEI.01.b	200.29 €
UHC.05	VEI.01.c	100	0.1900	0.02	A-B-C	D	E	F	UHC.05 VEI.01.c	221.29 €
UHC.05	VEI.02.a	80	0.2356	0.04	A-B-C				UHC.05 VEI.02.a	143.90 €
UHC.05	VEI.02.b	100	0.1940	0.03	A-B-C	D			UHC.05 VEI.02.b	153.15 €
UHC.05	VEI.02.c	120	0.1649	0.02	A-B-C	D	E	F	UHC.05 VEI.02.c	161.68 €

Figure 14. Extract of the spreadsheets for the verification of the analyzed solutions.

3. Results

Based on simulations carried out to identify standardized energy efficiency solutions applicable to the upper closures that characterize the national residential building stock, an abacus of the most efficient standardized retrofit solutions has been developed. This abacus facilitated the formulation and definition of performance sheets aimed at providing a broad range of solutions for the renovation of existing roof structures, including parametric construction costs. The simulation results for standardized solutions, as detailed in the performance sheets, allowed the creation of a matrix serving as a valuable tool for identifying the most suitable retrofit solutions for a specific type of upper closure characterizing the building subject to energy retrofit interventions.

3.1. Abacus of Standardized Sustainable Insulation Solutions

The simulations conducted on the most common upper horizontal closures in the national residential building stock, along with assessments related to the sustainability and costs of the solutions, have led to the identification of 13 non-ventilated solutions at the extrados, 4 ventilated solutions, and 4 solutions applicable to the intrados, mostly prefabricated. An abacus has been developed for the 21 proposed insulation solutions, as detailed below:

- Performance indicators related to the materials constituting the stratigraphy (thickness, conductivity, specific heat, and density);
- PEI as an environmental impact indicator for the insulation material;
- Percentage of disassembly of the insulation system;
- Minimum total percentage of recycled or recovered content of the insulation system;
- Most EUR/m^2 resulting from the price analysis for each simulated thickness.
- Installation methods.

The abacus of the identified solutions is presented as an example (Figure 15).

The comprehensive abacus is presented in Supplement S3.

3.2. Performance Sheets of the Identified Solutions

Once the insulation solutions described in the chart were identified through simulations conducted on the most prevalent upper horizontal closures in the national residential building stock, which are the most distinctive, performance data sheets were developed. These sheets allow for a comparison of the performance parameters achievable with the application of the identified solutions on the same type of upper horizontal closure they are applicable to. The proposed solutions contribute to reducing the energy consumption of existing buildings, with reference to both winter and summer air conditioning. Each performance data sheet includes the following information:

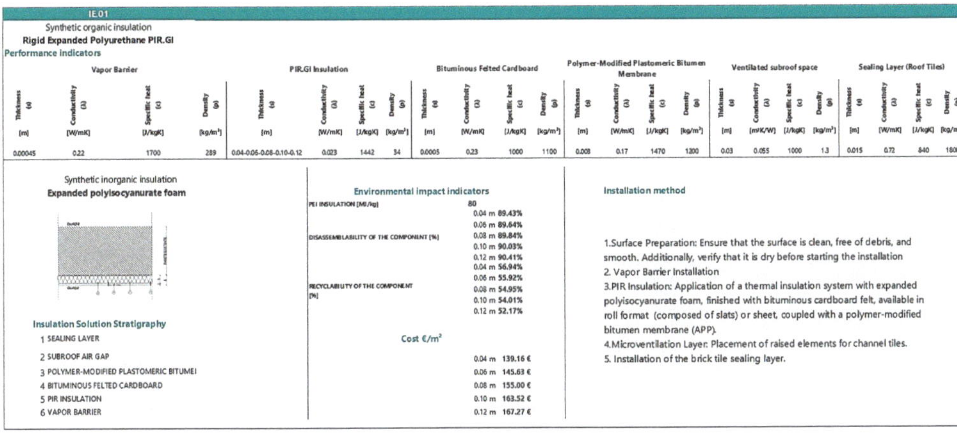

Figure 15. Abacus of the most performing solutions—IE01.

- UHC code of the existing upper horizontal closure, the time during which it was most widely used, and the stratigraphy to which the proposed solution is applicable;
- Code of the insulation solution (IE = insulation at the non-ventilated or micro-ventilated extrados, VEI = ventilated external insulation, II = intrados insulation);
- Type, thickness, and conductivity of the insulation material of the proposed solution, PEI value, percentage of disassembly, and percentage of prefabrication;
- Drawing of the roof stratigraphy and the applied insulation solution with a description of its components;
- Thermophysical properties of the building components of UHC and the proposed insulation solution;
- Results of the performance values achievable in winter and summer by applying the insulation solutions;
- Verification of the absence of surface condensation (isotherms);
- Verification of the absence of interstitial condensation (Glaser diagram—critical month);
- Radial chart to compare values related to steady-state U-value, periodic Y-value, PEI, percentage of disassembly, and prefabrication percentage of the proposed solution;
- Verification of compliance with the limit values specified by current regulations in winter and summer conditions.

The thermophysical properties of the building components of the upper horizontal closures and the proposed insulation solutions were studied to assess their ability to reduce the transmission of thermal energy, preventing its dissipation to the external environment in winter and its entry into the conditioned environment in summer. The studied parameters reported in each performance data sheet include the following: the thickness (s) of each material, thermal conductivity (λ), thermal resistance (R), specific heat (c), density (ρ), and surface mass (Ms).

The thermal transmission parameters of the studied opaque envelope components, which define their ability to control the flow of heat due to a temperature difference between the internal and external environment, are as follows: steady-state thermal transmittance (U) to verify performance in winter conditions according to the limit values specified by Annex E of the Eco-bonus Requirements Decree GU 5 October 2020 (calculated according to UNI EN ISO 6946); periodic thermal transmittance (Yie) to verify performance in summer conditions according to the limit values specified by the Minimum Requirements Decree 26 June 2015 (calculated according to UNI EN ISO 13786:2008 Thermal performance of building components—Dynamic thermal characteristics—Calculation methods and subsequent updates); phase shift (Φ) of the thermal wave verified according to qualitative parameters specified by UNI EN ISO 13786:2008; coefficient of attenuation (f_a) of the thermal wave verified according to qualitative parameters specified by UNI EN ISO 13786:2008; areic thermal capacity calculated according to UNI EN ISO 13786:2008, represents the ability of a building component to accumulate heat from the internal side of the closure.

An opaque envelope characterized by high heat accumulation capacity allows for the reduction in unwanted temperature fluctuations in both summer and winter periods, contributing to reducing energy consumption for air conditioning. Low values of the attenuation factor f_a, together with high values of periodic thermal transmittance, areic thermal capacity, and the phase shift Φ of the thermal wave, denote the good behavior of opaque closures in attenuating the effects of external summer thermal stresses. The analysis and performance verification of existing and optimized structures were conducted using energy performance certificate software. Simulations with different insulation thicknesses were performed on each type of roof from the matrix to define sustainable standardized insulation solutions. The matrix and performance data sheets contain the results of the simulations conducted in climate zone E, in the municipality of San Didero (Province of Turin), because this climatic zone hosts the most significant national residential building stock, as shown in Table 9 [16].

Table 9. Number of residential buildings in 2018 by climate zone.

Climate Zone	Number of Buildings	m^2
A	5217	170,118,357
B	710,079	
C	2,737,222	615,486,151
D	2,896,204	734,707,925
E	5,340,672	1,383,758,265
F	731,009	145,735,486

The thermo-hygrometric verification aims to ensure that the following phenomena do not occur:

- Interstitial condensation between the layers that make up the upper horizontal closure, as the occurrence of such a condition causes material degradation, especially of the insulating material, compromising the thermal performance of the roof;
- Surface condensation can lead to a conducive environment for mold and fungal growth.
- The calculation for verifying interstitial condensation is defined by the European Standard EN 13788 (Glaser diagram). The profiles of temperatures and water vapor pressures (saturated and actual) within the roof have been calculated: if the actual vapor pressure (Pe) reaches or exceeds that of the saturated vapor pressure (Ps), condensation will occur in the wall. Each data sheet includes the Glaser verification for the most critical month. To facilitate the comparison between the proposed insulation systems, ranges have been defined for each indicator based on the limit values specified by current regulations. A score from 1 to 5 has been assigned to each range, where 1 corresponds to the best performance.
- The data identified for the ranges of each indicator are reported in Table 10.

Table 10. Range of indicators for insulation solutions comparison.

Classification	Steady-State Thermal Transmittance (U) (W/m^2K)	Periodic Thermal Transmittance (Yie) (W/m^2K)	PEI (MJ/kg)	Disassembly (%)
1	$U \leq 0.19$	$Yie \leq 0.04$	$PEI \leq 10$	Disa. = 100
2	$0.19 < U \leq 0.20$	$0.04 < Yie \leq 0.06$	$10 < PEI \leq 30$	50 < disa. ≤ 75
3	$0.20 < U \leq 0.22$	$0.6 < Yie \leq 0.10$	$30 < PEI \leq 40$	25 < disa. ≤ 50
4	$0.22 < U < 0.27$	$0.10 < Yie \leq 0.17$	$40 < PEI \leq 100$	0 < disa. ≤ 25
5	$U = 0.27$	$Yie > 0.18$	$PEI > 100$	Disa. = 0

An example of the performance sheets is provided in Figure 16.

UHC.11_IE.01.A.c

UPPER HORIZONTAL CLOSURE UHC.11

Used from 1991 to 2005

Total thickness: 370mm

1 : River pebbles
2 : Waterproofing membrane
3 : Insulation panel
4 : Concrete screed
5 : Cement mortar
6 : Reinforced concrete
7 : Lateral-cementitious floor
8 : Internal plaster

U 0.57 W/mqk
YIE 0.08 W/mqk

Insulating material
Expanded polyurethane
λ : 0.023
Thickness 80 mm
PEI : 80
DISA 94
PREF 70

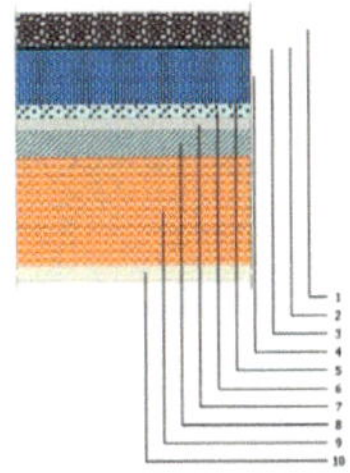

	Description of the stratigraphic solution	Thickness
	EXTERNAL	(mm)
1	River pebbles	50
2	Polymer-modified bitumen plastomeric m	4
3	Bituminous cardboard felt	1
4	Rigid expanded polyurethane panel	80
5	Vapor barrier	1
6	Ordinary screed	20
7	Cement mortar	20
8	Reinforced concrete casting	40
9	Lattice concrete block slab	160
	INTERNAL	(mm)
	Spessore tot:	396

THERMOPHYSICAL PROPERTIES OF BUILDING COMPONENTS

LAYER	DESCRIPTION	s [mm]	λ [W/mK]	C [W/m²K]	M [kg/m²]	c [J/kgK]	R [m²K/W]
EXT	Adduttanza	0		25.00			0.0400
1	River pebbles	50	0.70	14.00	75.00	1,000	0.0714
2	Polymer-modified bitumen plastomeric r	4	0.17	42.50	4.80	1,470	0.0235
3	Bituminous cardboard felt	1	0.23	460.00	0.55	1,000	0.0022
4	Rigid expanded polyurethane panel	80	0.023	0.29	2.72	1,442	3.478
5	Vapor barrier	1	0.22	440.00	0.14	1,700	0.0023
6	Ordinary screed	20	1.06	53.00	40.00	1,000	0.0189
7	Cement mortar	20	1.4	70.00	40.00	1,000	0.0143
8	Reinforced concrete casting	40	1.91	47.75	96.00	1,000	0.0209
9	Lattice concrete block slab	160		3.70	144.00	1,000	0.2700
10	Internal plaster	20	0.7	35.00	28.00	1,000	0.0286
INT	Adduttanza	0		10.00			0.1000

RESULTS

TOTAL TRANSMITTANCE	W/m²K	0.2439
SURFACE MASS	kg/m²	403.21
PHASE SHIFT	ore	11.42
ATTENUATION FACTOR		0.12
AREAL HEAT CAPACITY	kJ/m²K	60.11
PERIODIC TRANSMITTANCE	W/m²K	0.03

THERMOHYGROMETRIC VERIFICATIONS

SURFACE CONDENSATION

Minimum temperature factor fRsi = 0.9386,
Maximum critical month temperature factor, fRsi,max = 0.7242,
Critical month = December,
Vapor concentration class = Medium,
Maximum allowable U value = 1.1032 W/m²K.

fRsi ≤ fRsi,max ✓

INTERSTITIAL CONDENSATION

Critical Month **December**

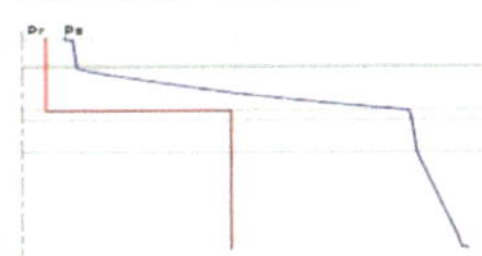

Pv ≤ Ps ✓

COMPARATIVE RADAR

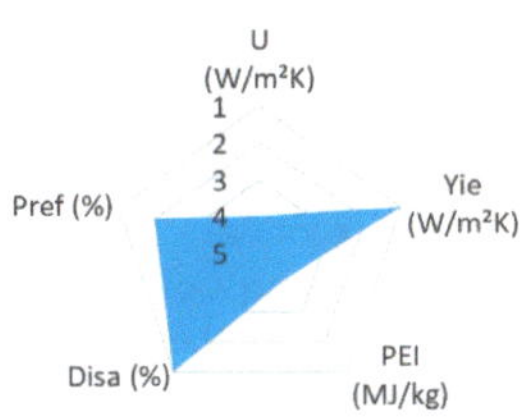

LIMIT VALUES VERIFICATION

Winter Mode

Allegato E (UNI EN ISO 6946)

A-B-C	≤	0.27 W/m2*K	☑
D	≤	0.22 W/m2*K	☐
E	≤	0.20 W/m2*K	☐
F	≤	0.19 W/m2*K	☐

ENEA
Agenzia nazionale per le nuove tecnologie, l'energia e lo sviluppo economico sostenibile

SAPIENZA
UNIVERSITÀ DI ROMA

Figure 16. Example of a performance sheet.

3.3. Matrix of Upper Closures and Their Corresponding Most Efficient Solutions

Following the development of the abacus for existing upper horizontal closures, the abacus for the most efficient solutions, and the performance sheets of the identified solutions for each roof, it was deemed appropriate to compile a matrix summarizing the results obtained from the simulations and already reported in the aforementioned sheets.

The building matrix is a scheme that represents possible configurations of building envelope types found within the national residential building stock. The matrix serves as a useful tool for identifying the most suitable retrofit solution for a specific roof type characterizing a building undergoing energy retrofit interventions. This tool is valuable for designers, guiding them towards the selection of the most suitable sustainable solution based on the characteristics of the building to be improved and the intervention location. The rows composing the matrix represent possible roof types, organized based on the construction era class, building typology, and the most prevalent construction type.

The sequence of different closure types follows the chronological evolution of construction techniques. The matrix is based on a system of sequential filters. Once the construction period of the building undergoing an insulation intervention is identified and cross-referenced with the building typology and construction type, the existing closure types of the opaque building envelope can be selected. At this juncture, transitioning to the delineation of geometric, performance, and morphological attributes, the categories of extant horizontal upper enclosures are recognized.

Each row of the matrix includes the following:

- Configuration number (ID);
- Construction era class (Class 1: up to 1900; Class 2: 1901–1920; Class 3: 1921–1945; Class 4: 1946–1960; Class 5: 1961–1975; Class 6: 1976–1990; Class 7: 1991–2005; Class 8: after 2005);
- Main building typologies, subdivided into isolated buildings (single-family and multi-family) and aggregated buildings (detached, tower, row, balcony, small building, and block);
- Construction types, subdivided into load-bearing masonry (stone, stone, and brick, brick and reinforced concrete) and framed (wood, reinforced concrete, and steel);
- Characteristics of the existing building envelope;
- Characteristics of the pre-existing upper horizontal closures, including the following:
- Thickness of the opaque building envelope;
- Steady-state thermal transmittance (W/m^2K);
- Morphological and geometrical characteristics;
- Accessibility level.
- Link to the sheet containing simulation outcomes with an indication of the performance sheets of solutions applicable to the specific opaque building envelope.

It is then possible to proceed by coupling the identified renovation solutions with different thicknesses of insulating material to the existing enclosure, determining the thermal transmittance (steady-state and periodic) of the entire stratigraphy. This tool is able to provide the climatic zones in which individual solutions can be adopted to meet the minimum requirements specified by current regulations, and a link is provided that allows viewing the descriptive and performance sheet resulting from the coupling of the insulation solution to the existing structure. The technical sheets are identified by combining the code of the upper horizontal closure UHC and the code of the retrofit solution IE, VEI, II, followed by a lowercase letter indicating the type of thickness of the solution.

An excerpt of the matrix is presented in Figure 17.

NATIONAL RESIDENTIAL BUILDING STOCK MATRIX																																					
	Construction Era Class								Building Type								Construction Type						Building Envelope Characteristics			Characteristics of Ante Operam Upper Horizontal Closures											Insulation Solutions
									Isolated Buildings		Multi-family Buildings						Load-bearing Masonry				Frame					Upper Horizontal Closures -Ante Operam		Morphological		Geometric		Accessibility					
ID	1. Up to 1900	2. From 1901 to 1920	3. From 1921 to 1945	4. From 1946 to 1960	5. From 1961 to 1975	6. From 1976 to 1990	7. From 1991 to 2005	After 2005	Single-family	Two-family	In line	Tower	Row	With balcony	Gallery	Block	Stone	Stone and Brick	Brick	Contrete	Wood	Reinforced Contrete	Upper Horizontal Closures	Vertical Closures	Lower Horizontal Closures	Thickness (cm)	Thermal Transmittance U = W/m2K	Continuous	Discontinuous	Horizontal plane 1-5%	Inclined plane >5%	Class A	Class B	Class C	Class D	Class E	SOLUTIONS SHEET LINK
1	x								x		x		x	x	x	x	x						UHC.01 - Pitched roof with wooden structure and boarding 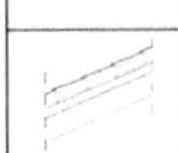	VC.01 - Plastered stone masonry (45 cm) 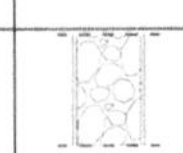	LHC.02 - Brick vaulted ceiling 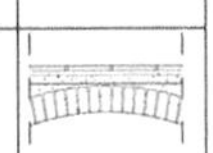	4.5	2.56		x		x	x	x				SC 01
2	x								x		x		x	x	x	x	x						UHC.01 - Pitched roof with wooden structure and boarding 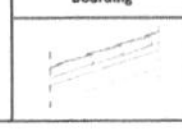	VC.01 - Plastered stone masonry (45 cm) 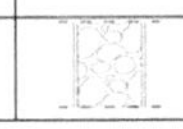	LHC.01 - Wooden floor with brick tiles 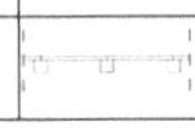	4.5	2.56		x		x	x	x				SC 01

Figure 17. Excerpt of the national residential building stock matrix.

4. Discussion

The age of the existing national residential building stock highlights the need to intervene in the building stock with energy efficiency measures, including thermally insulating the building envelope [46]. This measure reduces heat loss in winter and heat gain in summer, enhancing climate control efficiency. Energy retrofit interventions aim to decrease energy consumption for heating and cooling while reducing greenhouse gas emissions and improving comfort. The promotion of sustainable construction goes beyond the efficiency of existing assets. This also means rethinking materials, components, and building systems more efficiently in terms of resource consumption. This involves the use of recycled and recyclable materials, streamlining installation methods, as well as designing prefabricated and dry-assemblable systems to maximize recycling or reuse of components at the time of decommissioning. This must be performed in compliance with the technical specifications provided by the minimum environmental criteria (C.A.M.) applicable to building components.

During this research, a wide range of insulation materials available on the market was examined, comparing them based on specific energy performance indicators, environmental impact, cost, and usage indications. This analysis allowed for the selection of twenty-one insulation solutions characterized by the best performance and limited environmental impact. To determine the thickness of the insulation material for each proposed technological solution, a methodology was developed in which the performance of insulation scenarios was precalculated. This guides the designer towards more efficient, prefabricated, and sustainable standardized solutions for the national residential building stock. Simulations were carried out for each solution with different thicknesses of insulating material on each type of upper horizontal closure present in the diagram. To ensure the applicability of the proposed insulation solutions, compliance with common criteria for all building components, such as the percentage of disassembly and the percentage of recovered or recycled material content, was verified. These criteria both reduce the need for raw materials and stimulate the waste valorization chain from construction and demolition.

An examination of insulation intervention costs facilitated an economic assessment of the viability of implementing the identified solutions, offering the designer an additional selection criterion. This process culminated in the creation of an optimized technological solutions diagram based on climatic, geometric/dimensional, and technical/performance parameters. The solutions were primarily chosen based on prefabrication criteria, providing guidance for designers involved in energy efficiency interventions within the residential real estate sector. Following the identification of the most efficient standardized insulation solutions, performance data for each solution pertaining to every type of upper horizontal closure in the national residential building stock were compiled through simulations. Subsequently, these performance data were consolidated into performance sheets for each solution and for every roof, facilitating the immediate identification of the most performance-oriented system based on climatic zones. These elaborations have given rise to the development of a matrix populated by the configurations of the most common opaque closures in the national residential building stock, with a description of existing roofs and a link that allows access to a summary of the simulation results subsequently detailed in the relevant performance sheet. Through the matrix and performance sheets, the designer can easily identify the case that best corresponds to the specific energy efficiency intervention based on the construction era, building type, construction type, climatic zone, and the most effective technological solutions according to the performance requirements to be met in both winter and summer. The matrix produced is an open system that can be implemented following the evolution of new components and systems. Dashboards have also been developed that provide a summary for each UHC:

- Description of the upper horizontal closure before work;
- The number and type of applicable solutions;
- The minimum number of applicable solutions in different climatic zones based on the results of energy simulations;
- The number of solutions by type of insulation (IE = non-ventilated or micro-ventilated insulation solutions, VEI= ventilated external insulation; II = underside insulation solutions);
- The minimum number of solutions by type of insulation applicable for each climatic zone;
- Classification of solutions by type of insulation based on cost per m^2.

The following is an example of a dashboard (Figure 18).

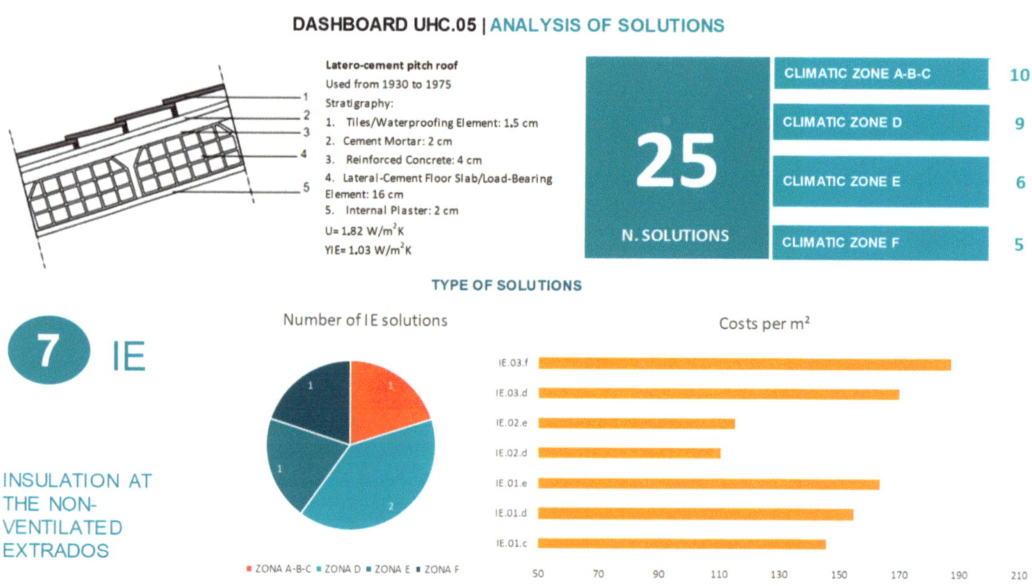

Figure 18. *Cont.*

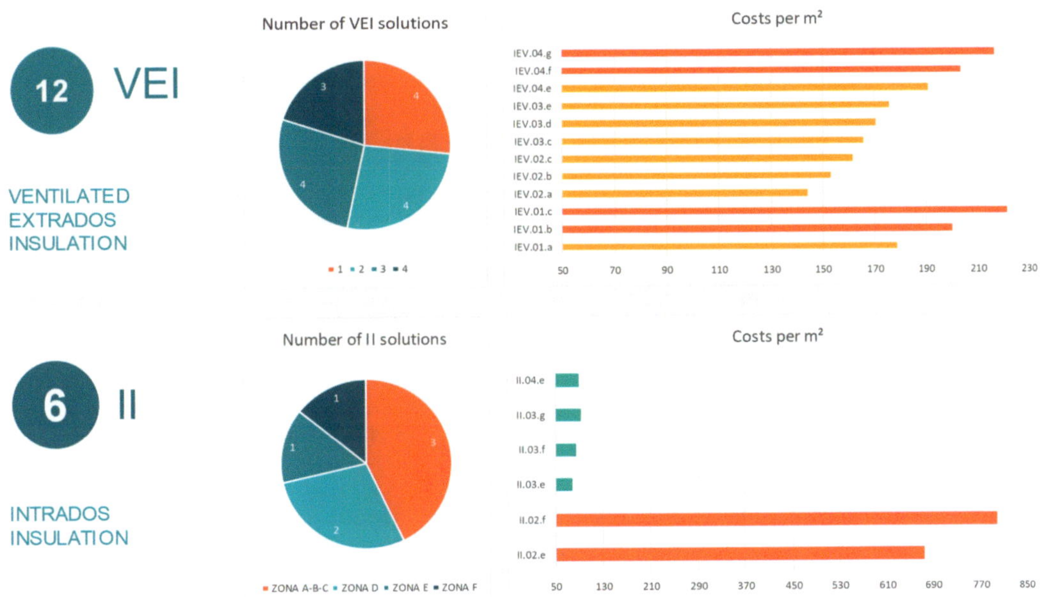

Figure 18. Example of a dashboard.

5. Conclusions and Future Developments

Defining eco-sustainable solutions is one of the most crucial challenges for the future, particularly in the context of energy retrofitting.

It is an innovation process that is affecting the entire construction supply chain and the renewal of existing building stock, as increasingly, plans are being implemented to align buildings with new criteria for energy efficiency and eco-sustainability.

The objective guiding the research presented here consists of attempting to identify and promote sustainable solutions capable of directing designers towards energy retrofit interventions for the upper closures of national residential buildings, primarily characterized by high levels of prefabrication and recycled content, in accordance with current minimum environmental criteria (CAM). The main limitations of this research are related to the absence of an evaluation of maintenance costs for the standardized thermal insulation solutions proposed. Furthermore, the roofing considered for defining the pre-calculated solutions are those most commonly found within the national residential building stock and may not include some types. Lastly, for a comprehensive energy analysis, it is necessary to consider all building envelope closures. Indeed, this study will be integrated in the second year of ENEA research with a prior focus on identifying standardized thermal insulation solutions for vertical closures—vertical perimeter walls [23].

The research outcomes will constitute the core for the development of a tool supporting energy efficiency interventions in residential buildings during renovation. The development in the second year of research will indeed focus on creating a tool to guide the designer in identifying scenarios that best correspond to the real situation of the property to be redeveloped, based on the construction era class, building type, construction type, and climatic zone. Utilizing the database of identified optimized proposals, the tool will enable the selection of the most effective technological solutions (standardized, energy-efficient, and sustainable) that ensure compliance with performance requirements set by current regulations in winter conditions, following Off-Site construction indications.

The tool will provide the opportunity to conduct a preliminary energy analysis capable of outlining the potentialities offered by a range of proposed retrofit interventions concerning the opaque building envelope. It will be a valuable instrument for the professional technician, who, during the selection of redevelopment options for a building, can efficiently assess, through preliminary energy performance simulations, the outcome of applying the different optimized solutions of the opaque building envelope previously selected through the dynamic matrix in terms of energy savings and economic feasibility. The calculation of energy demand will encompass both heating and cooling requirements. The tool's calculations will cover both the building envelope and the most common plant types.

The research findings underscore the effectiveness of adopting the described methodology in enhancing the percentage improvement of steady-state thermal transmittance (W/m^2K). Notably, a substantial average improvement of over 66% is observed in upper horizontal closures, ranging from a minimum of 34% to a maximum of 92%. This indicates a substantial enhancement in energy efficiency achievable through targeted interventions. The graph visually represents the distribution of improvement percentages in steady-state thermal transmittance across the existing roofs of the national building stock utilizing optimized, standardized, and sustainable solutions. It is evident that the percentage improvement in transmittance is particularly higher in cases where the existing roof system belongs to an older construction era class, characterized by significantly higher pre-renovation transmittance values. These values necessitate post-renovation adjustments to comply with prevailing regulatory standards. This observation underscores the need for tailored interventions to address specific building characteristics and regulatory requirements, ensuring optimal energy efficiency outcomes (Figure 19).

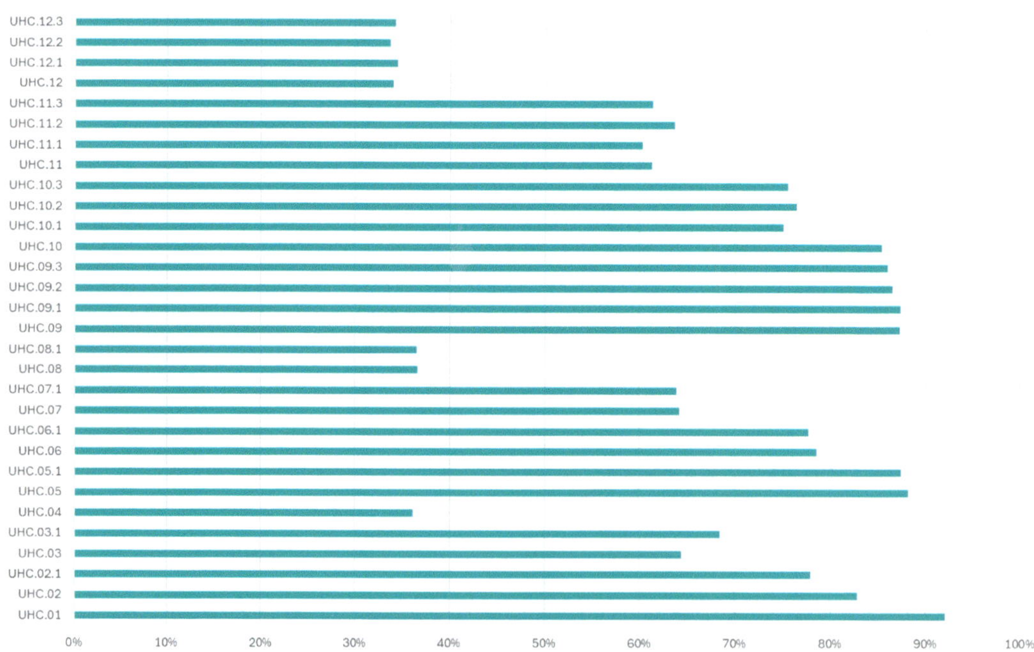

Figure 19. Steady-state thermal transmittance improvement (%).

The tool will allow for the identification of the economic savings achievable with the application of the proposed insulation solutions and the net present value (NPV). It will inform the user about any incentives available for the implementation of the simulated intervention.

The user can download performance sheets containing the stratifications of applicable solutions on vertical closures (perimeter walls) and upper horizontal closures (roofs), thermophysical properties, simulation results, thermo-hygrometric verifications, and sustainability indicators. When there is a need to intervene in a property, the tool will enable following:

- Creating an energy model of the building subject to redevelopment interventions in the pre-intervention situation using the matrix of existing vertical and horizontal closures that characterize the national residential building stock.
- Evaluating the energy performance of the real building in its current state through synthetic indicators.
- Guiding the choice using databases of identified optimized solutions for both perimeter walls and roofs.
- Constructing a post-intervention building model.
- Evaluating the outcome of adopting the identified solutions in terms of technical/economic feasibility downstream of the simulated scenarios.

Using the tool, it will also be possible to create a "ranking" of priority for energy redevelopment actions by comparing different solutions.

In conclusion, the new European directives represent a significant step towards a more sustainable and energy-efficient Europe, laying the groundwork for a comprehensive transformation of the construction sector towards climate neutrality. By adopting ambitious measures and innovative tools, such as the one proposed in this article, it will be possible to enhance the durability and overall performance of existing buildings and drastically reduce greenhouse gas emissions and energy consumption in the construction sector by 2030.

Supplementary Materials: The following supporting information can be downloaded at: https://www.mdpi.com/article/10.3390/su16093850/s1, Supplement S1: Abacus of existing upper horizontal closures; Supplement S2: Study of insulating materials for the redevelopment of existing upper closure types; Supplement S3: Abacus of insulation solutions for upper horizontal closures (UHCs).

Author Contributions: Conceptualization, C.R. and E.P.; methodology, E.P.; validation, E.P., C.Z. and C.R.; formal analysis, E.P. and C.Z.; investigation, E.P. and C.Z.; resources, E.P.; data curation, E.P. and C.Z.; writing—original draft preparation, E.P., C.R. and C.Z.; writing—review and editing, E.P., C.R. and C.Z.; visualization, E.P., C.R. and C.Z.; supervision, E.P. and C.R.; project administration, E.P.; funding acquisition, E.P. and C.R. All authors have read and agreed to the published version of the manuscript.

Funding: This research was funded by MINISTRY OF ECONOMIC DEVELOPMENT, in the program for "Research of the Electricity System" in cooperation with ENEA on project "High-efficiency buildings for the energy transition", 2022–2024. Grant number: I53C22003050001.

Institutional Review Board Statement: Not applicable.

Informed Consent Statement: Not applicable.

Data Availability Statement: Data are contained within the article Supplementary Material.

Acknowledgments: The authors would like to express their gratitude to Fabrizio Cumo for his valuable support.

Conflicts of Interest: The authors declare no conflicts of interest.

References

1. Hafez, F.S.; Sa'di, B.; Safa-Gamal, M.; Taufiq-Yap, Y.H.; Alrifaey, M.; Seyedmahmoudian, M.; Stojcevski, A.; Horan, B.; Mekhilef, S. Energy Efficiency in Sustainable Buildings: A Systematic Review with Taxonomy, Challenges, Motivations, Methodological Aspects, Recommendations, and Pathways for Future Research. *Energy Strategy Rev.* **2023**, *45*, 101013. [CrossRef]
2. Shi, X.; Tian, Z.; Chen, W.; Si, B.; Jin, X. A review on building energy efficient design optimization from the perspective of architects. *Renew. Sustain. Energy Rev.* **2016**, *65*, 872–884. [CrossRef]

3. Alhazmi, H.; Alduwais, A.K.; Tabbakh, T.; Aljamlani, S.; Alkahlan, B.; Kurdi, A. Environmental Performance of Residential Buildings: A Life Cycle Assessment Study in Saudi Arabia. *Sustainability* **2021**, *13*, 3542. [CrossRef]
4. Hariram, N.P.; Mekha, K.B.; Suganthan, V.; Sudhakar, K. Sustainalism: An Integrated Socio-Economic-Environmental Model to Address Sustainable Development and Sustainability. *Sustainability* **2023**, *15*, 10682. [CrossRef]
5. Economidou, M.; Ringel, M.; Valentova, M.; Castellazzi, L.; Zancanella, P.; Zangheri, P.; Serrenho, T.; Paci, D.; Bertoldi, P. Strategic energy and climate policy planning: Lessons learned from European energy efficiency policies. *Energy Policy* **2022**, *171*, 113225. [CrossRef]
6. Economidou, M.; Todeschi, V.; Bertoldi, P.; D'Agostino, D.; Zangheri, P.; Castellazzi, L. Review of 50 years of EU energy efficiency policies for buildings. *Energy Build.* **2020**, *225*, 110322. [CrossRef]
7. Tsemekidi-Tzeiranaki, S.; Labanca, N.; Cuniberti, B.; Toleikyte, A.; Zangheri, P.; Bertoldi, P. *Analysis of the Annual Reports 2018 under the Energy Efficiency Directive, EUR 29667 EN*; Publications Office of the European Union: Luxembourg, 2019; ISBN 978-92-76-00173-7. [CrossRef]
8. European Commission. Focus on Energy Efficiency in Buildings. 2020. Available online: https://commission.europa.eu/news/focus-energy-efficiency-buildings-2020-02-17_en (accessed on 21 April 2024).
9. Almeida, D.V.; Kolinjivadi, V.; Ferrando, T.; Roy, B.; Herrera, H.; Vecchione Gonçalves, M.; Van Hecken, G. The "Greening" of Empire: The European Green Deal as the EU first agenda. *Political Geogr.* **2023**, *105*, 102925. [CrossRef]
10. A Renovation Wave for Europe—Greening Our Buildings, Creating Jobs, Improving Lives. 2020. Available online: https://eur-lex.europa.eu/legal-content/EN/TXT/?qid=1603122220757&uri=CELEX:52020DC0662 (accessed on 16 February 2024).
11. von Platten, J.; de Fine Licht, K.; Mangold, M.; Mjörnell, K. Renovating on Unequal Premises: A Normative Framework for a Just Renovation Wave in Swedish Multifamily Housing. *Energies* **2021**, *14*, 6054. [CrossRef]
12. European Commission. Renovation Wave. Available online: https://energy.ec.europa.eu/topics/energy-efficiency/energy-efficient-buildings/renovation-wave_en (accessed on 16 February 2024).
13. European Commission. Proposal for a Directive of the European Parliament and of the Council on the Energy Performance of Buildings (Recast) COM/2021/802 Final. Available online: https://eur-lex.europa.eu/legal-content/EN/TXT/?uri=celex:52021PC0802 (accessed on 22 April 2024).
14. Abrahamsen, F.E.; Ruud, S.G.; Gebremedhin, A. Assessing Efficiency and Environmental Performance of a Nearly Zero-Energy University Building's Energy System in Norway. *Buildings* **2023**, *13*, 169. [CrossRef]
15. Ruggieri, G.; Andreolli, F.; Zangheri, P. A Policy Roadmap for the Energy Renovation of the Residential and Educational Building Stock in Italy. *Energies* **2023**, *16*, 1319. [CrossRef]
16. Ministry for Ecological Transition. Strategy for Energy Retrofitting of National Building Stock. Available online: https://energy.ec.europa.eu/system/files/2021-12/2020_ltrs_italy_-_en.pdf (accessed on 22 April 2024).
17. Agostinelli, S. Deep Renovation. Criteri di Efficientamento Energetico degli Edifici; Project "ENERSELVES" Interreg Europe Horizon. ISBN: 979-12-200-5959-6. 2020. Available online: https://www.researchgate.net/publication/355169433_DEEP_%20RENOVATION_Criteri_di_efficientamento_energetico_degli_edifici (accessed on 28 June 2023).
18. Mavromatidis, L.E.; Bykalyuk, A.; Lequaya, H. Development of polynomial regression models for composite dynamic envelopes thermal performance forecasting. *Appl. Energy* **2013**, *104*, 379–391. [CrossRef]
19. Sala Lizarraga, J.M.P.; Picallo-Perez, A. 12—Design and optimization of the envelope and thermal installations of buildings. In *Exergy Analysis and Thermoeconomics of Buildings. Design and Analysis for Sustainable Energy Systems*; Butterworth-Heinemann: Oxford, UK, 2020; pp. 911–1005.
20. Paolino, G. Il sistema tetto. In *Progettazione, Comportamento e Realizzazione Delle Coperture Degli edifici*; Maggioli Editore: Santarcangelo di Romagna (RN), Italy, 2013.
21. Zhenjun, M.; Cooper, P.; Daly, D.; Ledo, L. Existing Building Retrofits: Methodology and State-of-theArt. *Energy Build.* **2012**, *55*, 889–902.
22. Lizana, J.; Barrios-Padura, A.; Molina-Huelva, M.; Chacartegui, R. Multi-Criteria Assessment for the Effective Decision Management in Residential Energy Retrofitting. *Energy Build.* **2016**, *129*, 284–307. [CrossRef]
23. Paraschiv Lizica, S.; Paraschiv, I.S.; Ion, V.I. Increasing the energy efficiency of buildings by thermal insulation. *Energy Procedia* **2017**, *128*, 393–399.
24. Cumo, F.; Giustini, F.; Pennacchia, E.; Romeo, C. Support Decision Tool for Sustainable Energy Requalification the Existing Residential Building Stock. The Case Study of Trevignano Romano. *Energies* **2021**, *14*, 74. [CrossRef]
25. Konstantinou, T. A Methodology to Support Decision-Making towards an Energy-Efficiency Conscious Design of Residential Building Envelope Retrofitting. *Buildings* **2015**, *5*, 1221–1241. [CrossRef]
26. Bianco, V.; Marmori, C. Modelling the deployment of energy efficiency measures for the residential sector. The case of Italy. *Sustain. Energy Technol. Assess.* **2022**, *49*, 101777. [CrossRef]
27. Ballarini, I.; Corrado, V. A New Methodology for Assessing the Energy Consumption of Building Stocks. *Energies* **2017**, *10*, 1102. [CrossRef]
28. Di Turi, S.; Stefanizzi, P. Energy analysis and refurbishment proposals for public housing in the city of Bari, Italy. *Energy Policy* **2015**, *79*, 58–71. [CrossRef]

29. Hüttler, W.; Bachner, D.; Hofer, G.; Kremp, M.; Trimmel, G.; Wall, I. Decision support tool for the innovative and sustainable renovation of historic buildings (HISTool). In *Energy Efficiency in Historic Buildings;* Broström, T., Nilsen, L., Carlsten, S., Eds.; Uppsala University, Department of Art History: Uppsala, Sweden, 2018; pp. 226–235.
30. Ibañez Iralde, N.S.; Pascual, J.; Salom, J. Energy retrofit of residential building clusters. A literature review of crossover recommended measures, policies instruments and allocated funds in Spain. *Energy Build.* 2021; 252, 111409.
31. Piras, G.; Muzi, F. Energy Transition: Semi-Automatic BIM Tool Approach for Elevating Sustainability in the Maputo Natural History Museum. *Energies* **2024**, *17*, 775. [CrossRef]
32. Alhammad, M.; Eames, M.; Vinai, R. Enhancing Building Energy Efficiency through Building Information Modeling (BIM) and Building Energy Modeling (BEM) Integration: A Systematic Review. *Buildings* **2024**, *14*, 581. [CrossRef]
33. Prada-Hernández, A.V.; Rojas-Quintero, J.S.; Vallejo-Borda, J.A.; Ponz-Tienda, J.L. Interoperability of building energy modelling (BEM) with building information modelling (BIM). In Proceedings of the SIBRAGEC ELAGEC, Sao Carlos, Brazil, 7–9 October 2015; pp. 519–526.
34. Farid Mohajer, M.; Aksamija, A. Integration of building energy modelling (BEM) and building information modelling (BIM): Workflows and case study. In Proceedings of the Building Technology Educator's Society Conference, Amherst, MA, USA, 17–22 June 2019.
35. Bastos Porsani, G.; de Lersundi, K.D.V.; Sánchez-Ostiz Gutiérrez, A.; Fernández Bandera, C. Interoperability between building information modelling (BIM) and building energy model (BEM). *Appl. Sci.* **2021**, *11*, 2167. [CrossRef]
36. Agostinelli, S.; Cumo, F.; Guidi, G.; Tomazzoli, C. Cyber-Physical Systems Improving Building Energy Management: Digital Twin and Artificial Intelligence. *Energies* **2021**, *14*, 2338. [CrossRef]
37. Jha, B.; Bhattacharjee, B. Tool for energy efficient building envelope retrofitting. In Proceedings of the Building Performance Analysis Conference and SimBuild Co-Organized by ASHRAE and IBPSA-USA, Chicago, IL, USA, 26–28 September 2018.
38. Belaïd, F.; Roubaud, D.; Galariotis, E. Features of residential energy consumption: Evidence from France using an innovative multilevel modelling approach. *Energy Policy* **2019**, *125*, 277–285. [CrossRef]
39. Action Plan for the Environmental Sustainability of Consumption in the Public Administration Sector (2023 edition). Available online: https://gpp.mite.gov.it/sites/default/files/2023-08/PAN_GPP.pdf (accessed on 21 February 2024).
40. Sicignano, E.; Di Ruocco, G.; Stabile, A. Quali—A Quantitative Environmental Assessment Method According to Italian CAM, for the Sustainable Design of Urban Neighbourhoods in Mediterranean Climatic Regions. *Sustainability* **2019**, *11*, 4603. [CrossRef]
41. Almusaed, A.; Almssad, A.; Alasadi, A.; Yitmen, I.; Al-Samaraee, S. Assessing the Role and Efficiency of Thermal Insulation by the "BIO-GREEN PANEL" in Enhancing Sustainability in a Built Environment. *Sustainability* **2023**, *15*, 10418. [CrossRef]
42. Clemente, C.; Piermattei, P. Parametri di valutazione e scelta dei materiali isolanti. In *Pluralità Tecnologica.Papers;* Clemente, C., Ed.; Rdesignpress: Roma, Italy, 2012; pp. 131–143.
43. D'Angola, A.; Pepe, R.; Scuderi, M. *Il Nuovo Conto Termico D.M. 16 Febbraio 2016. Con 16 Esempi di Calcolo;* MaggioliEditore: Santarcangelo di Romagna (RN), Italy, 2016.
44. Ministry of Economic Development. Decree August 6, 2020—Technical Requirements for Accessing Tax Deductions for Energy Retrofitting of Buildings—So-Called Ecobonus. Available online: https://www.gazzettaufficiale.it/do/atto/serie_generale/caricaPdf?cdimg=20A0539400500010110001&dgu=2020-10-05&art.dataPubblicazioneGazzetta=2020-10-05&art.codiceRedazionale=20A05394&art.num=1&art.tiposerie=SG (accessed on 26 April 2024).
45. *UNI EN ISO 13786:2008;* Thermal Performance of Building Components—Dynamic Thermal Characteristics—Calculation Methods. International Organization for Standardization: Geneva, Switzerland, 2008. Available online: https://store.uni.com/uni-en-iso-13786-2008 (accessed on 26 April 2024).
46. Laaroussi, Y.; Bahrar, M.; Zavrl, E.; El Mankibi, M.; Stritih, U. New qualitative approach based on data analysis of European building stock and retrofit market. *Sustain. Cities Soc.* **2020**, *63*, 102452. [CrossRef]

sustainability

MDPI

Article

Frictional Behavior of Chestnut (*Castanea sativa* Mill.) Sawn Timber for Carpentry and Mechanical Joints in Service Class 2

José Ramón Villar-García [1], Manuel Moya Ignacio [1], Pablo Vidal-López [2] and Desirée Rodríguez-Robles [2,*]

[1] Forest Research Group, Department of Forest and Agricultural Engineering, Universitary Center of Plasencia, University of Extremadura, Av. Virgen del Puerto 2, 10600 Plasencia, Spain; jrvillar@unex.es (J.R.V.-G.); manuelmi@unex.es (M.M.I.)

[2] Mechanical and Fluid Engineering Research Group, Department of Forest and Agricultural Engineering, School of Agricultural Engineering, University of Extremadura, Av. Adolfo Suarez s/n, 06071 Badajoz, Spain; pvidal@unex.es

* Correspondence: desireerodriguez@unex.es

Abstract: Wood is poised to become a material of choice for future construction. When appropriately managed, it is a renewable material with unique mechanical properties. Thus, there has been a growing demand for hardwoods, including *Castanea sativa* Mill., the focal point of this investigation, for structural applications. Albeit in a limited capacity, Eurocode 5-2 offers friction coefficients for softwoods, but it falls short for hardwoods. These coefficients play a critical role in numerical simulations involving friction, enabling the optimization of joints and, by extension, the overall structural integrity. Test samples were evaluated at 15% and 18% moisture content (Service Class 2) for various orientations of timber-to-timber and timber-to-steel friction. The results provide an experimental database for numerical simulations and highlight the influence of moisture on the stick–slip phenomenon, which was absent for the timber-to-timber tests, as well as on the rising friction values. At 18%, the static and kinetic coefficients were 0.70 and 0.48 for timber-to-timber and 0.5 and 0.50 for timber-to-steel. The increase was around 50% for timber-to-timber friction and over 170% for timber-to-steel pairs. Moreover, the findings proved a relationship between both coefficients and the validity of the linear estimation approach within the 12–18% moisture commonly applied to softwoods.

Keywords: friction coefficient; tribology; mechanical properties; contact simulation; Eurocode 5

Citation: Villar-García, J.R.; Moya Ignacio, M.; Vidal-López, P.; Rodríguez-Robles, D. Frictional Behavior of Chestnut (*Castanea sativa* Mill.) Sawn Timber for Carpentry and Mechanical Joints in Service Class 2. *Sustainability* **2024**, *16*, 3886. https://doi.org/10.3390/su16103886

Academic Editors: Uroš Klanšek and Tomaž Žula

Received: 8 March 2024
Revised: 2 May 2024
Accepted: 5 May 2024
Published: 7 May 2024

1. Introduction

Historically, wood has been a traditional and widely used material in construction due to its abundance, ease of use, and adequate mechanical properties. However, as technological advancements made steel and concrete not only more accessible but also cost-effective, these materials began to be perceived as superior alternatives due to their modern aesthetic, enhanced durability, and significantly improved fire resistance. In turn, the prominence of wood in the construction industry diminished as it was relegated to small-scale or less demanding structural applications due to concerns regarding instability, fire safety, decay, and sound transmission [1].

Currently, with the construction sector widely recognized as a major contributor to environmental degradation due to substantial material and energy consumption, greenhouse gas emissions, and waste generation, wood is experiencing a resurgence as a sustainable construction material. The favorable mechanical properties relative to its weight, the enhancement of its durability through innovative treatments, and the advent of new engineered timber products, e.g., glued laminated timber (glulam), cross-laminated timber (CLT), and laminated veneer lumber (LVL), are some of the driving factors in its resurgence besides the pursuit of sustainable development. In this regard, beyond its inherent sustainability, the use of wood has a crucial role in addressing climate change concerns due to its significantly lower embodied energy [2] and reduced CO_2 emissions [3,4] while simultaneously acting as a carbon sink of approximately 1.5 t of CO_2 per m^3 of wood [5]. As a renewable resource originating from responsibly managed forests, wood further alleviates the pressures of raw material scarcity, highlighting its multifaceted contribution to environmental conservation.

For this investigation, chestnut wood (*Castanea sativa*, Mill.) was selected as this deciduous species covers more than 2.5 million hectares in Europe around the Mediterranean region, with 89% concentrated in France, Italy, Spain, Portugal, and Switzerland, in decreasing order of importance, as shown in Figure 1 [6]. Several research works have underscored its ecological relevance as a source of support for a wide variety of flora and fauna [7,8] and the European Council has included "9260 *Castanea sativa* woods'" in Annex 1 of the Habitats Directive [9]. Commercially, chestnut is valued both for fruit and non-wood products as well as timber. For instance, in Spain, the average total volume (with bark) of chestnut stands harvested in 2021 reached 97,878 m^3 [10], mostly from the north provinces (Galicia, Asturias, Navarre, and Catalonia), but also arising from the center and south of the country (Figure 1). Chestnut wood is valued for its appearance and strength; it is particularly appreciated for external use due to its natural protection against decay [11,12], and it possesses a vast tradition of use for both structural and non-structural purposes in construction (beams, joists, and traditional grain stores), woodworking, furniture, flooring, fine veneer, general joinery, and poles) [11]. Nowadays, sustainability concerns have spurred a new interest in its use. In this regard, Carbone et al. [13], who investigated the market competitiveness of laminated chestnut timber products, forecasted a bright future for this type of wood while indicating the need for a targeted chestnut wood policy to significantly bolster its market penetration and growth.

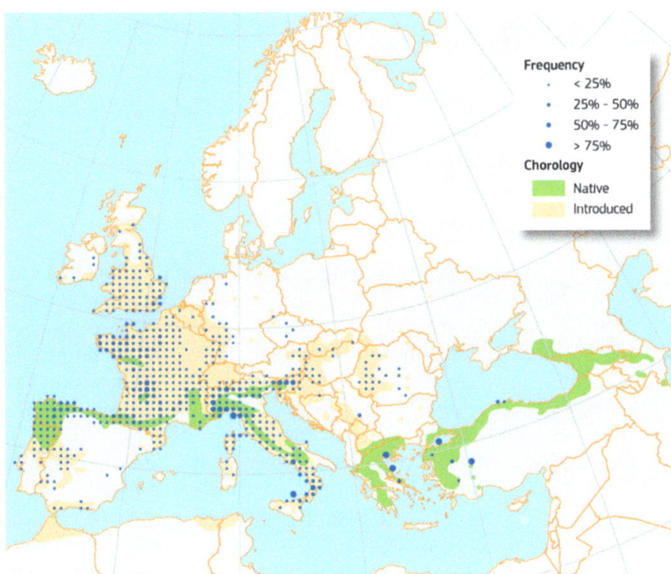

Figure 1. Frequency and chorology map of the distribution of *Castanea sativa* in Europe [6].

In structural timber engineering, the friction properties of wood, which are the focus of this study, hold significant relevance, particularly in the designing of joints and supports. The friction coefficient between wooden parts or between wood and metal connectors significantly influences the magnitude and manner of force transmission [14–20]. For instance, in step joints [14] and reverse joints [18], load distribution varies across contact faces depending on the coefficient of friction. For tie-rafter connections in trusses, a higher friction coefficient increases force transmission in areas of greater contact while reducing it in areas of lesser contact, thus decreasing the shear stress at the tie end and increasing the compression oblique to the grain in the rafter. For dowel joints [15], a higher friction coefficient leads to a more uniform stress distribution around the perimeter, which reduces joint slippage. In tapered tenon joints [16,17], the friction coefficient affects the forces on the frontal and lateral faces differently, with a higher friction coefficient reducing the contact forces on the front face. For connections with dowel-type fasteners and nut–washer fixings [19], the pre-tensioning creates an initial axial load that improves the friction effect at the wood–wood interfaces and the load distribution between faces. For timber connections with metal fasteners [20], the friction coefficient impacts the distribution of loads transmitted directly between pieces, whether through metal–wood or wood–wood contacts.

Thus, the understanding of this parameter is crucial for the analysis and simulation of both carpentry joints and mechanical connections. As with most mechanical properties of wood, friction also varies with the moisture content reached by the specimen in balance with the relative humidity and temperature of its surrounding environment. Consequently, Eurocode 5 [21] incorporates this effect in design by establishing three service classes reflecting the environmental conditions (i.e., temperature and relative humidity of the surrounding air) to which the wood will be exposed and its eventual equilibrium moisture content:

- Service Class 1: corresponds to conditions (20 °C and 65% relative humidity) where the average moisture content in most softwoods remains below 12%;
- Service Class 2: corresponds to conditions (20 °C and 85% relative humidity) where the average moisture content in most softwoods remains below 20%;
- Service Class 3: corresponds to conditions where the average moisture content in most softwoods exceeds 20%.

It should be noted that although Eurocode 5 [21] identifies service classes for softwoods, the temperature and relative humidity conditions describing the different service classes and moisture contents are also applicable to hardwoods such as chestnut. In this regard, there are international standards that define service classes applicable to both softwood and hardwood. For instance, the National Design Specification for Wood Construction [22] issued by ANSI defines two service conditions: "dry" (with up to 16% moisture content for laminated wood and CLT and 19% for sawn wood) and "wet" (for moisture contents exceeding these levels). Likewise, the Canadian standard for engineering design in wood [23] specifies a "dry" service condition, where the average equilibrium moisture content of solid wood over a year is 15% or less and does not exceed 19%, whereas the "wet" service condition encompasses all conditions that do not meet the dry criteria.

Therefore, the standards used to characterize the mechanical properties of wood stipulate testing at a specific moisture level, commonly 12%. Then, subsequent adjustments are made in calculations through the use of coefficients based on the intended service class. However, there is no European standard regarding the experimental determination of friction coefficients, but conversely, it is referenced in Table 6.1 of Eurocode 5-2 [24] for conifer timber in the context of stress-laminated decks. Specifically, values for the static friction coefficient are provided at moisture contents of $\leq$12% and $\geq$16%, with the provision that values within this range can be linearly interpolated.

Although several researchers [25–27] have commented on the linear variation in properties with moisture contents from 8% to 20%, or until fiber saturation is reached, limited research has explored the relationship between moisture content variations and friction, with investigations predominantly centered at the 12% equilibrium moisture content. Among those that do consider or provide insights on moisture content, the following studies (Table 1) are noteworthy.

Table 1. Noteworthy fiction coefficients from the literature review.

Test	Static Friction Coefficient	Kinetic Friction Coefficient	Moisture Content	References
Timber-to-timber	0.25 to 0.7	0.15 to 0.4	Dry	Argüelles et al. [26,28]
Timber-to-timber	0.5 to 0.71	0.3 to 0.65	From 11.25% to 20% at different wood sections (tangential, diagonal, and radial)	Fu et al. [29]
Timber-to-timber	0.36 to 052	0.25 to 0.34	12% at different orientation of the contact surfaces	Villar-García et al. [30]
Timber-to-timber	0.44 to 0.51	0.33 to 0.39	12% at different orientation of the contact surfaces	Villar-García et al. [31]
Timber-to-steel	- -	0.1 to 0.3 0.4 to 0.64	From 10% to 14% At fiber saturation	McKenzie et al. [32]
Timber-to-steel	- - -	0.3 to 0.5 0.5 to 0.7 0.7 to 0.9	Dry Intermediate moisture Close to saturation	Glass and Zelinka [27]
Timber-to-steel	0.156 to 0.238 0.121 to 0.176 0.280 to 0.344	- - -	12% at different fiber directions Oven-dried at different fiber directions Saturated at different fiber directions	Dorn et al. [33]
Timber-to-steel	0.16 to 0.21	0.15 to 0.18	12% at different orientations of the contact surfaces	Villar-García et al. [31]

For varying moisture content values, Argüelles et al. [26,28] reported values for the static friction coefficient ranging from 0.25 to 0.7 and for the kinetic friction coefficient within the 0.15 to 0.4 range. The coefficients increased with the moisture content of the timber-to-timber testing specimen up to saturation and remained constant beyond that point. This effect was also noticed by Glass and Zelinka [27], who reported that the coefficients continuously increase until fiber saturation is reached. Then, the values stabilize until water is present on the surface, triggering a decrease in the coefficients due to the lubricating effect. Although for beech timber, Fu et al. [29] examined the influence of both the moisture content and wood section (i.e., tangential, diagonal, and radial) on the static and kinetic friction coefficients. Both values increased with the moisture content within the 5–30% range, but greater moisture contents are responsible for marginal increases. For the different orientations of the contact surfaces, the authors reported static friction coefficients ranging from 0.5 to 0.71 and kinetic friction coefficients ranging from 0.3 to 0.65 at 11.25% and 20% moisture levels, respectively.

Regarding timber-to-steel friction, there are a limited number of studies, predominantly focused on dynamic assessments. McKenzie et al. [32] performed an extensive examination of the kinetic friction coefficients of numerous wood species against rough and smoot steel surfaces, although chestnut was not included in the investigation. For smooth surfaces, which are common in timber connections, the study reported kinetic friction coefficients ranging from 0.1 to 0.3 for moisture contents between 10% and 14%, depending on the speed of sliding. For moisture levels at fiber saturation, the values ranged from 0.4 to 0.64 for increasing sliding speeds. Moreover, based on the figures describing the dynamic friction included in the research, it could be inferred that the static friction values were only slightly higher than those reported for the kinetic friction.

Similarly, Glass and Zelinka [27] noticed that the kinetic friction coefficient for smooth timber in contact with hard, smooth surfaces, such as steel, can vary from 0.3 to 0.5 in dry specimens, from 0.5 to 0.7 at intermediate moisture contents, and from 0.7 to 0.9 when approaching saturation. Despite the distinct properties compared to sawn timber, it is worth mentioning the study on the friction behavior of microlaminated *Picea abies* against steel carried out by Dorn et al. [33]. The authors recorded static friction coefficient values ranging between 0.16 and 0.24 at a 12% moisture content. For oven-dried specimens, these values remained mostly constant. However, for saturated specimens, the static friction coefficient increased between 74% to 123% for tests parallel to the grain and between 82% and 182% for tests perpendicular to the grain.

This research focuses on the study of both static and kinetic friction coefficients of chestnut timber. Through an enhanced understanding of friction, the aim is to expand the use of *Castanea sativa* for structural designs involving frictional forces, promoting construction sustainability by encouraging the use of less exploited materials, which entails a diversification in the range of species used in construction and thus alleviates the demand for more commonly exploited ones. Examples of targeted applications include stressed plate bridges and walkways, timber trusses with carpentry joints, and constructions with mechanical timber-to-steel connections. The experimental program takes into account the orthotropic nature of wood by assessing different wood orientations involving both the wooden frictional pairs as well as against a steel plate. Moreover, the influence of the moisture content was considered by carrying out tests at 15% and 18% (i.e., Service Class 2 conditions). The results arising from the experimental program would provide a comprehensive database to be used as an input for precise engineering calculations, such as those carried out in numerical simulations, that would allow for a more accurate volumetric optimization of this natural resource. Additionally, in combination with previous findings by the authors on timber-to-timber and timber-to-steel tests at a 12% moisture content [30,31], this program would be used to validate the interpolation approach suggested for softwoods within the 12–18% moisture content range for hardwoods.

2. Materials and Methods

Test samples of 105 × 50 × 25 mm were prepared from Spanish chestnut (*Castanea sativa* Mill.) with a density of 670 kg/m^3 (12% moisture content). Since the variation in moisture content changes the frictional properties of wood, the tests were carried out at two moisture contents: firstly, at 18% moisture content, which represents Service Class 2 according to Eurocode 5 [21] (e.g., structures under cover but open to the air, canopies, covered pergolas, walkways, and bridges that are either covered or protected by a wear layer, as well as indoor and enclosed swimming pools [21,25,26]), then at 15% moisture content (i.e., an intermediate value to the 12% moisture content used to represent the conditions of Service Class 1 established in Eurocode 5 [21]). Thus, one set of specimens was stored in a condition room with a constant temperature of 20 °C and a relative humidity of 85% to ensure hygroscopic equilibrium and the desired moisture content of 18%. Conversely, for conditioning to a humidity of 15%, a temperature of 38 °C and a humidity of 80% were set [27]. The moisture levels were checked immediately before carrying out the tests using a hygrometer and afterwards via oven drying according to EN 13183-1 [34].

In the absence of a specific European standard test for determining the friction coefficient of wood and drawing upon the general recommendations provided by the American standard ASTM G115-10 [35], the authors developed and validated a test procedure based on a direct shear machine [36]. The proposed method adapts common geotechnical equipment to facilitate the placement and contact of the surfaces to be tested (i.e., specimens were positioned in the device by their largest surface area, ensuring that sliding occurred along the longest dimension), thereby facilitating both accurate experimental conditions as well as the application and recording of the necessary variables. Firstly, this method allows for the application of a normal load (N) to the upper face of the specimen through a distribution plate connected to a load bridge and counter-balance device while controlling the sliding speed. Similar to other research works [30,31,37,38], this study employed a 0.5 MPa load and an 8 mm·min^{-1} speed to simulate conditions encountered in practice while also effectively preventing the occurrence of inertial forces. Moreover, this method enables the measurement of both displacement and the necessary force (F) required to produce sliding by means of an LVDT (Linear Variable Differential Transformer) displacement sensor and load cell sensor, respectively. Therefore, the coefficient of friction (μ) is determined according to Equation (1):

$$F = \mu \times N, \quad (1)$$

Here, the proportionality constant is the friction coefficient, designated as either the static friction coefficient (μ_s) or kinetic friction coefficient (μ_k), contingent upon whether it pertains to the value at the precise moment just before sliding commences or during the ongoing relative displacement of the solids or the surfaces under examination.

Two separate experimental series were executed to evaluate the frictional behavior between pairs of materials: one set examined timber-to-timber interactions while the other focused on timber-to-steel contacts. Moreover, to simulate the conditions of surfaces that are designed to come into contact within the joint assembly, the influence of both the orthotropic nature of wood as well as the different roughness across the cutting planes was considered. As such, three distinct orthogonal axes were considered: longitudinal -L- (parallel to the fiber or grain, i.e., the axis of the tree), radial -R- (perpendicular to the grain in the radial direction and normal to the growth rings), and tangential -T- (perpendicular to the grain but tangent to the growth rings), as shown in Figure 2.

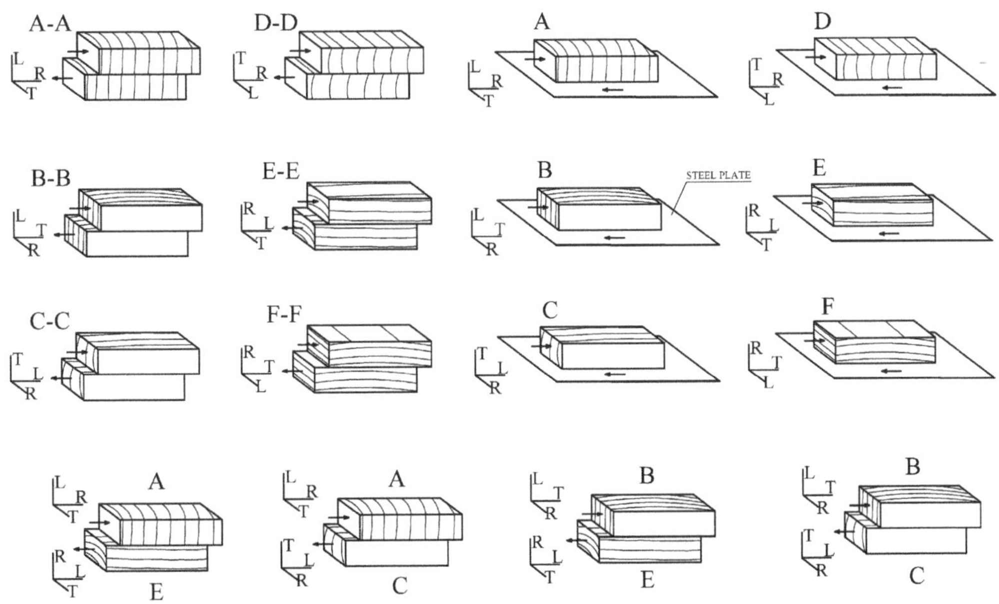

Figure 2. Timber-to-timber and timber-to-steel friction planes for the varying anatomical directions (L, R, and T) of the specimen of wood and their respective sliding directions.

Consequently, the three possible friction planes and their two respective directions of slippage were evaluated (Figure 2), ensuring a comprehensive analysis of frictional behavior under varied conditions:

- Transverse plane (perpendicular to the fiber):

1. (A) predominant direction of radial sliding (sliding parallel to the radius of the growth rings);
2. (B) predominant direction of tangential sliding to the growth rings;

- Radial plane (defined by the axis of the three and a radius of the trunk):

1. (C) sliding direction parallel to the fiber (i.e., radial surfaces);

2. (D) sliding direction perpendicular to the fiber;

- Tangential plane (tangent to the growth rings):

1. (E) sliding direction parallel to the fiber (i.e., tangential surfaces);
2. (F) sliding direction perpendicular to the fiber.

Therefore, Figure 2 presents the array of friction pairs that reflect combinations frequently encountered in structural connections. On the one hand, timber-to-timber tests could be divided among surfaces with identical orientations: A-A, B-B, C-C, D-D, E-E, and F-F, and tests between surfaces of differing orientations: A-C, A-E, B-C, and B-E. On the other hand, timber-to-steel tests were designed as A-S, B-S, C-S, D-S, E-S, and F-S, with S indicating the steel plate. Thus, the experimental program reached a total of over 400 tests and ultimately offered significant insights into frictional behavior.

3. Results and Discussion

3.1. Timber-to-Timber Tests with Identical Orientations

Figure 3 showcases the most illustrative graphs depicting the variation of the friction coefficient relative to the displacement for tests involving surfaces of identical orientations and friction directions under a controlled moisture content of 18%.

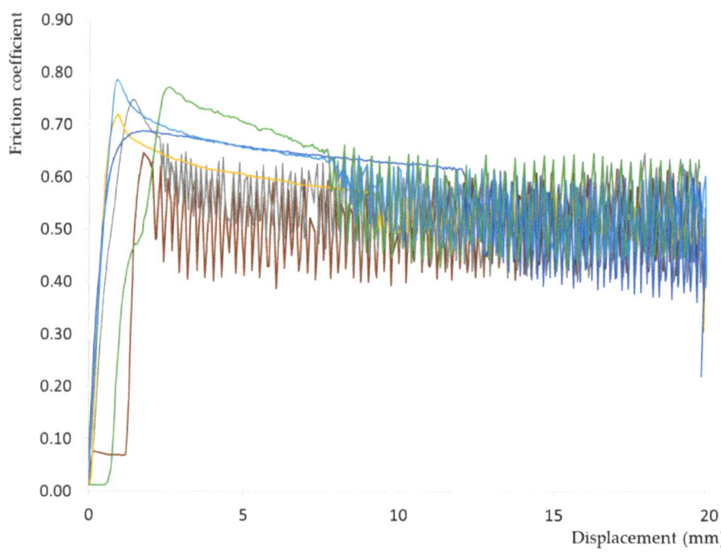

Figure 3. Representative examples (— A-A; — B-B; — C-C; — D-D; — E-E; — F-F) of the friction coefficient variation for sections with the same orientation in both specimens at a moisture content of 18%.

The oscillations observed in Figure 3 are illustrative of the stick–slip phenomenon. However, the fluctuation manifested with reduced intensity compared to similar tests conducted at lower moisture levels [30,31]. This reduction aligns with findings by [29,39], highlighting that higher moisture weakens the stick–slip behavior between the wood surfaces. The differences between the frictional force–displacement curves of dry and wet surfaces were also observed by Fu et al. [29], who attributed them to the softening of the fibers and the decreased amplitude of the rough peaks, which led to a weakened stick–slip motion in the 5–30% moisture range. As described by Möhler and Herröder [40] in friction scenarios A and C, the sliding motion occurs continuously across the friction path and is characterized by a parabolic decrease in the horizontal force, at least in an initial segment.

Table 2 shows the friction coefficients from various sawn specimens and frictional directions grouped by friction pairs with identical orientations and a moisture content of 18%. Both the mean value derived from the 15 tests performed for each specific pairing and the coefficient of variation (CoV) are indicated to highlight the average performance and the variability within each set.

Table 2. Friction coefficients between wood surfaces of identical orientation at 18% moisture content.

Mean (CoV %)	A-A	B-B	C-C	D-D	E-E	F-F
μ_s	0.67 (15.3)	0.71 (11.4)	0.68 (14.4)	0.78 (8.2)	0.63 (13.9)	0.73 (9.9)
μ_k	0.42 (4.8)	0.47 (12.7)	0.49 (12.9)	0.56 (16.7)	0.46 (29.3)	0.54 (24.6)

Based on the comparison between the results presented in Table 2 and those obtained for these same orientations in a previous work [30], it becomes noticeable that moisture content significantly impacts both static and kinetic friction, overshadowing the effects of the testing orientation. This finding aligns with observations made by [29]. Nonetheless, knowledge of the specific friction values for different wood orientations can significantly enhance decision making during joint construction. Then, it would allow for more favorable designs by tailoring the cut of wood, notches, and contact interfaces to optimize frictional force transmission between the components. By strategically exploiting the orthotropic nature of wood, such as by rotating the R and T axes of the beams, the distribution of stresses could be improved.

On average, disregarding orientation, the static friction coefficient stands at 0.70 and the kinetic friction coefficient at 0.48. Such values represent roughly 50% increases compared to those measured at 12% moisture content. These findings are consistent with those reported by Argüelles et al. [26] (i.e., a 0.7 static friction coefficient and 0.4 kinetic friction coefficient) and Fu et al. [29] at 20% moisture content (μ_s = 0.5–0.71; μ_k = 0.3–0.65). In this experimental program, the average coefficients of variation for the different orientation test series were 7.6% for static friction and 10% for kinetic friction. Notably, the CoV for each friction pair significantly decreased by about 15% compared to the 12% moisture tests, suggesting that increased moisture on the contact surfaces leads to less variability in friction.

3.2. *Timber-to-Timber Tests with Different Orientations*

Figure 4 presents some representative examples that capture the fluctuation of the friction coefficient as a function of displacement, focusing on experiments that involve surfaces with different orientations and sliding directions, conducted at a moisture content of 18%.

Similar observations apply to Figure 4 regarding the stick–slip behavior of the tested specimens. The performance of the friction pairs demonstrates a consistent relationship between the displacement and friction coefficient, closely aligning with the patterns noted in scenarios of identical orientation between wood surfaces (Figure 3).

Table 3 compiles the friction coefficients from various sawn specimens and frictional directions, grouped by friction pairs with identical orientations and a moisture content of 18%. A trend consistent with the earlier discussion is observed as values exhibit a notable increase compared to those at 12% moisture content [31]. Specifically, there is a 42% surge in the static friction coefficient, averaging at 0.67, and a 30% rise in the kinetic friction coefficient, averaging at 0.47. Nevertheless, the increment is less pronounced than the increase observed for samples with identical orientations, as recorded in Table 2.

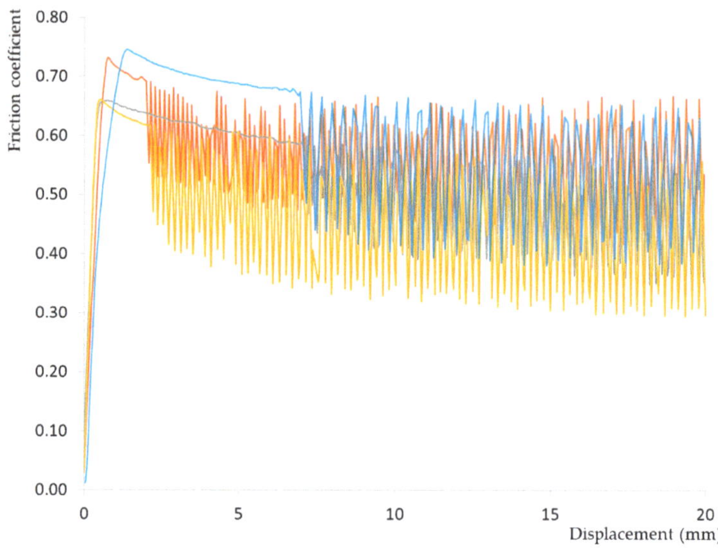

Figure 4. Representative examples (— A-C; — A-E; — B-C; — B-E) of the friction coefficient variation for specimens with different orientations at a moisture content of 18%.

Table 3. Friction coefficients between wood surfaces of different orientations at 18% moisture content.

Mean (CoV %)	A-C	A-E	B-C	B-E
μ_s	0.70 (18.1)	0.65 (15.6)	0.64 (9.9)	0.70 (10.3)
μ_k	0.48 (25.7)	0.45 (13.6)	0.43 (14.3)	0.50 (20.7)

The overall average values, both static and kinetic, are remarkably similar to those obtained for the same orientation at 18% moisture content and align with the referenced literature from the previous section. The lack of significant variance for any specific pair could respond to a homogenizing effect of increased moisture levels. Notably, the A-C and B-E pairings continue to register the greatest friction values, a pattern consistent with observations at 12% moisture. However, no direct correlation is observed between the highest values in Table 3 and the superior frictional values arising from the friction of wood surfaces of identical orientations (Table 2). Regarding the coefficient of variation, the values decreased compared to the 12% moisture level for each tested friction pair, mirroring the trend observed for pairs of identical orientation. Nonetheless, the CoV values remained in the same range as those obtained for 18% moisture content for wood surfaces of identical orientation.

It is worth noting that the average static friction coefficient value ($\mu_s = 0.69$) significantly exceeds those specified in Eurocode 5-2 [24]. For the calculation of stress-laminated deck plates consisting of sawn softwood at a moisture content greater than 16%, the design values established for the static friction coefficient are 0.45 for scenarios perpendicular to grain and 0.35 for scenarios parallel to grain. Nevertheless, this comparison should take into account the case-specificity differences regarding the type of wood and moisture content (i.e., specimens derived from *Castanea sativa*, a deciduous tree, conditioned at 18% moisture content). Moreover, it should be acknowledged that the values stipulated by Eurocode 5-2 [24] serve as design guidelines, factoring in safety margins to ensure structural integrity.

Thus, the proposed values in Eurocode 5-2 [24] are deliberately conservative since greater friction coefficient values are beneficial for the outcomes of the engineering calculations.

3.3. Timber-to-Steel Tests

Figure 5 showcases representative friction cases of the tests comparing the timber specimens at 18% moisture content and the steel plate, focusing on experiments that involved different fiber orientations relative to the sliding direction.

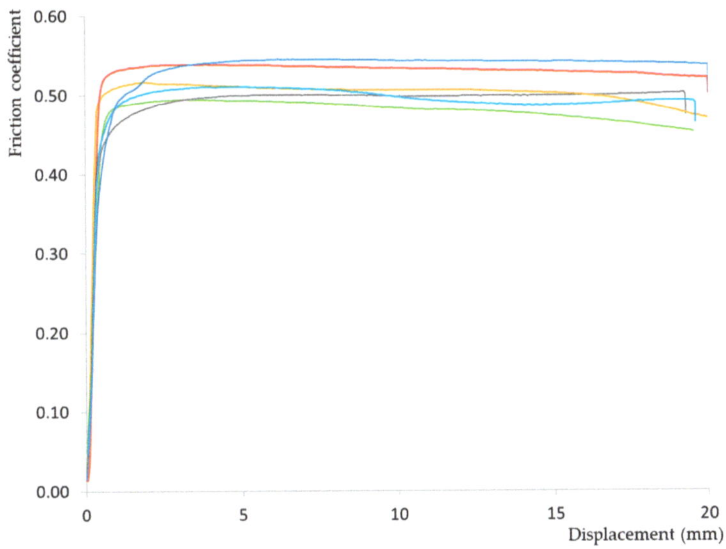

Figure 5. Representative examples (— A-S; —B-S; — D-S; — C-S; — E-S; - F-S) of the friction coefficient variation between the timber specimens at 18% moisture content and the steel plate.

Conversely to previous timber-to-timber test series, Figure 5 illustrates the absence of the stick–slip phenomenon, corroborating findings from other studies [29,41]. As also noted by those researchers, the increase in moisture does not introduce a pronounced inflection at the beginning of displacement. The shape of the obtained curves (Figure 5) is similar to the type B classification proposed by Möhler and Herröder [40] in which the frictional force exhibits a flat parabolic shape, indicative of a friction trajectory that either slightly decreases or, in certain instances, remains constant after reaching the peak load. Notably, in some instances, the value of friction marginally increases shortly after the sliding begins.

Table 4 details the mean values for both static and kinetic friction coefficients, accompanied by the coefficient of variation from tests involving the interaction between a steel plate and a wood specimen conditioned at 18% moisture content and sawn to exhibit a specific orientation.

Table 4. Friction coefficients involving a wood surface at 18% moisture content and the steel plate.

Mean (CoV %)	A-S	B-S	C-S	D-S	E-S	F-S
μ_s	0.48 (2.5)	0.49 (6.1)	0.55 (4.6)	0.53 (3.2)	0.54 (4.9)	0.52 (4.4)
μ_k	0.45 (7.2)	0.47 (7.2)	0.53 (7.2)	0.52 (3.1)	0.53 (5.2)	0.50 (5.3)

Analysis of the data in the test series for the different wood orientations reveals an average static friction coefficient of 0.52 and a kinetic friction coefficient of 0.50, with coefficients of variation of 6% and 7%, respectively. Two main insights emerge from these findings.

On the one hand, there is a substantial increase in both static and kinetic friction coefficients compared to steel–wood pairs at 12% moisture content [31]. For instance, the average static coefficient exhibits a 173% rise and the kinetic coefficient a 194% surge, which closely places the kinetic value on par with the static coefficient. The greater increase in the kinetic coefficients is indicative of a more pronounced effect of moisture that facilitated the lifting of the wood fibers during sliding interactions [29]. These substantial increases are in line with those documented by Dorn et al. [33], who conducted tests on wood against steel ranging from oven-dried to fully saturated specimens. Moreover, the obtained values fall within the range specified by Glass and Zelinka [27] for the friction of wood against hard and smooth surfaces at intermediate moisture (μ_k = 0.5–0.7) and are consistent with the findings reported by McKenzie et al. [32] of μ_k = 0.4–0.64. Nonetheless, it is worth noting that this increase significantly exceeds that observed for the same moisture variation in the timber-to-timber tests, suggesting that when friction occurs against a very smooth surface, such as steel, the moisture content of the wood has a significantly major role in the friction coefficient.

On the other hand, the CoV values within each orientation are considerably lower compared to those obtained for wood specimens at 12% moisture content. Such a reduction in variability is attributed to both the increased moisture at the contact surface and the homogenizing effect of steel (i.e., the limited roughness) in the wood–steel friction dynamics.

3.4. Correlation between μ_k and μ_s

For each friction specimen pairing within the timber-to-timber test series, Figure 6a illustrates the relationship between the static friction coefficient (μ_s) and the kinetic friction coefficient (μ_k). Similarly, Figure 6b displays the average values for each friction combination. The μ_k/μ_s ratio for surfaces of identical orientation averaged 0.72, similar to the values obtained at 12% moisture content, which indicated no significant change in their relationship. For surfaces of different orientations, the μ_k/μ_s ratio was 0.69, yielding a value comparable to that of surfaces with identical orientation at an 18% moisture level. This similarity suggests that the orientation of wood surfaces does not markedly affect the relationship between static and kinetic friction coefficients under the same moisture conditions.

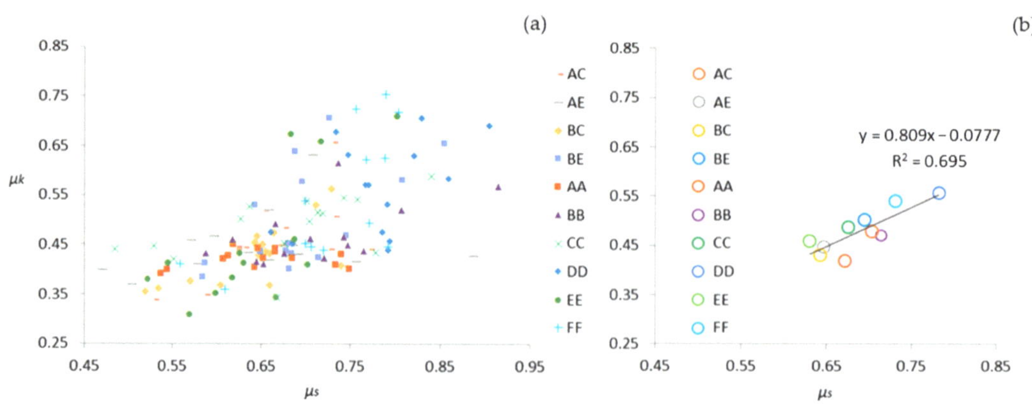

Figure 6. Relationship between the values of μ_k and μ_s for the different timber-to-timber friction pairs (**a**), as well as the mean value for each group, denoted by a circle in the corresponding color (**b**).

Although no strong relationship emerged from the entire dataset, the analysis of the average values (Figure 6b) allowed for an acceptable correlation ($R^2 = 0.70$) between static and kinetic friction coefficients (Equation (2)).

$$\mu_k = 0.809\mu_s - 0.0777 \ (R^2 = 0.70), \tag{2}$$

Additionally, the consideration of specific friction orientations allowed for improved correlations such as those shown in Equations (3) and (4) for the A-C and E-E friction pairs, respectively.

$$\text{For the friction pair A-C: } \mu_k = 0.98\mu_s - 0.19 \ (R^2 = 0.76), \tag{3}$$

$$\text{For the friction pair E-E: } \mu_k = 0.89\mu_s - 0.035 \ (R^2 = 0.74), \tag{4}$$

A similar approach for the relationship of the static and kinetic friction coefficients of the timber-to-steel tests is followed in Figure 7a,b. The μ_k/μ_s ratio averages 0.97, which numerically captures the behavior depicted in Figure 5 (i.e., a flat parabolic curve with a minimal reduction in the coefficient value during sliding). In this case, a notable difference is observed in the ratio compared to the values obtained at 12% moisture, which had an average of 0.83, indicating a further reduction in the differences between static and kinetic values. The greater proximity to unity reflects the absence of the initial inflection point in the registered friction behavior. This phenomenon was also noted by Fu et al. [29], who observed that the difference between μ_s and μ_k decreases with higher moisture contents.

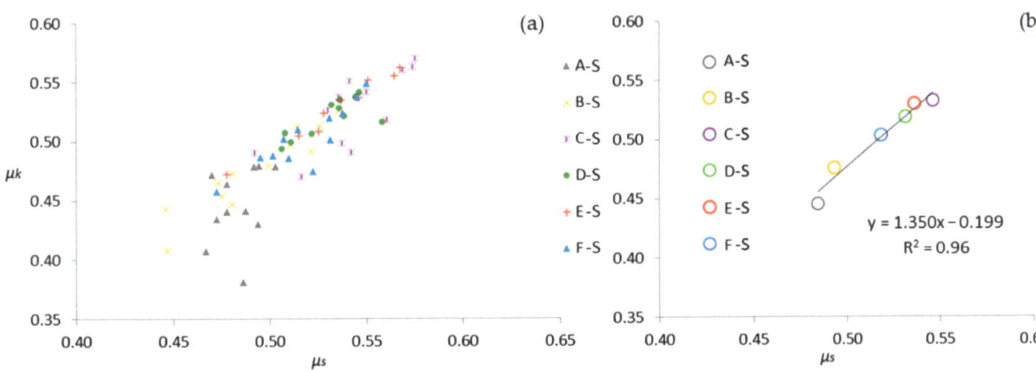

Figure 7. Relationship between the values of μ_k and μ_s for the different timber-to-steel friction pairs (**a**), as well as the mean value for each group, denoted by a circle in the corresponding color (**b**).

From the average coefficients for the different timber-to-steel tests, Equation (5) shows the relationship between static and kinetic friction. The robustness of the correlation ($R^2 = 0.96$) allows for a highly reliable prediction of the kinetic coefficient from a known static coefficient and vice versa. Moreover, the specific friction pair combinations also display strong correlations between both coefficients. It should be noted that the high degree of correlation was also identified for the 12% moisture content [31] pointing to a generalization of this observation across the entire studied moisture spectrum, as further detailed in subsequent discussions.

$$\mu_k = 1.350\mu_s - 0.199 \ (R^2 = 0.96), \tag{5}$$

3.5. Influence of Moisture Content on Friction Coefficients

To evaluate the validity of the linear coefficient–moisture relationships, the experimental program included a targeted series of tests at an intermediate moisture level of 15% while maintaining all other test parameters at constant levels. The average value and coefficient of variation from 10 determinations within each friction pair combination (i.e., between wood surfaces of identical orientation, wood surfaces of different orientations, and wood and steel) of static and kinetic coefficients are displayed in Tables 5–7.

Table 5. Friction coefficients between wood surfaces of identical orientation at 15% moisture content.

Mean (CoV %)	A-A	B-B	C-C	D-D	E-E	F-F
μ_s	0.59 (6.3)	0.61 (5.2)	0.51 (33.1)	0.69 (31.1)	0.48 (7.5)	0.70 (7.7)
μ_k	0.37 (17.7)	0.33 (11.8)	0.37 (28.3)	0.47 (26.3)	0.37 (27.1)	0.43 (6.4)

Table 6. Friction coefficients between wood surfaces of different orientations at 15% moisture content.

Mean (CoV %)	A-C	A-E	B-C	B-E
μ_s	0.56 (32.0)	0.57 (20.9)	0.51 (26.5)	0.56 (17.4)
μ_k	0.44 (16.2)	0.39 (31.7)	0.40 (26.3)	0.41 (25.9)

Table 7. Friction coefficients between a wood surface at 15% moisture content and the steel plate.

Mean (CoV %)	A-S	B-S	C-S	D-S	E-S	F-S
μ_s	0.33 (7.3)	0.34 (17.2)	0.36 (10.6)	0.35 (2.9)	0.33 (8.8)	0.37 (17.6)
μ_k	0.31 (8.1)	0.31 (5.5)	0.32 (10.2)	0.34 (5.7)	0.32 (5.4)	0.32 (15.9)

Consistent with previous observations, the CoV resembles more closely the results from the 12% moisture tests due to a lower moisture-induced homogenizing effect on the frictional behavior at this intermediate level. Nevertheless, taking into account the inherent variability of wood as a natural material, these CoV values are deemed acceptable, especially when considering those reported in the literature.

Both the static and kinetic friction coefficients fall within the range of those arising from specimens conditioned at 12% and 18% moisture contents, which is in accordance with the known dependence behavior between the moisture content and the mechanical properties of wood. Certainly, Eurocode 5-2 [24] and several researchers [25–27] accept that intermediate friction coefficients could be determined through linear interpolation. Therefore, taking into account the static and kinetic coefficient results obtained by the authors at 12% [30,31] and 18% moisture contents, all possible linear regressions were determined. Figures 8 and 9 show these linear relationships as dotted lines colored according to each friction pair combination of the timber-to-timber and timber-to-steel tests. Moreover, to evaluate the precision of the interpolation method, the corresponding experimental results at 15% moisture content are also included in Figures 8 and 9.

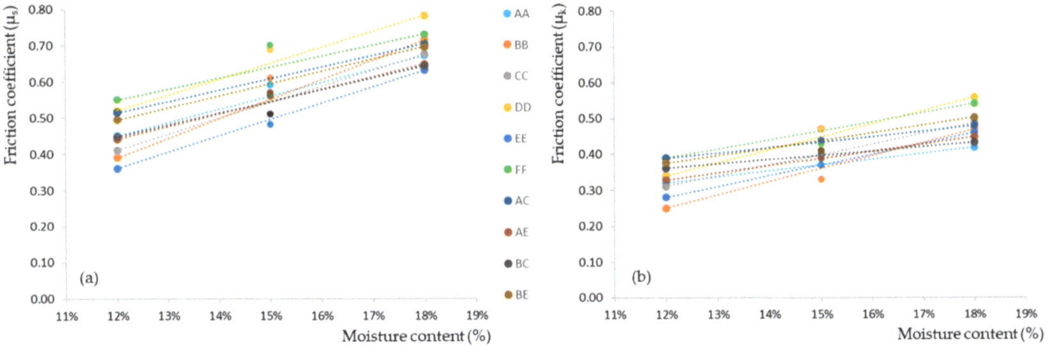

Figure 8. For each group of timber-to-timber tests, average static (**a**) and kinetic (**b**) friction coefficient values at moisture contents of 12% (from [30,31]), 15%, and 18%, as well as the linear regression between the two extreme values of the studied range, are shown.

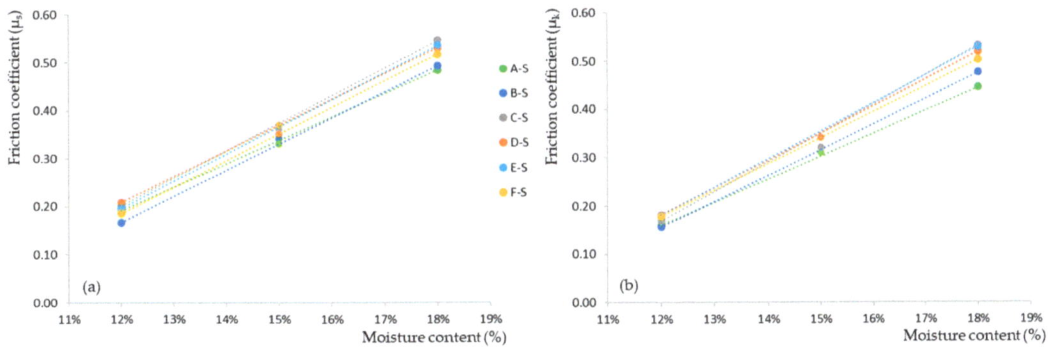

Figure 9. For each group of timber-to-steel tests, average static (**a**) and kinetic (**b**) friction coefficient values at moisture contents of 12% (from [31]), 15%, and 18%, as well as the linear regression between the two extreme values of the studied range, are shown.

Aside from a few exceptions, the slopes of the linear regressions are similar for each type of friction coefficient displayed in the different figures, which is especially apparent in timber-to-steel friction cases. This observation underscores the robustness of the linear estimation approach across different materials and conditions. Moreover, Table 8 presents the interpolated friction coefficients at the 15% moisture content from each linear regression (i.e., dotted lines) in Figures 8 and 9. Although, in most cases, the accuracy of the linear regression compared to the experimental value is evident from the figures, the observed error compared to the average experimental result at the same moisture content is also reported in the table.

The observed errors (Table 8), particularly in scenarios involving timber-to-steel friction, are consistently lower than the coefficients of variation recorded across all experimental tests carried out at 15% moisture content. This finding highlights the precision of the linear estimation approach within the 12–18% moisture range, but also confirms its applicability to hardwoods like the sawn chestnut (*Castanea sativa* Mill.). Therefore, the method that was originally limited to the friction coefficient of softwoods in stress-laminated decks as per Eurocode 5-2 [24] proved to be significantly effective in enhancing the predictability of the frictional behavior of this particular hardwood species (i.e., chestnut), which previously lacked specific and comprehensive friction coefficient data or prior testing for linear estimation accuracy.

Table 8. For each studied scenario (friction coefficients involving wood surfaces of identical orientation, wood surfaces of different orientations, and wood and steel), the values of the static and kinetic friction coefficients resulting from the linear interpolation and the percentage of error relative to the experimental values at a 15% moisture content are shown.

Interpolated value (error %)	**A-A**	**B-B**	**C-C**	**D-D**	**E-E**	**F-F**
μ_s	0.56 (−5.0%)	0.55 (−9.5%)	0.54 (6.4%)	0.65 (−5.7%)	0.50 (3.2%)	0.64 (−8.6%)
μ_k	0.37 (−0.1%)	0.36 (9.2%)	0.40 (7.7%)	0.45 (−4.5%)	0.37 (0.0%)	0.47 (8.2%)
Interpolated value (error %)	**A-C**	**A-E**	**B-C**	**B-E**		
μ_s	0.61 (8.7%)	0.54 (−4.5%)	0.54 (6.8%)	0.59 (6%)		
μ_k	0.43 (−1.2%)	0.39 (−0.4%)	0.40 (−0.9%)	0.44 (7%)		
Interpolated value (error %)	**A-S**	**B-S**	**C-S**	**D-S**	**E-S**	**F-S**
μ_s	0.34 (2%)	0.33 (−3%)	0.37 (4%)	0.37 (6%)	0.37 (11%)	0.35 (−5%)
μ_k	0.30 (−2%)	0.32 (2%)	0.35 (10%)	0.35 (3%)	0.35 (11%)	0.34 (6%)

4. Conclusions

As current knowledge regarding the friction properties of hardwood is limited, this investigation studied both the static and kinetic friction behaviors of sawn chestnut timber. The understanding of this parameter is crucial for the analysis and simulation of both carpentry joints and mechanical connections; thus, the friction behaviors involving both wood samples at identical and different orientations as well as wood samples against a steel plate were considered. Moreover, since moisture content plays a major role in this property, tests were carried out at 15% and 18%, providing insights into the wood's performance under Service Class 2, a common scenario in wooden structures, and allowing comparison with the 12% moisture content representing Service Class 1.

For timber-to-timber tests, a reduction in the stick–slip phenomenon, up to its almost-disappearance in some initial phases of tests, was observed due to the increased moisture. However, a clear initial peak was still noticed, albeit one less pronounced than at 12% moisture content, and higher μ_k/μ_s ratios were determined. For timber-to-steel tests, there was a complete absence of the stick–slip phenomenon reported at 12% moisture content determinations. It was also noticed the lack of any peak at the onset of sliding and either the maintenance of or slight increase in the friction coefficient once relative motion commenced, which resulted in a higher μ_k/μ_s ratio of 0.97.

Both static (μ_s) and dynamic (μ_k) coefficients exhibited increased values compared to those at 12% moisture content (Service Class 1). Although the results were in line with those found by other researchers, given the limited literature available on wood friction at moisture contents exceeding the 12% value associated with standard testing, the direct comparison of the results was challenging, particularly for hardwood and chestnut. For the 18% tests, the average values were μ_s = 0.68 and μ_k = 0.47 for timber-to-timber tests and μ_s = 0.52 and μ_k = 0.5 for timber-to-steel tests. The increase was around 50% for timber-to-timber friction pairs and over 170% for timber-to-steel friction pairs compared to the 12% moisture content. Moreover, it was proven that these new data points could be used in the same manner as the linear interpolation outlined in Eurocode 5-2 [24] for softwoods. In this regard, the study confirmed the accuracy of this approach by comparing each interpolated value with the corresponding experimental result at the intermediate moisture content of 15%.

Author Contributions: Conceptualization, J.R.V.-G., P.V.-L. and D.R.-R.; methodology, J.R.V.-G.; validation, formal analysis, investigation, funding acquisition, J.R.V.-G., M.M.I., P.V.-L. and D.R.-R.; writing—original draft preparation, J.R.V.-G. and D.R.-R.; writing—review and editing, supervision, J.R.V.-G., P.V.-L. and D.R.-R. All authors have read and agreed to the published version of the manuscript.

Funding: This research was funded by Junta de Extremadura and by the European Regional Development Fund of the European Union through grants GR21163 and GR21091.

Institutional Review Board Statement: Not applicable.

Informed Consent Statement: Not applicable.

Data Availability Statement: Data is contained within the article.

Acknowledgments: The administrative and technical support from the Forest Research Group and the Mechanical and Fluid Engineering Research Group of the University of Extremadura is gratefully acknowledged.

Conflicts of Interest: The authors declare no conflicts of interest.

References

1. Roos, A.; Woxblom, L.; McCluskey, D. The Influence of Architects and Structural Engineers on Timber in Construction–Perceptions and Roles. *Silva Fenn.* **2010**, *44*, 871–884. [CrossRef]
2. Minunno, R.; O'Grady, T.; Morrison, G.M.; Gruner, R.L. Investigating the Embodied Energy and Carbon of Buildings: A Systematic Literature Review and Meta-Analysis of Life Cycle Assessments. *Renew. Sustain. Energy Rev.* **2021**, *143*, 110935. [CrossRef]
3. Tonelli, C.; Grimaudo, M. Timber Buildings and Thermal Inertia: Open Scientific Problems for Summer Behavior in Mediterranean Climate. *Energy Build.* **2014**, *83*, 89–95. [CrossRef]
4. Martínez-Alonso, C.; Berdasco, L. Carbon Footprint of Sawn Timber Products of *Castanea Sativa* Mill. in the North of Spain. *J. Clean. Prod.* **2015**, *102*, 127–135. [CrossRef]
5. Arriaga, F.; Wang, X.; Íñiguez-González, G.; Llana, D.F.; Esteban, M.; Niemz, P. Mechanical Properties of Wood: A Review. *Forests* **2023**, *14*, 1202. [CrossRef]
6. Conedera, M.; Tinner, W.; Krebs, P.; de Rigo, D.; Caudullo, G. Castanea Sativa in Europe: Distribution, Habitat, Usage and Threats. In *European Atlas of Forest Tree Species*; Publication Office of the European Union: Luxembourg, 2016; p. e0125e0+.
7. Rodríguez-Guitián, M.; Rigueiro, A.; Real, C.; Blanco, J.; Ferreiro da Costa, J. El Habitat" 9269 Bosques de *Castanea Sativa*" En El Extremo Noroccidental Iberico: Primeros Datos Sobre La Variabilidad Floristica de Los" Soutos". *Bull. Société d' Hist. Nat. Toulouse* **2005**, *141*, 75–81.
8. González-Varo, J.P.; López-Bao, J.V.; Guitián, J. Presence and Abundance of the Eurasian Nuthatch *Sitta Europaea* in Relation to the Size, Isolation and the Intensity of Management of Chestnut Woodlands in the NW Iberian Peninsula. *Landsc. Ecol.* **2008**, *23*, 79–89. [CrossRef]
9. Council Directive 92/43/EEC On the Conservation of Natural Habitats and of Wild Fauna and Flora. Available online: http://data.europa.eu/eli/dir/1992/43/2013-07-01 (accessed on 7 March 2024).
10. MITECO. *Forest Statistics Yearbook 2021*; Spnanish Ministry for Ecological Transition and Demographic Challenge: Madrid, Spain, 2022.
11. Vega, A.; Dieste, A.; Guaita, M.; Majada, J.; Baño, V. Modelling of the Mechanical Properties of Castanea Sativa Mill. Structural Timber by a Combination of Non-Destructive Variables and Visual Grading Parameters. *Eur. J. Wood Prod.* **2012**, *70*, 839–844. [CrossRef]
12. Beccaro, G.; Alma, A.; Bounous, G.; Gomes-Laranjo, J. (Eds.) *The Chestnut Handbook: Crop & Forest Management*; CRC Press: Boca Raton, FL, USA; London, UK; New York, NY, USA, 2020; ISBN 978-0-429-44560-6.
13. Carbone, F.; Moroni, S.; Mattioli, W.; Mazzocchi, F.; Romagnoli, M.; Portoghesi, L. Competitiveness and Competitive Advantages of Chestnut Timber Laminated Products. *Ann. For. Sci.* **2020**, *77*, 51. [CrossRef]
14. Villar, J.R.; Guaita, M.; Vidal, P.; Arriaga, F. Analysis of the Stress State at the Cogging Joint in Timber Structures. *Biosyst. Eng.* **2007**, *96*, 79–90. [CrossRef]
15. Sjödin, J.; Serrano, E.; Enquist, B. An Experimental and Numerical Study of the Effect of Friction in Single Dowel Joints. *Holz. Roh. Werkst.* **2008**, *66*, 363–372. [CrossRef]
16. Koch, H.; Eisenhut, L.; Seim, W. Multi-Mode Failure of Form-Fitting Timber Connections—Experimental and Numerical Studies on the Tapered Tenon Joint. *Eng. Struct.* **2013**, *48*, 727–738. [CrossRef]

17. Aira, J.R.; Íñiguez-González, G.; Guaita, M.; Arriaga, F. Load Carrying Capacity of Halved and Tabled Tenoned Timber Scarf Joint. *Mater. Struct.* **2016**, *49*, 5343–5355. [CrossRef]
18. Villar-García, J.R.; Crespo, J.; Moya, M.; Guaita, M. Experimental and Numerical Studies of the Stress State at the Reverse Step Joint in Heavy Timber Trusses. *Mater. Struct.* **2018**, *51*, 17. [CrossRef]
19. Domínguez, M.; Fueyo, J.G.; Villarino, A.; Anton, N. Structural Timber Connections with Dowel-Type Fasteners and Nut-Washer Fixings: Mechanical Characterization and Contribution to the Rope Effect. *Materials* **2021**, *15*, 242. [CrossRef] [PubMed]
20. Fonseca, E.M.M.; Leite, P.A.S.; Silva, L.D.S.; Silva, V.S.B.; Lopes, H.M. Parametric Study of Three Types of Timber Connections with Metal Fasteners Using Eurocode 5. *Appl. Sci.* **2022**, *12*, 1701. [CrossRef]
21. *UNE-EN 1995-1-1*; Eurocódigo 5: Proyecto de Estructuras de Madera. Parte 1-1: Reglas Generales y Reglas Para Edificación. AENOR: Madrid, Spain, 2016.
22. American Wood Council. *The 2024 National Design Specification (NDS) for Wood Construction*; American Wood Council: Leesburg, VA, USA, 2023.
23. *CSA O86*; Engineering Design in Wood. Canadian Standards Association: Toronto, ON, Canada, 2019.
24. *EN 1995-2*; Eurocode 5: Design of Timber Structures—Part 2: Bridges. CEN: Brussels, Belgium, 2016.
25. Blass, H.; Aune, P.; Choo, B.; Görlacher, R.; Grifiths, D. *Timber Engineering. STEP 1: Basis of Design, Material Properties, Structural Components and Joints*; Centrum Hout: Almere, The Netherlands, 1995.
26. Argüelles, R.; Arriaga, F.; Esteban, M.; Íñiguez, G.; Argüelles Bustillo, R. *Timber Structures. Basis for Calculation [Estructuras de madera. Bases de cálculo]*; AITIM. Technical Research Association of the Wood and Cork Industries: Madrid, Spain, 2013; ISBN 978-84-87381-44-7.
27. Glass, S.V.; Zelinka, S.L. Chapter 4: Moisture Relations and Physical Properties of Wood. In *Wood handbook: Wood as an Engineering Material*; General Technical Report FPL-GTR-282; US Department of Agriculture, Forest Service, Forest Products Laboratory: Madison, WI, USA, 2010.
28. Argüelles, R.; Arriaga, F.; Esteban, M.; Íñiguez, G.; Argüelles Bustillo, R. *Timber Structures. Joints [Estructuras de madera. Uniones]*; AITIM. Technical Research Association of the Wood and Cork Industries: Madrid, Spain, 2015; ISBN 978-84-87381-44-7.
29. Fu, W.; Guan, H.; Chen, B. Investigation on the Influence of Moisture Content and Wood Section on the Frictional Properties of Beech Wood Surface. *Tribol. Trans.* **2021**, *64*, 830–840. [CrossRef]
30. Villar-García, J.R.; Vidal-López, P.; Corbacho, A.J.; Moya, M. Determination of the Friction Coefficients of Chestnut (*Castanea Sativa* Mill.) Sawn Timber. *Int. Agrophys.* **2020**, *34*, 65–77. [CrossRef] [PubMed]
31. Villar-García, J.R.; Vidal-López, P.; Rodríguez-Robles, D.; Moya Ignacio, M. Friction Coefficients of Chestnut (*Castanea Sativa* Mill.) Sawn Timber for Numerical Simulation of Timber Joints. *Forests* **2022**, *13*, 1078. [CrossRef]
32. McKenzie, W.M.; Karpovich, H. The Frictional Behaviour of Wood. *Wood Sci. Technol.* **1968**, *2*, 139–152. [CrossRef]
33. Dorn, M.; Habrová, K.; Koubek, R.; Serrano, E. Determination of Coefficients of Friction for Laminated Veneer Lumber on Steel under High Pressure Loads. *Friction* **2021**, *9*, 367–379. [CrossRef]
34. *EN 13183-1*; Moisture Content of a Piece of Sawn Timber—Part 1: Determination by Oven Dry Method. CEN: Brussels, Belgium, 2002.
35. *ASTM G115-10*; Standard Guide for Measuring and Reporting Friction Coefficients. ASTM International: West Conshohocken, PA, USA, 2018.
36. Villar-García, J.R.; Vidal-López, P.; Moya Ignacio, M. Device to Perform Friction Tests between Solid Bodies. Utility Model U 201932027. 2020. Available online: https://patentscope.wipo.int/search/en/detail.jsf?docId=ES289318649 (accessed on 7 March 2024).
37. Crespo, J.; Regueira, R.; Soilán, A.; Díez, M.R.; Guaita, M. Methodology to Determine the Coefficients of Both Static and Dynamic Friction Apply to Different Species of Wood. In Proceedings of the 1st Ibero-Latin American Congress on Wood in Construction (CIMAD), Coimbra, Portugal, 7–9 June 2011.
38. Aira, J.R.; Arriaga, F.; Íñiguez-González, G.; Crespo, J. Static and Kinetic Friction Coefficients of Scots Pine (*Pinus Sylvestris* L.), Parallel and Perpendicular to Grain Direction. *Mater. Construcción* **2014**, *64*, e030. [CrossRef]
39. Berman, A.; Ducker, W.; Israelachvili, J. Experimental and Theoretical Investigations of Stick-Slip Friction Mechanisms. In *Physics of Sliding Friction*; Persson, B.N.J., Tosatti, E., Eds.; NATO ASI Series; Springer: Dordrecht, The Netherlands, 1996; pp. 51–67. ISBN 978-94-015-8705-1.
40. Möhler, K.; Herröder, W. Range of the coefficient of friction of spruce wood rough from sawing [Obere und untere Reibbeiwerte von sägerauhem Fichtenholz]. *Holz. Roh. Werkstoff* **1979**, *37*, 27–32. [CrossRef]
41. Xu, M.; Li, L.; Wang, M.; Luo, B. Effects of Surface Roughness and Wood Grain on the Friction Coefficient of Wooden Materials for Wood–Wood Frictional Pair. *Tribol. Trans.* **2014**, *57*, 871–878. [CrossRef]

MDPI AG
Grosspeteranlage 5
4052 Basel
Switzerland
Tel.: +41 61 683 77 34

Sustainability Editorial Office
E-mail: sustainability@mdpi.com
www.mdpi.com/journal/sustainability

www.ingramcontent.com/pod-product-compliance
Lightning Source LLC
Chambersburg PA
CBHW041406210926
51850CB00013B/3196
* 9 7 8 3 7 2 5 8 2 3 3 8 3 *